45th Vibration Institute Annual Training Conference 2022 and Society for Machinery Failure Prevention Technology Annual Conference 2022

Savannah, Georgia, USA
1-5 August 2022

ISBN: 978-1-7138-6057-0

Printed from e-media with permission by:

Curran Associates, Inc.
57 Morehouse Lane
Red Hook, NY 12571

Some format issues inherent in the e-media version may also appear in this print version.

Vibration Institute
2625 Butterfield Road
Suite 128N
Oak Brook, IL 60523-3415
USA

Phone: (630) 654-2254
Fax: (630) 654-2271

information@vi-institute.org

Additional copies of this publication are available from:

Curran Associates, Inc.
57 Morehouse Lane
Red Hook, NY 12571 USA
Phone: 845-758-0400
Fax: 845-758-2633
Email: curran@proceedings.com
Web: www.proceedings.com

VIATC
VIBRATION INSTITUTE ANNUAL TRAINING CONFERENCE & EXPO
2022

MFPT
ANNUAL CONFERENCE

AUGUST 1 - 5 | SAVANNAH, GA

CONFERENCE PROCEEDINGS

45th Annual Training Conference and Expo

Savannah, Georgia

www.vi-institute.org | 630-654-2254

TABLE OF CONTENTS

PAPERS:

TABLE OF CONTENTS

PAPERS (continued):

THE PRACTICAL APPLICATION OF SIGNAL PROCESSING

Author: Nelson L. Baxter P.E.

THE PRACTICAL APPLICATION OF SIGNAL PROCESSING

Nelson L. Baxter P.E.
ABM Technical Services

Nelson Baxter is owner of ABM Technical Services. He has 40 years in troubleshooting mechanical and structural vibration problems. Nelson is a member of the Board of Directors and teaches the Vibration Institute AVA course.

ABSTRACT:
Signal processing considerations are often thought of as technical issues that only people who design spectrum analyzers have to deal with. This session's purpose is to show how digital signal processing can significantly alter the data an analyst acquires. If analysts do not understand how digital signal processing affects their data, then there is the potential for their analysis to be faulty. The main focus of the talk will be to show signal processing demonstrations that illustrate the critical nature of the data acquisition decisions made by analysts.

KEY WORDS:
Aliasing; FFT; A-D Convertors; Demodulation; Windows

INTRODUCTION:

This paper is dedicated to Jack Frarey with whom I taught the Vibration Institute advanced course. Without the knowledge passed on by Jack, this paper would not have been possible. Jack's spent countless hours proving out concepts and often finding that things that had been accepted for years were in fact neither true nor accurate.

Digital signal processing has brought forth many positive changes to the field of vibration analysis. Dedicated DSP chips have made it possible to perform spectrum analysis, with small portable devices. For the most part, the conversion of analog signals to a series of numbers, that can be stored and processed into spectra has been a good thing. However, as with most things there are other aspects of this conversion that produce unexpected side effects.

THE FFT PROCESS:

Without going into a great deal of detail, the FFT process consists of:
1: Sampling analog signals with an A/D converter to produce a series of discrete numbers that represent the time signal from the transducer and then storing that block of sampled data in a buffer.

2: The next step is to multiply that block of numbers by a series of numbers that represent sines and cosines at each analysis bin's frequency. If the series of numbers that represent the analog signal have frequency content that matches the frequencies of the multiplying sines and cosines, then for each frequency where there is a match, there is a value generated which represents that frequency's amplitude. That output is the value that is seen for a particular frequency in the vibration spectrum. The key point here is that the FFT is a batch process. It works great if the signals being analyzed are sinusoidal, of constant amplitude, are of a constant frequency and if the signals start and end with zero during the period of the block of data that is being processed. Unfortunately, in the real world, such signals are rare.

What happens if instead of sine waves, the signals are impacts? What happens if during the time in which the block of data was obtained, the amplitude varies?
What happens if during the acquisition period the frequency changes? What happens if the signal does not start and end with zero for the block of data that will be analyzed? As it turns out, these are all valid questions, because they can greatly influence the spectral outputs.

ANTI-ALIASING FILTERS

Whenever data is sampled by the A/D converter there must not be any signals present which are greater than half the sampling rate. Another way of looking at it is that two samples must be taken for each cycle of the highest frequency sine wave that is being sampled. The figure below shows what happens if the sampling frequency is too low.

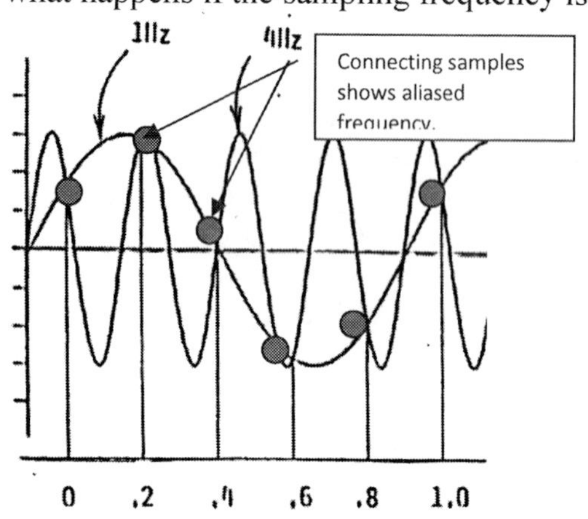

The previous plot shows what happens if the sampling rate is not at least twice the frequency of highest frequency that is present. Since the above sample rate is too low for the higher frequency that is present, an aliased frequency is produced. From a field analyst's view point there are two questions that must be considered. The first is, what happens if the anti-aliasing filters are not turned on? (It should be noted that some analyzers that are on the market at the time this article was written do allow these filters to be shut off.)

What occurs is that frequencies that do not exist show up in the spectrum. Any analyst can understand how confusing this would be. For the previous plot, when the FFT is performed, there would appear in the spectrum a 1 HZ signal that does not actually exist. Imagine trying to figure out what mechanical component these frequencies come from. It would be a long search indeed. The solution to the problem is to check out your software and make sure the anti-aliasing filters are on. During a recent Level IV course, the instructors were testing out one of the most popular models of spectrum analyzer and discovered that there was a small check box that had to be checked before the filters were turned on.

The second question that occurs is what is the effect on the data if the anti-aliasing filters are turned on? The answer is that any signals above Fmax will be significantly attenuated. This means that if you have your analyzer set at an Fmax of say 2000 Hz, then if there is a I inch/second gear mesh frequency at 4000 Hz, for a 120 db/octave anti-aliasing filter, then the signal will be attenuated by 120 db or 1 million to 1.

The field solution to this problem is to first set your Fmax high to make sure there is nothing present, then following that test set it at a lower frequency. Otherwise, you would be unaware of the higher frequency. Because of the presence of anti-aliasing filters, an interesting effect can occur which many analysts have probably seen. The situation is as follows. The analyst is looking at a spectrum and it has a spectral component with a magnitude of .1 inches/second at 60 HZ. That is the only frequency that is present, and it should result in a time wave signal of .1 G. However, when the time domain is displayed, the peak G level is 2 G's. In other words, what is seen on the analyzer does not make sense. This situation is shown below.

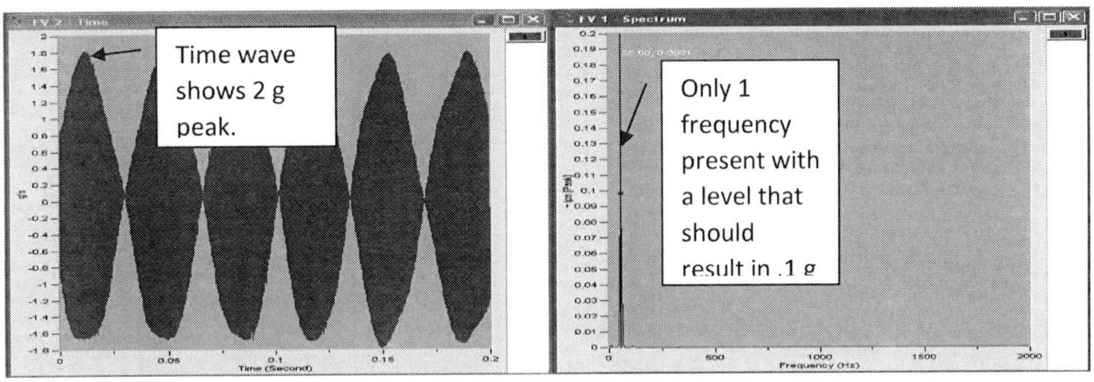

What is actually present.

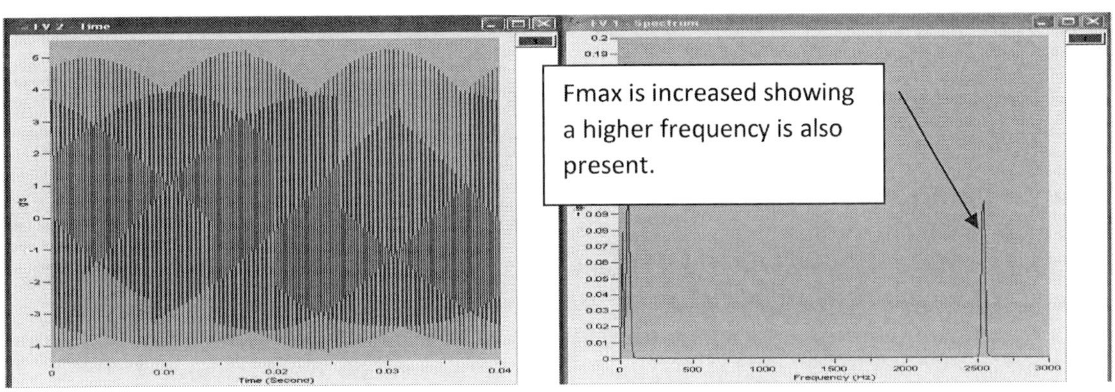

What has occurred in the above plots is quite interesting. Even though the frequency plot only shows signals up to, in this case 2000 HZ, the time plot obviously sees beyond the Fmax frequency. In this case, the time plot is seeing a rotor bar frequency at 2520 Hz. The reason that the data above Fmax can be seen in the time plot is that the anti-aliasing filters are not brick wall filters, so the data above the Fmax, even though it has been attenuated, is none the less still present. The thing to note here is that if there is a G level in the time plot that just does not add up, based upon what is seen in the spectrum, then the first step is to increase the Fmax to see if a higher frequency is coming through.

A TO D CONVERTER ISSUES:
The A/D converter is the device that samples the analog signals produced by the vibration pickups and converts them into a series of numbers from which the FFT can be calculated. A/D converters have a given range that they operate in, for instance +/- 1 volt.

A/D converters are specified as to their ability to break the signal into parts. A 12-bit A/D converter can break a signal into 2 to the 12th discrete parts or 4096 discrete amplitude values. Since the signal can be either positive or negative, that means that there can be 2048 discrete positive amplitude values and 2048 discrete negative amplitude values. Presently most devices have 16-bit A/D converters and the newest devices have 24 bit A/D converters. The larger the bit count, the more parts that a signal can be broken up into. A way to visualize this is the look at a sine wave as a series of stair steps. The higher the bit count, the more individual steps there will be. A higher number of steps mean that we can see small signals better in the presence of large ones. This is called dynamic range. It is found by taking the log of the number of steps and then multiplying it by 20. Example: For a 12-bit A/D converter, the number of amplitude steps is 2048, the log of 2048 is 3.311 so if you multiply this by 20, the result is approximately 66 dB. All this is fine and well. but since this is a paper on the practical application aspects of signal processing, what relevance does it have for the average user?

PRACTICAL A/D CONVERTER EXAMPLE:
An analyst is monitoring a cooling tower gearbox and he would like to look at the unbalance of the fan and the tooth mesh of the gearbox. The fan runs 2Hz (120 CPM).
The pinion gear has 18 teeth and runs 30 HZ for a tooth mesh of 540 HZ. Can the analyst with an instrument with a 12-bit A/D converter detect a 5 mil unbalance of the tower in the presence of .6 inches per second of the tooth mesh frequency? A 5 mil signal at 2 HZ is equal to .001 g's whereas. .6 inches/second at the 540 Hz tooth mesh frequency is 5.24 G's. The ratio of the two signals is 5240. In terms of G's this means that the tooth mesh signal is 5240 times as large as the unbalance signal.

Since a 12-bit A/D converter can only divide the signal up into 2048 positive and negative parts, it is therefore not possible to see the slow speed fan unbalance in the presence of the tooth mesh. In this situation, what then are the options that the analyst has?

1: Take two sets of readings. Take a first reading with a low Fmax that will let the anti-aliasing filters eliminate the tooth mesh.
The analyzer can then auto range and amplify the unbalance signal so the A/D converter can read it. The second reading with the higher Fmax can then be used to see the tooth mesh vibration.
2: Use analog integration prior to sampling the data. If the signal is integrated into velocity, then the 2 HZ 5 mil signal would be equal to .031 inches/second. When the .6 in/second tooth mesh vibration is divided by the .0126 signal, the ratio is 19.3 to one which can easily be handled by the 12 bit A/D converter.
3: Use an instrument with a higher count A/D converter. A 16 bit A/D converter can break a signal up into 32768 bits, so it would have enough amplitude discrimination to see both the low level 2Hz signal and the 540 Hz tooth mesh signal at the same time.

Note that we do not want to go to using the last bit of the A/D converter to measure our signal. The plot below shows what happens when we start using the last few bits.

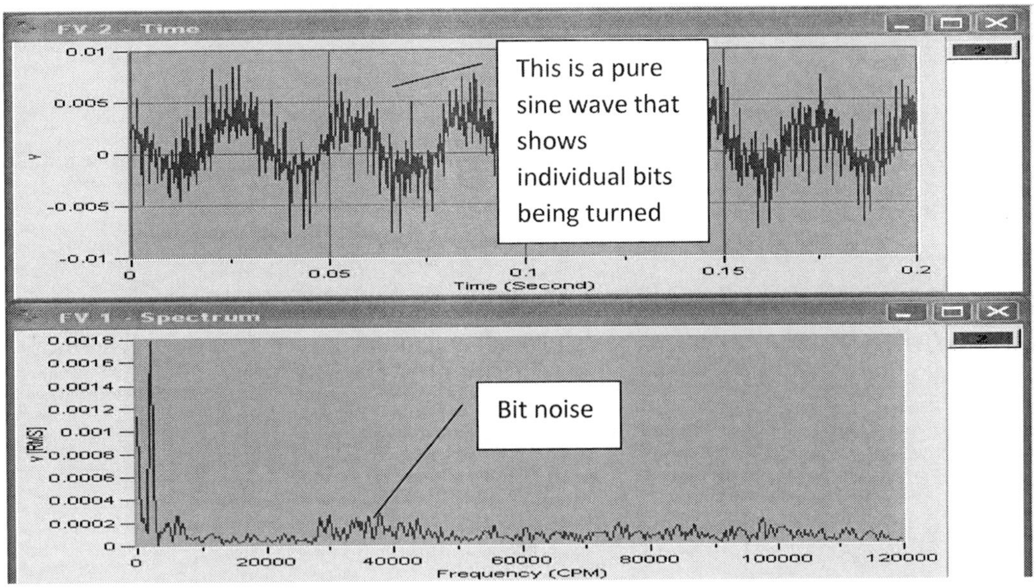

The above plot illustrates what happens when the auto range is shut off and a low-level signal is sampled by the A/D converter. A slight increase causes the next higher bit to fire and then a decrease causes a step drop down to the next lower bit. The FFT of this is a noisy signal with very poor dynamic range.

PROBLEMS ASSOCIATED WITH THE FFT BEING A BATCH PROCESS:

Thus far, problems with digital sampling, anti-aliasing issues and A/D converter dynamic range concerns have been discussed. As has been illustrated, digital sampling can cause frequencies to appear that do not exist. Anti-aliasing filters can cut out frequencies above the Fmax, but even though the spectrum does not show anything, frequencies just above the Fmax may still appear in the waveform. The lack of resolution in the A/D converter can mean that small signals that are present do not appear and bit noise that does not actually exist may appear. If all that does not make one start to wonder, then here comes the most disturbing part of it all. Unless signals are constant in both frequency and amplitude and fall within the center of the bin, then the signatures produced by the FFT can be in significant error. The following problems will be illustrated in this section of the talk.

1: If the time block does not contain data for the entire time period, then the amplitude, while appearing correct in the time display, will be off in the frequency display.
2: If the amplitude varies during the sample period, then the values will be in error.
3: If the frequency varies during the period of the time block that is being FFTd, then serious errors will result.
4: Spectra produced by Impacts, like those present from a defective bearing, do not produce spectra that are representative of their actual amplitudes.

The plot below shows the effect of not having the buffer completely full of the signal to be analyzed. As can be seen in the time plot on the left the amplitude of the sine wave is .3. The spectrum however only shows half that amount.

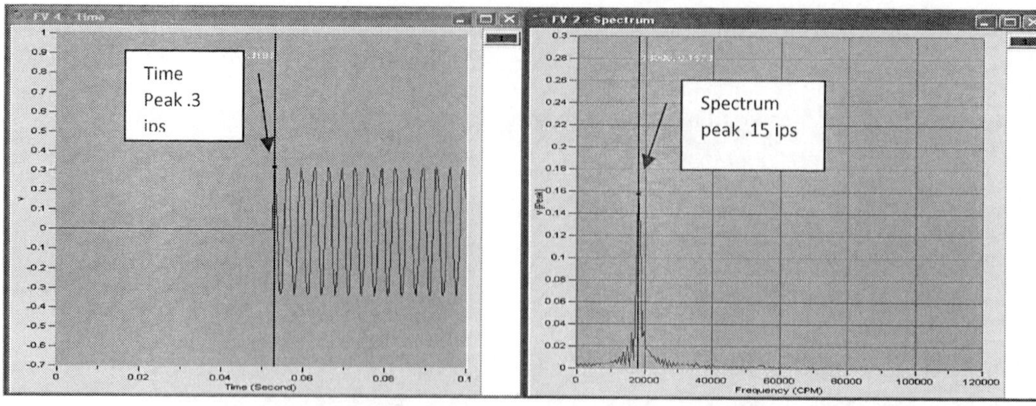

This discrepancy between the peak values in the time display and the spectrum can be even more if the Hanning Window is applied to the data in the partially filled time block.

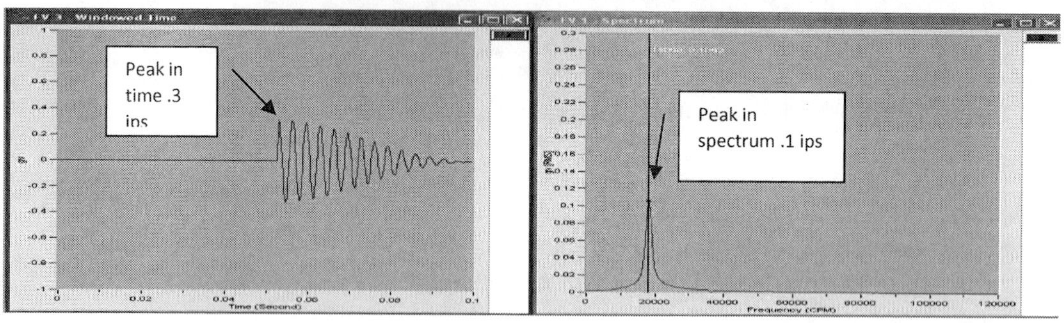

The above illustration shows that the FFT calculation is not correct unless the whole buffer is filled with the data at a particular frequency. Again, this is an interesting fact, but how does that affect an analyst's field measurements.

The following plot to the right is a peak hold plot taken from the coast down of a fan. It shows a peak value of 27.4 mils at 1029 cpm. This data was taken at 400 lines of resolution and 50% overlap. The time for each block of data is 2 seconds.

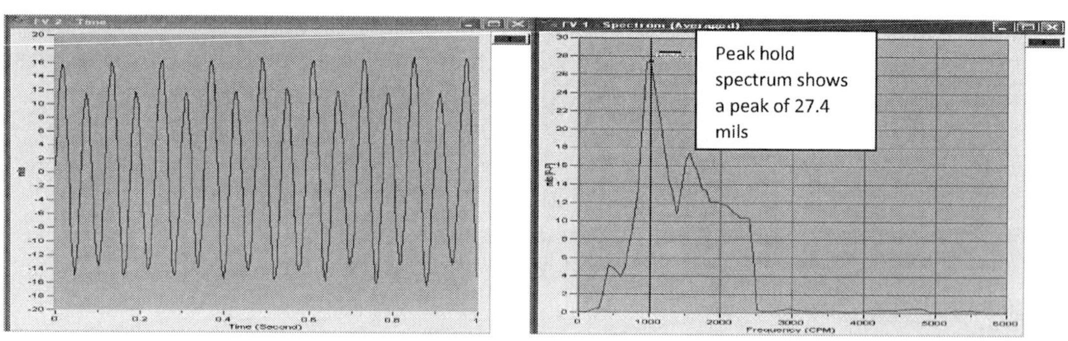

The analyst thinks to himself, I want better resolution so I can determine the critical speed better so instead of using 400 lines of resolution, he or she uses 1600 lines of resolution.

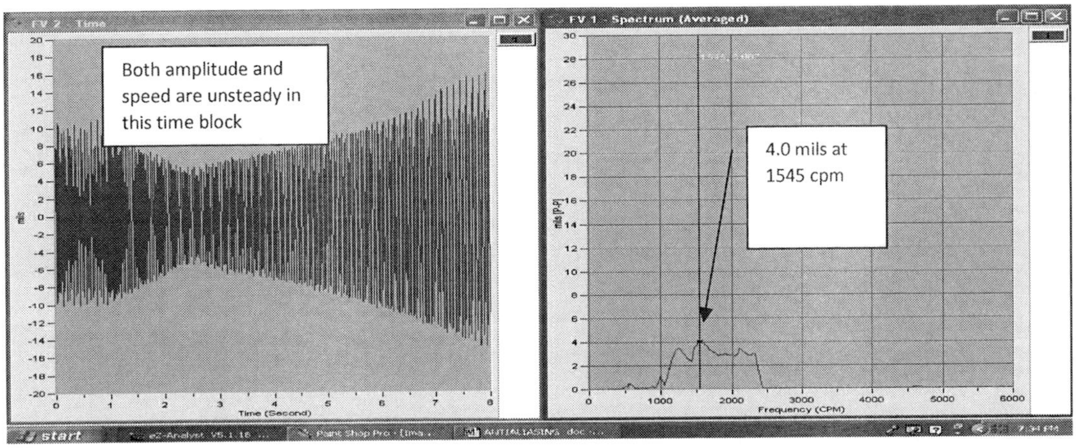

In the case of the 400 line spectrum, the peak in response was over 27 mils and the critical speed was 1029 cpm. In the case of the 1600 line peak hold plot, which was taken with 90% overlap, the peak value was 4.0 mils and the frequency of the highest peak was 1545 cpm. That is quite a difference. By looking at the peak response in the time plot, it can be determined that the 400 line spectra shows the peak very near to its actual value and is therefore correct. Imagine if you are monitoring the coast down of a compressor and you want to know what level the vibration was on the proximity probes. If the choice was made to go for high frequency resolution, then the answer could be off by a large factor. Another ironic twist is that by using high resolution in the pursuit of finding the correct critical speed, instead of helping to more accurately find the critical speed, the decision to use the high resolution setting actually made matters worse. What causes the significant errors seen in this example?

The answer as mentioned before is that the FFT is a batch process. For it to be accurate, the amplitude and the frequency need to be constant for the length of the FFT time period. If the time data in the above two plots is examined, it can be seen that for the 2 second period that it took to get the data for the 400-line spectrum, the amplitude and frequency were relatively constant. However, for the eight second time period that was required to get the data for a 1600-line spectrum, both the amplitude and the frequency varied significantly. The variation in the frequency meant that the energy was spread over several bins and thus the peak at any one frequency was incorrect. This is an excellent example of how the choices we make as analysts can significantly alter the results. For this example, the peak amplitude displayed was off by a factor of 7:1.

Now that all faith in the FFT has been put into question, lets totally destroy any confidence that the analyst might have in it in regards to accurately determining the peak values of non-sinusoidal signals. Vibration spectra have been used for years to diagnose bearing defects. The commonly taught method always begins with using the geometry of the bearings to determine the fault frequencies. According to the classical teaching, the next step is to look for these defect frequencies in the spectra. As it turns out seeing the actual first order of the defect frequencies often does not happen. This does not mean that spectra cannot be used to find band bearings, it is however, often the case that, rather than showing up at the fundamental frequencies in the spectra, the calculated defect frequencies show up as higher harmonics or modulation frequencies of a structural natural frequency or an accelerometer natural frequency. Why is that the case?

The plot below is typical of what an analyst might see from a bad bearing.

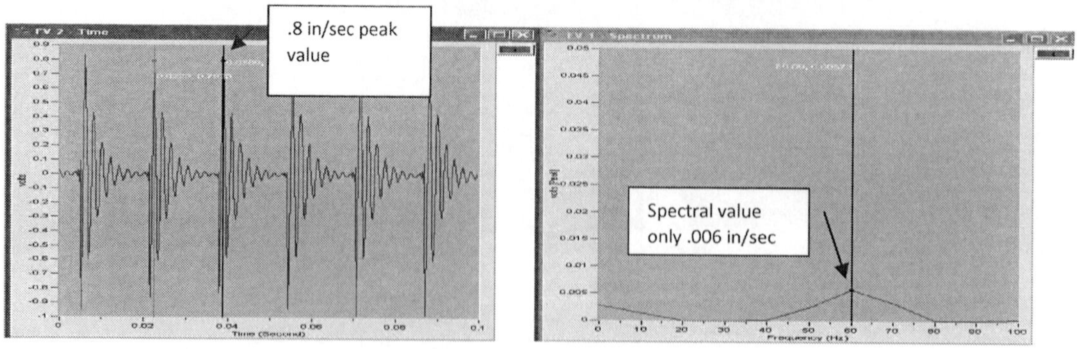

The actual defect frequencies are represented by the impact peaks. As can be seen, the level of the impacts is in this case .8. Using delta cursors, the time between the spacing is .0166 seconds or 60 HZ. The spectrum to the right shows the level of energy in the 60 HZ bin. It has a peak value of just .006. This means that, in this instance, the spectral value of the peak at 60 HZ is 133 times smaller than the peak value seen in the time plot. What causes this discrepancy? The answer is again that the FFT is a batch process. If you count the number of pulses in the time plot, it is six. There are 512 time samples, but only six are associated with the fundamental defect frequency, the rest are associated with the response of a structural natural frequency to the impacts. When the FFT is calculated by multiplying the pulses by the sines and cosines associated with each frequency bin, there are just too few samples associated with the fundamental defect frequency bin.

DEMODULATION

We have all heard of demodulation and been told that it does a lot better job of detecting impacts. Let's see why. The plot below shows what happens when we run the above signal into a capacitor and let it drain out through a resister. As can be seen, the amount of energy at the impact frequency is much larger. This is because the area under the impact curve is now much greater and when we multiply by the sine and cosine functions of the FFT, the resulting amount of energy that is at the impact frequency bin is much larger. The spectrum from the demodulated time wave form therefore contains a much higher level of energy associated with the fundamental defect frequency.

Demodulated Wave and its FFT

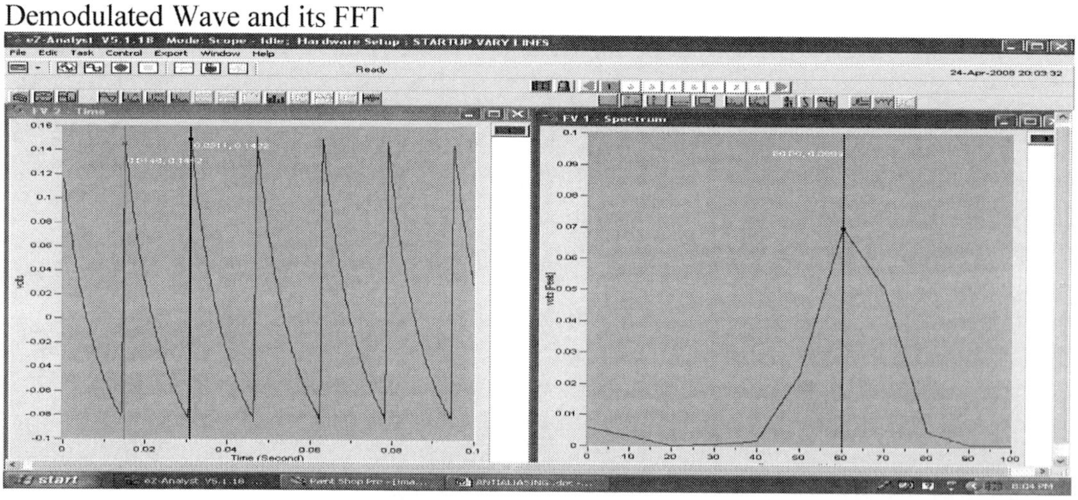

The following are two ways to demodulate a signal, both of which increase the amount of the fundamental impact energy. Both convert a narrow spike into a slowly decaying wave, thereby increasing the energy under the curve.

Method 1- Hardware Demodulation. This method was alluded to in the preceding paragraphs.

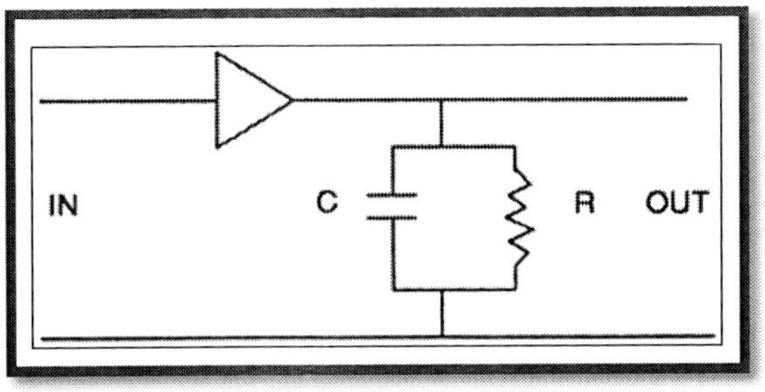

What happens in the above circuit is that the rectifier preserves only the positive parts of the signal. This positive potential then charges the capacitor. Once charged, the electrons drain off through the resister.

The net result is that the narrow spike, for instance from a bearing defect impact, is now represented by a sharp rise and a slow decay type of response. As noted before, this results in an increase in the area of the curve and a much larger amount of energy in the FFT bin of the defect frequency. The red line in the following figure shows the raw signal, the blue line is the enveloped response.

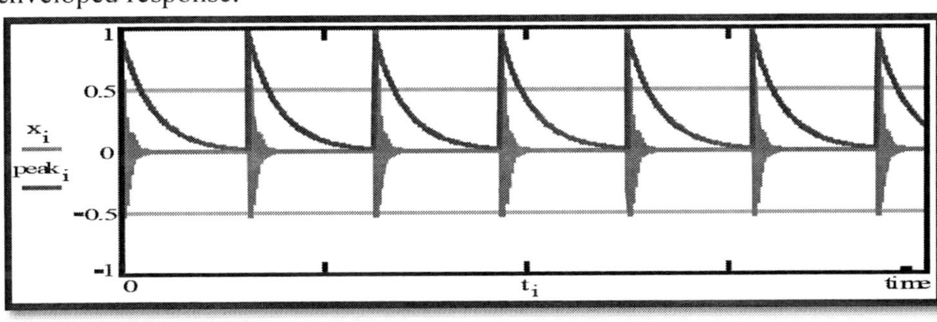

One question that often arises is: What significance are the harmonics that appear in a demodulated spectrum? The answer is none. The harmonics are a function of the rise and decay constants that depend upon the capacitor and the resister that are chosen for the demodulator.

Fast Decay from small resistance

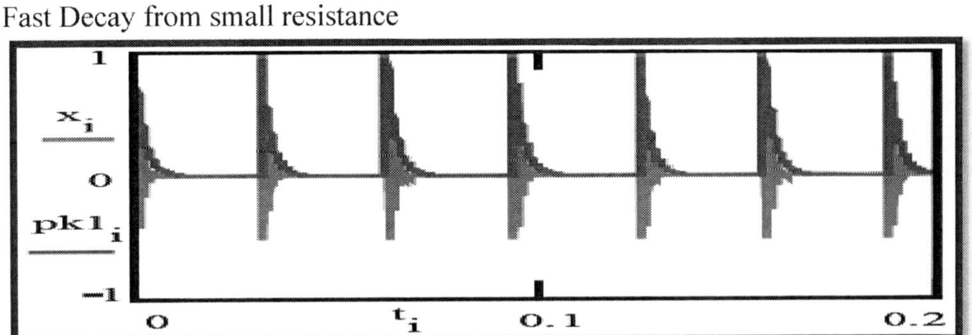

Spectrum from fast decay

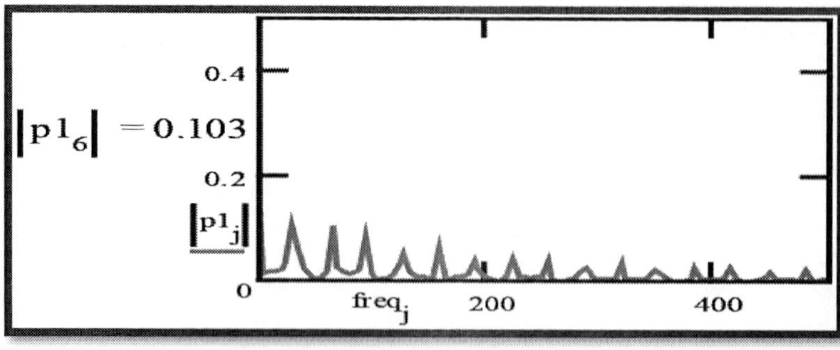

Slow decay from higher resistance

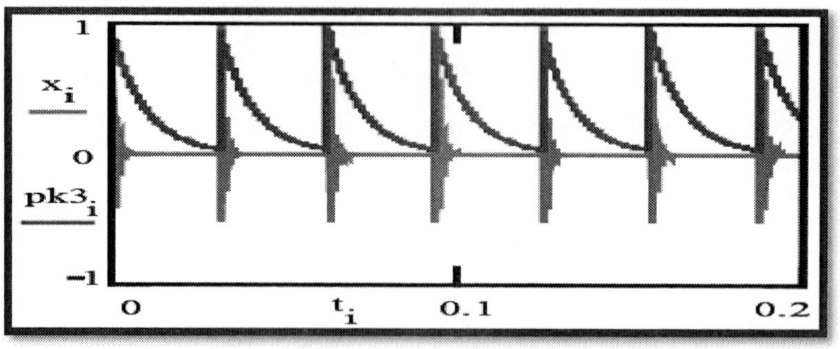

Spectrum from slower decay

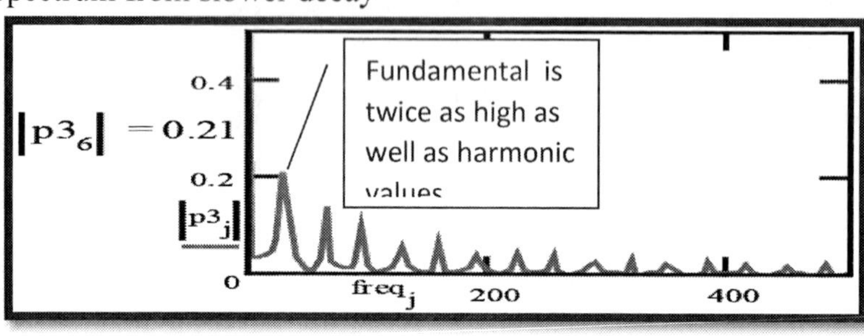

The practical aspect of all of this is that because the response is dependent upon the circuitry utilized to do the demodulation, it is not possible to accurately determine the severity of an impacting problem from a demodulated spectrum or compare readings from different test equipment manufacturers. Ref- From AVA notes produced by Jack Frarey.

Method 2-The second method of demodulation to be discussed is via the use of a software demodulator. The Hilbert transform is a method of demodulating a signal utilizing a software approach.

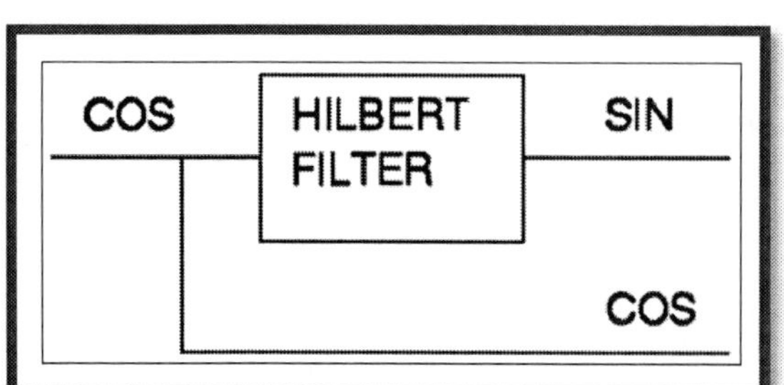

The Hilbert filter in the diagram above shifts the original signal by -90 degrees. This Hilbert shifted time wave is calculated by producing an IFFT of the phase shifted spectral components. We often forget that when an FFT is produced that each spectral component has both amplitude and phase. This is an instance where that knowledge is put to use. Once the phase shifted time wave is created, what is referred to as the analytic signal can be synthesized. The analytic signal which represents the envelope is made by taking the square root of the sum of the squares of the original signal and the phase shifted signal. Note that the squaring function serves the same purpose as the rectifier of the hardware demodulator.

Original Time Wave

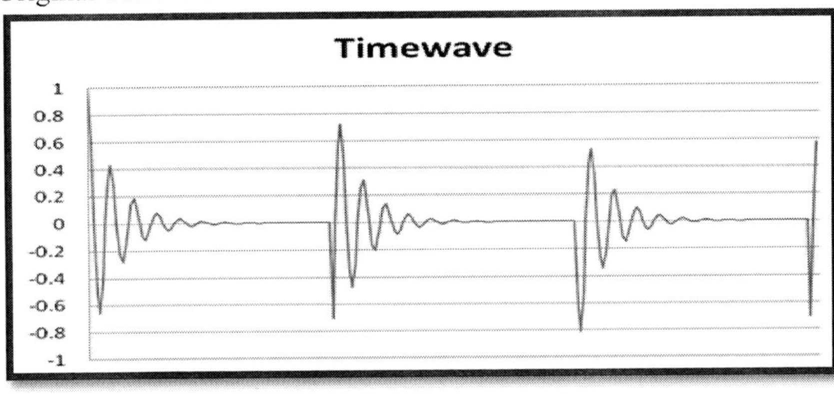

IFFT from -90 Degree (Hilbert) shifted spectral components

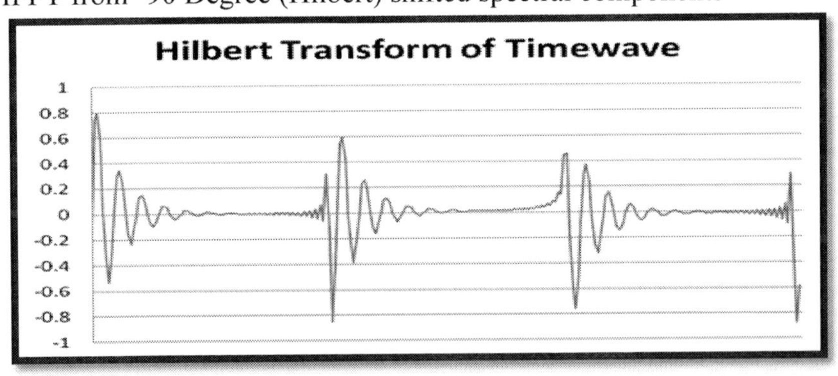

Original signal(blue) overlaid with Enveloped signal (Red).

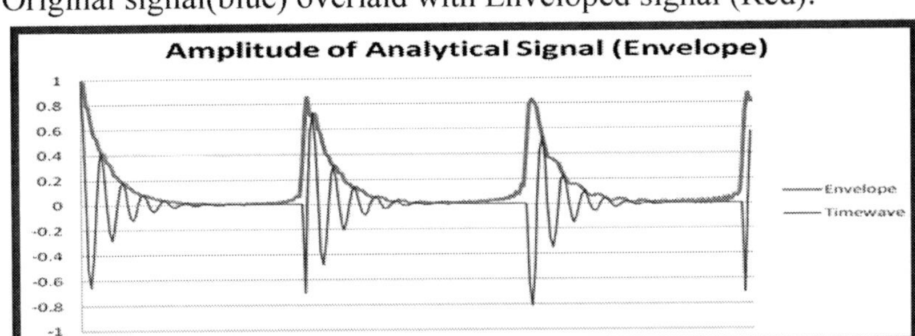

Note that the above software demodulation looks just like we charged a capacitor and drained it through a resister.

The advantages are:

1:The carrier signal does not have to be known.

2: There is no need to adjust any filters because the transform adjusts to the data set.

3: No additional signal processing is required.

REF- Hilbert Transform section from AVA Notes prepared by Dave Geswein.

EFFECTS OF WINDOWS

Windows are used to eliminate discontinuities in the data. Discontinuities result in the generation of shirts around the frequencies of interest and result in a significant loss of dynamic range. Windows solve this problem by multiplying the time block by a function that ensures that the beginning and end of the time block smoothly starts and ends with zero. This approach thus ensures that there will not be any discontinuities in the calculations. The figure that follows was taken from the Vibration Institute's level IV notes. It shows an example of the discontinuities that occur without windowing on non bin centered data.

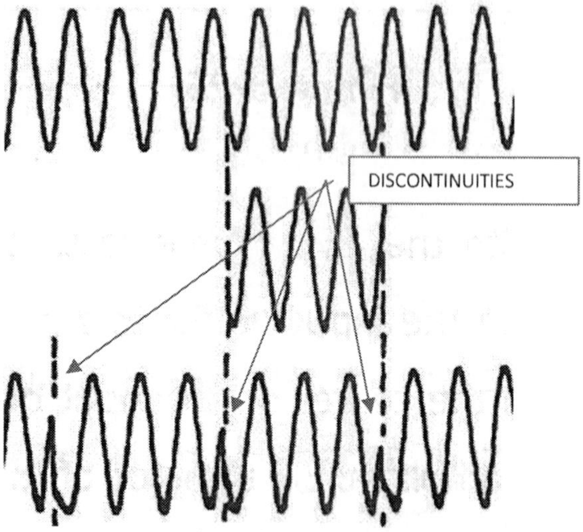

As mentioned earlier the windowing functions smooth out these discontinuities. Each window or lack thereof has an effect on the accuracy of the FFT process. It turns out that all the windows are accurate when applied to sine waves if the data is bin centered. However, lets take a look at what happens if the data does not fall into the center of a bin.

WHEN DATA IS BIN CENTERED, THEN ALL WINDOWS GIVE CORRECT VALUE

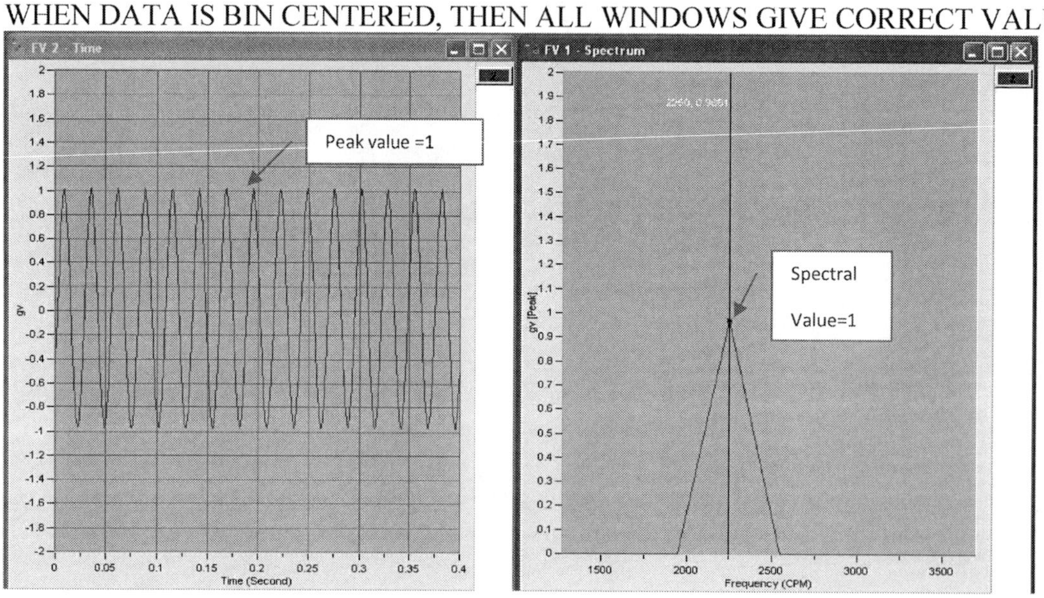

FOR RECTANGULAR WEIGHTING OF A SIGNAL BETWEEN BINS, THE ERROR CAN BE UP TO 37%

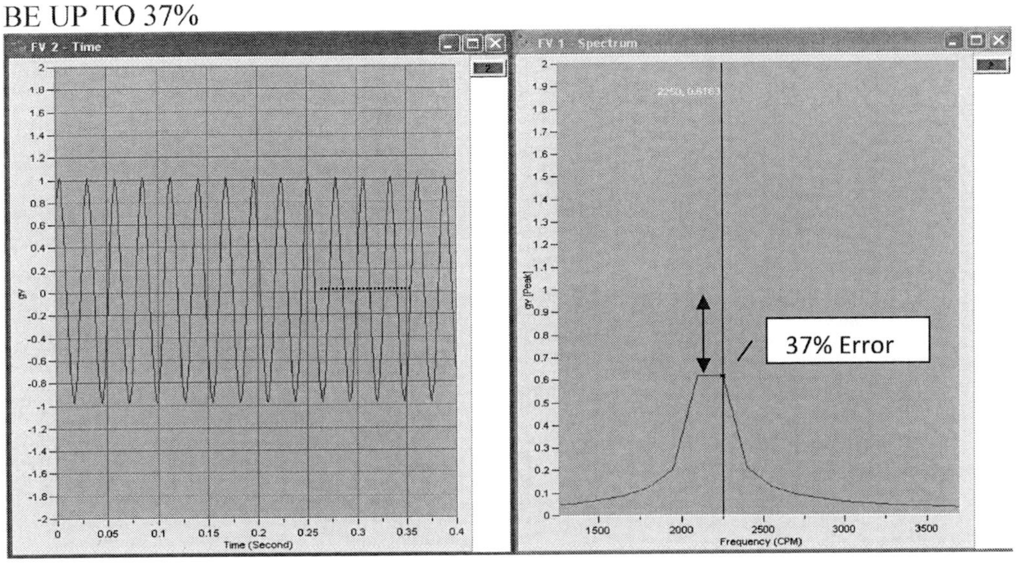

HANNING WEIGHTING ERROR CAN BE UP TO 17%

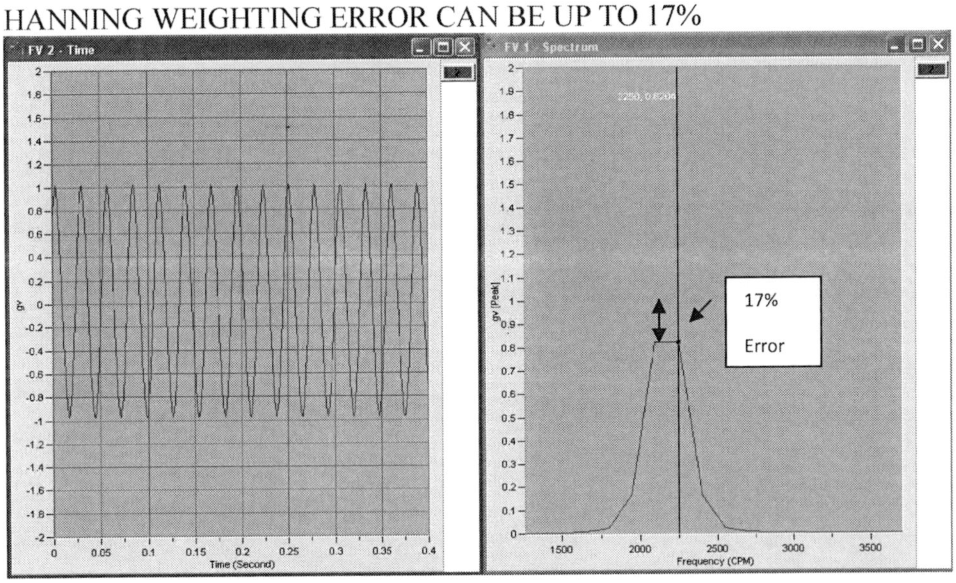

The above plots show that the lack of a window can result in a 37% amplitude error and even with a Hanning Window, the error can be up to 17%. The solution for the analyst when the Hanning window is applied is to use the interpolation feature of his analyzer if it contains one. These features use the ratio of the amplitude of the adjoining bins to determine the exact position of the signal in a bin. They then correct for any Hanning induced error and also accurately determine the actual frequency. This correction is a win-win situation and should be used if available.

FLATOP WINDOW BETWEEN BINS ERROR 2%

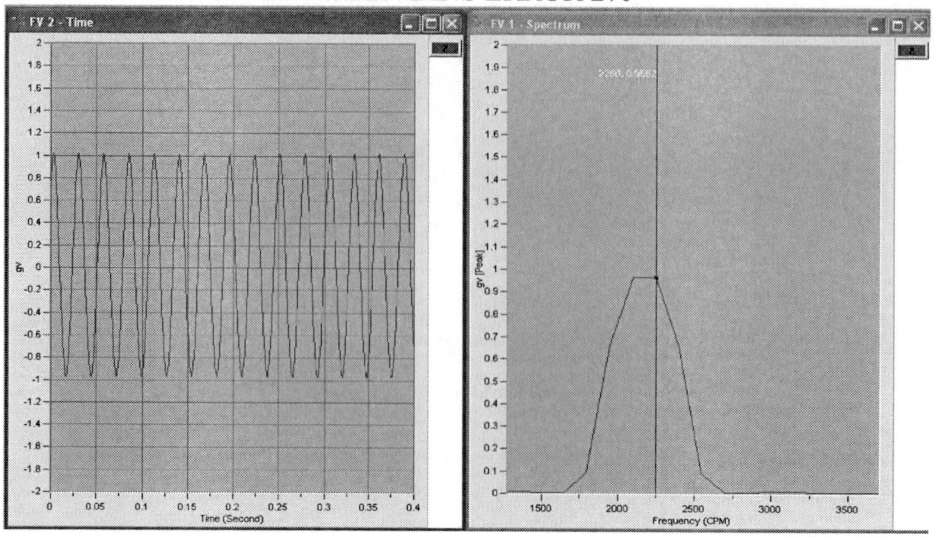

FALSE RESOLUTION

The windowing functions can have other strange effects. The first plot below shows a beat where there is one beat in the time window. This is the signal before any windowing has been applied. There is one beat in the time block, so it is not possible to separate out the two frequencies.

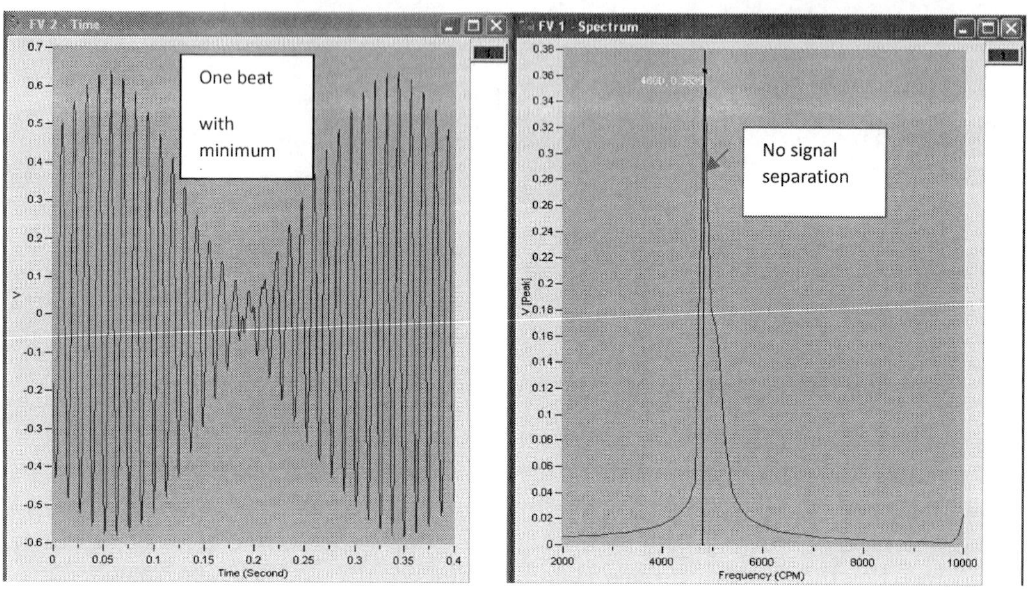

The data below shows what happens when the Hanning Window is applied to this same beat, where the minimum occurs at the center of the time block. If the beat is at a minimum at the center of the time block, then when the Hanning Weighting Factor is applied, it makes one beat per time period appear to be two beats per time period. Frequencies appear at 4800 and 5100 cpm.

Hanning Window modulates signal making it appear there are two discrete signals

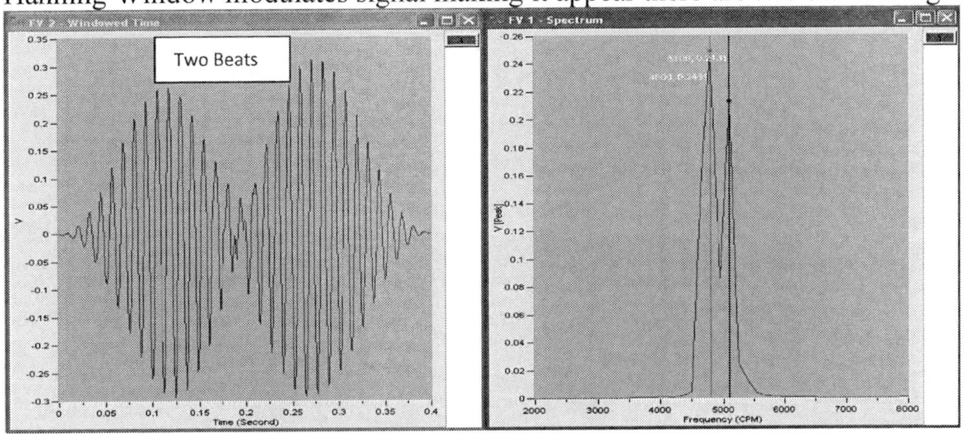

In the above instance there was originally only one beat and it would be impossible to separate the two frequencies. However, after the Hanning window was applied, it made it appear as if two beats occurred in the time and window and the spectrum shows the presence of two frequencies that do not actually exist. Now look at what happens if we apply the flat top window to the same data. If the beat is at the minimum in the time block, then the Flat Top window makes one beat per window appear as four beats per window. The spectrum now shows beats at 4650 and 5250 CPM

Flattop window modulates signal causing four beats to appear in time plot.

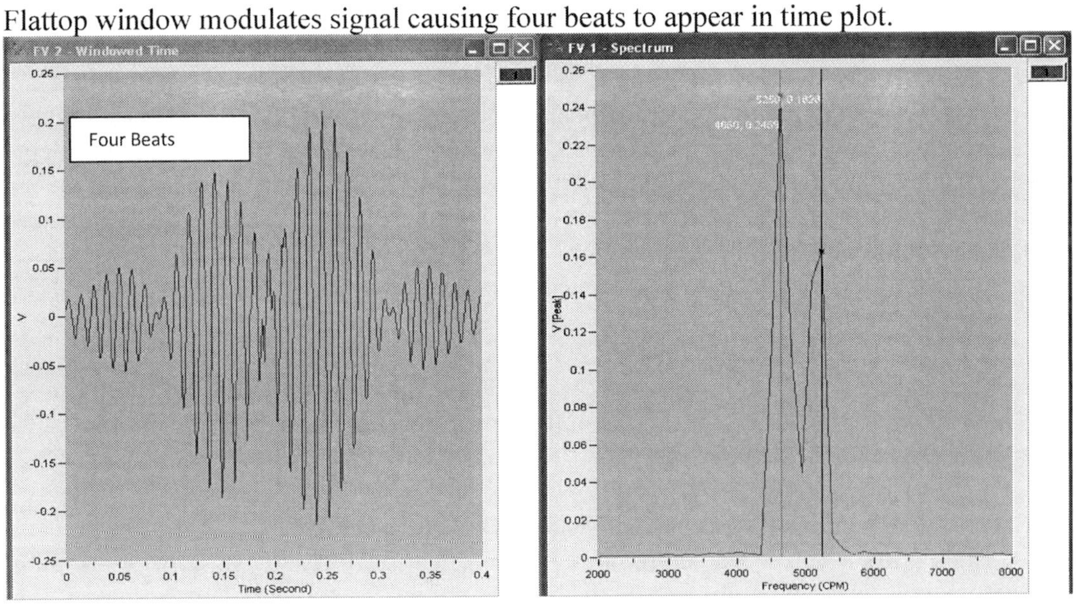

What is the truth? The only way to really know is to increase the resolution so there truly is more than 1 beat in the time window. The spectra produced by the effects of the windows gave a false sense that the signals were being separated when in fact they were not.

Spectrum and time plot with higher frequency resolution

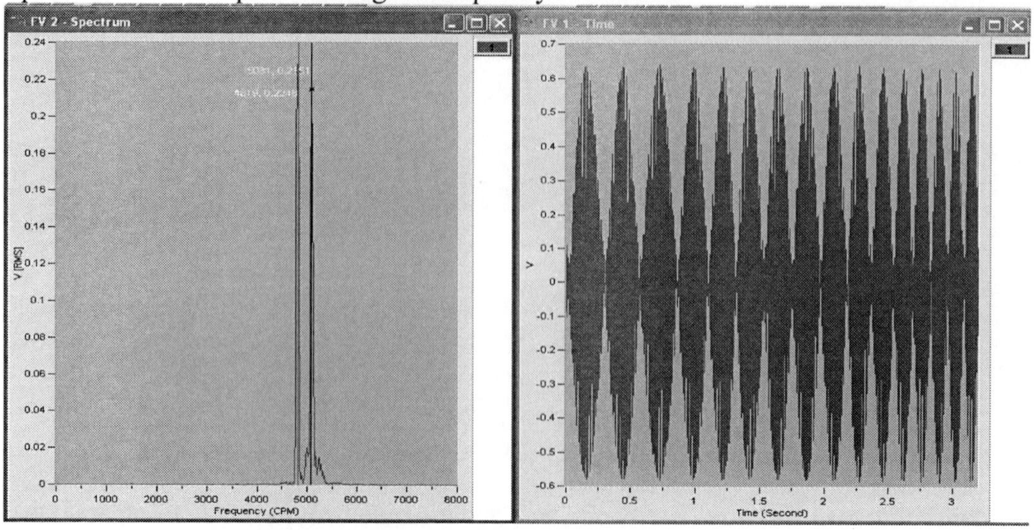

The plot above shows what needs to be done to determine the actual separation. What is needed is to take high enough resolution data to allow for several beats to occur in the time block. This allows for the true separation in the frequencies to be calculated.

The Hanning window indicated that the frequencies were 4800 and 5100 CPM or a separation or 300 CPM, which is two bins of separation. The Flat Top Window indicated that the beat frequencies occurred at 4650 CPM and 5250 CPM or 600 CPM which is equivalent to four bins of separation. The reason for the two bin separation when the Hanning Window is used is that the shape of the Hanning window produces two beats where there was in reality just one. In the same manner, the Flat Top Window produces four beats where there was only one. The problem here is that the windows are appearing to be able to separate out the two frequencies when in fact they cannot. What is the lesson for the analyst. The lesson is to never believe the FFT when there is a small separation between the frequencies. The way to solve this problem is to increase the resolution so that there is clear separation between the two signals.

WINDOWING EFFECTS ON IMPACT TEST DATA RESULTS:
Due to the transient nature of impact testing data, windows can have a significant effect on the test results. The following plots illustrate the amount of error that can result by choosing the wrong window for use in impact testing.

RESPONSE WITHOUT AND WITH EXPONENTIAL WEIGHTING:
The Log Dec calculation was used to compute the amplification factor from a response first without the exponential decay window then with the decay window applied. The Amplification factors calculated from the Log Dec. Were-Without decay window Q= 46.6 With decay window applied Q=12.4 Note that the exponential window in this case changed the amplification factor by 4:1. Analyst note- take off exponential window prior to calculating amplification factor.

Time based impact data without and with and exponential window applied

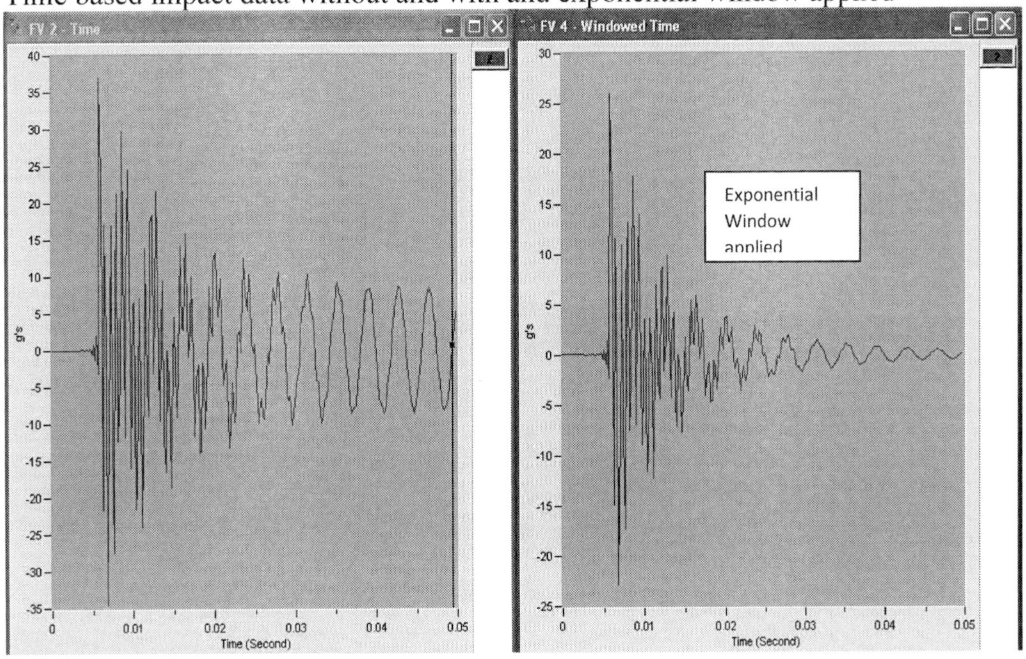

EFFECT OF USING THE HANNING WINDOW ON THE IMPACT PULSE

The next plot shows the pulse from an instrumented hammer with and without Hanning weighting applied. With no window on the pulse height is 15.9 lbf. With the Hanning window on the pulse amplitude is 2.6 lbf. The effect of the window causes an 84% reduction in the pulse height.

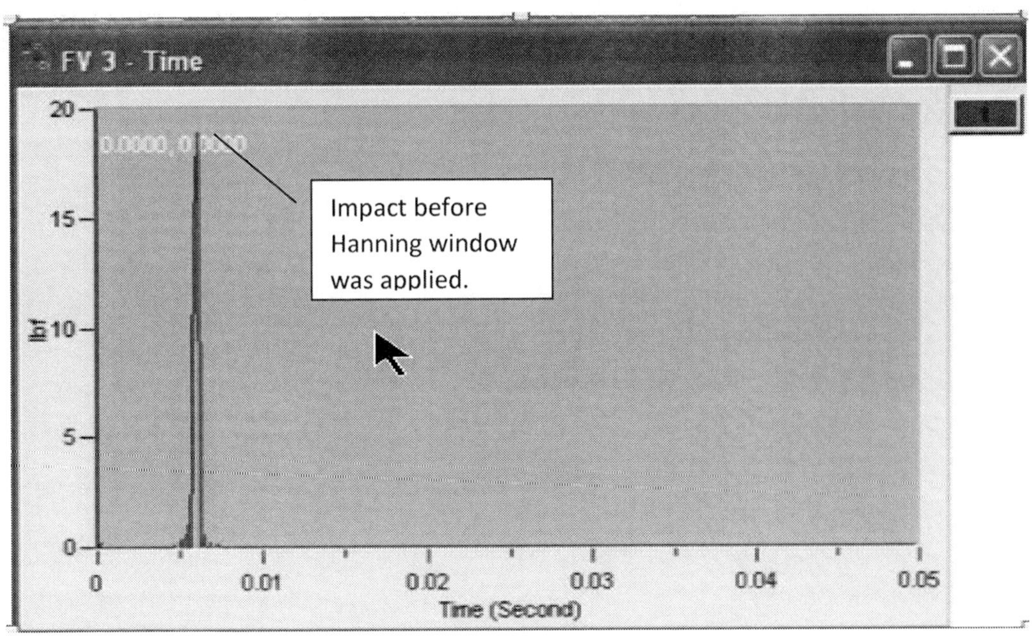

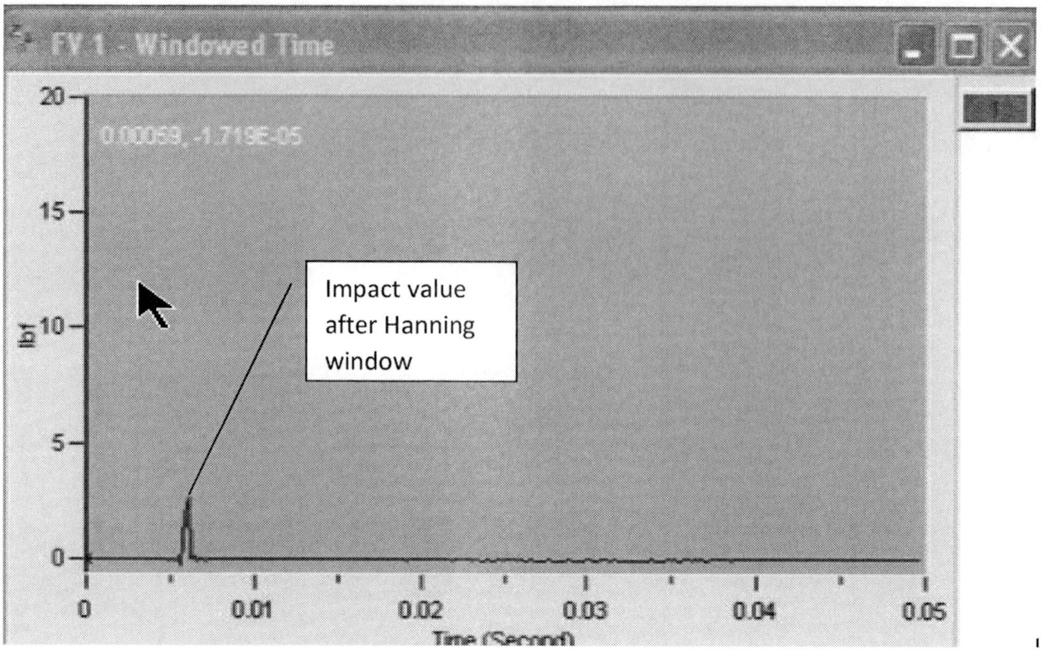

The next plot shows what happens when the Hanning window is applied to the response channel.

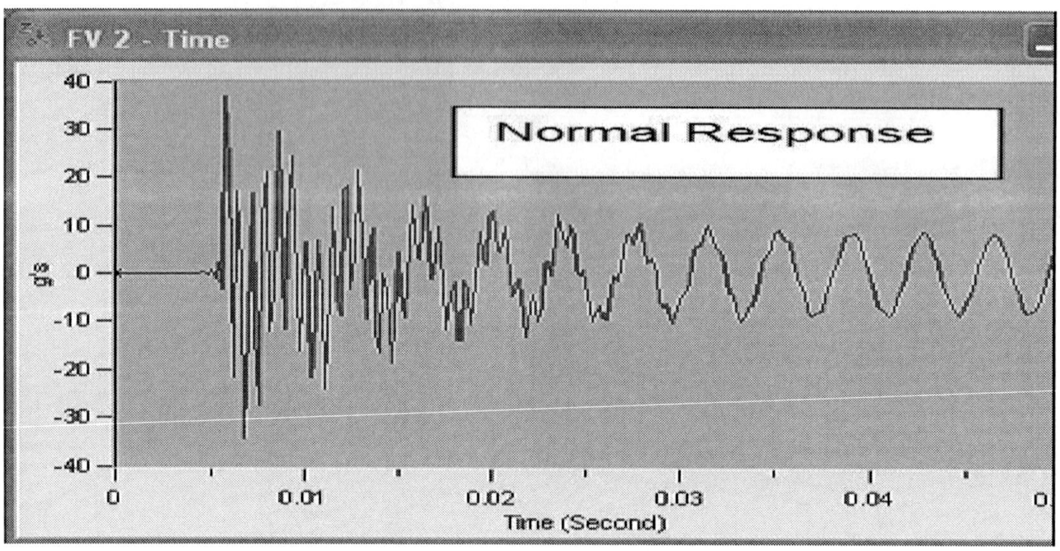

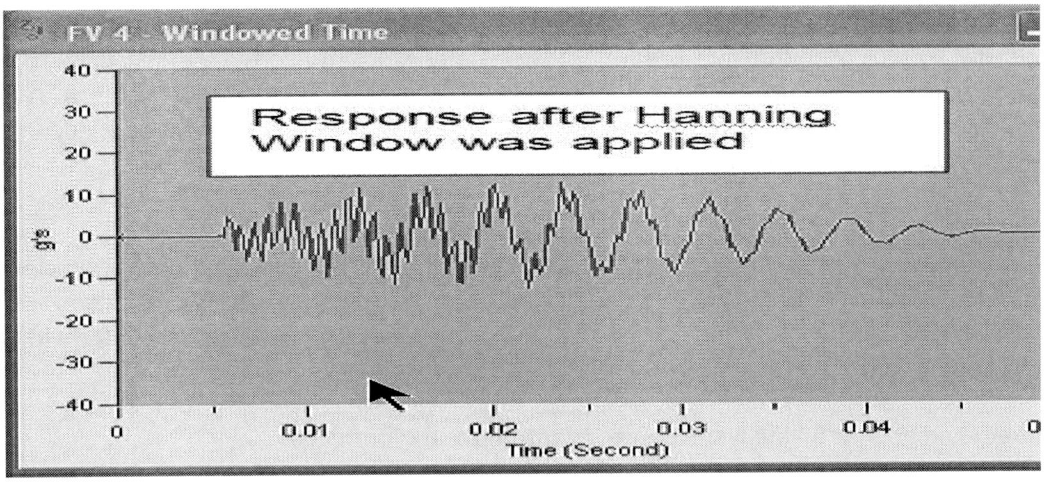

The above plots and the plots on the previous page show how the Hanning Window effects impact data. Since the transfer function is calculated by dividing the impulse power spectrum by the response spectrum, it is fairly obvious that the results will be significantly influenced by the application of the various windows. The list below shows how much the transfer function results can vary because of the application of different windows.

Transfer function window altered results:

WINDOW APPLIED	260 Hz	1400 Hz
No window	27.5 lb/g	26.21 lb/g
Hanning	120.6 lb/g	100.8 lb/g
Exponential	7.1 lb/g	13.0 lb/g

Note the huge discrepancy in results that occur by just changing the windows when performing impact tests.

NEGATIVE ALIASING

The final topic is negative aliasing. Negative aliasing is a strange thing that happens when the modulation frequency or one of its harmonics exceeds the carrier frequency.
This can happen when, for instance, there is nonlinear amplitude modulation or the presence of frequency modulation. The plot below shows a negative aliased frequency.

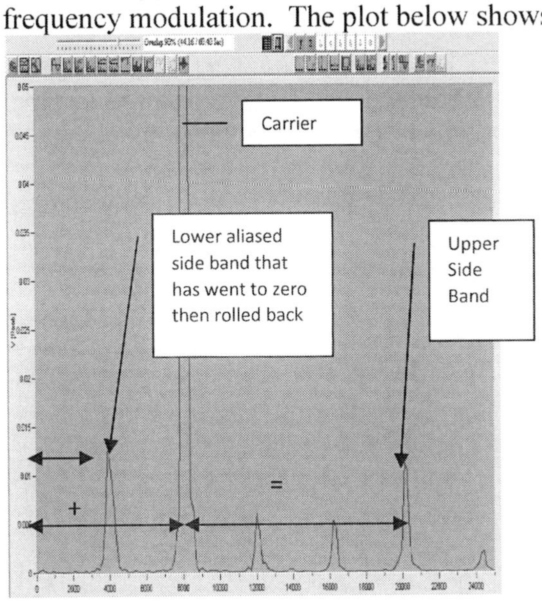

To an analyst, negative aliasing should be quite disturbing, because it results in frequencies showing up in the spectrum that do not actually exist. There is no protection against negative aliased frequencies. The only practical thing an analyst can do is be aware of their existence. As noted above, they are most likely to appear if you have a large family of sidebands like what occurs in non-linear amplitude modulation or whenever there is frequency modulation present.

CONCLUSION:
Since this is a practical application paper, lets summarize what we have learned and how we can apply this knowledge to our field testing. The important thing is to realize what is happening in each case and then to correct for it or at least be aware that the data may not be representing what is actually occurring.

Processing issues and actions an analyst can take to minimize their effects:

ANTI-ALIASING- Most devices are protected against the production of aliased frequencies that result from sampling high frequency signals too slowly, but some are not or allow the user to turn off the feature.
Action- Make sure you know how to turn the filters on, if your unit does not already do so automatically.
Action- If you see higher G levels in the time plot than cannot be explained by what is seen in the spectrum, then increase the Fmax to see if there is energy coming through from the region where the anti-aliasing filters have not taken full affect.

A/D CONVERTER ISSUES
Action-Be wary of turning off the auto range feature or this could result in under loading the A/D converter, resulting in bit noise and poor dynamic range. The opposite situation could be that the A/D converter could be overloaded resulting in clipping of the signal. In that case non existent harmonics and the introduction of a DC component that will drive integrated signals to very high levels could occur.

Action- If you are trying to see small signals in the presence of a large one, calculate the dynamic range needed to determine if the instrument's A/D converter is up to the task. If it will not, then consider analog integration ahead of the A/D converter.

FFT BATCH PROCESS PROBLEMS
If the signal is not constant in the block of data that the FFT is being processed upon, then serious amplitude errors will result. Changes in frequency are particularly bad because that causes the energy to bleed into the adjacent bins and the spectral amplitudes are therefore meaningless.

1: Action- When doing coast down or start up tests, use the minimum number of lines of resolution possible. This reduces the length of the time window, thereby minimizing the amount of variation in amplitude and frequency that will occur in the FFT calculation.

2: Action- Realize that if impacts are present that the spectral amplitude at the impact frequency does not represent the peak value seen in the time plot. In order to better determine the impact frequency, consider using a demodulation technique. Always view the raw time domain in order to determine the severity of impacts. Remember that the level of the fundamental impact frequency and its harmonics are significantly affected by the rise time and decay constants that come from the choice of the capacitor and resister values.

WINDOW ISSUES

ACTION- For steady signals, in order to avoid errors in amplitude and dynamic range problems on non bin centered data, a window needs to be applied. The most practical thing to do is use the Hanning Window and then utilize the interpolation feature available in most software packages to correct for its amplitude errors while at the same time greatly increase the ability to determine the frequency accurately. Flat top windows are accurate, but they have other issues such as less frequency discrimination and increased noise that render them less acceptable.

Action- Since windows effect the bandwidth and can produce false resolution, always use enough lines of resolution to clearly separate out the spectral frequencies.

Action- For impact testing, do not use the Hanning window unless the signal is delayed past the area where it is attenuated. As was seen in this paper, the Hanning window can drastically affect the transfer function results. If the exponential window is used to reduce the ring down of the response signal, it must be taken off prior to doing any damping and amplification factor calculations. It the case that was illustrated, the amplification factor was changed by a factor of 4:1.

NEGATIVE ALIASING- As stated above, this can occur when either the modulator or one of its side bands is of a higher frequency than the carrier. There is no action that can be taken to eliminate it. The analyst just needs to be aware of it when trying to explain strange frequencies.

FINAL THOUGHT

It is easy to ignore the signal processing issues listed in this paper, but as can be seen with the variation in the peak values on the coast down, the effect on the damping and transfer function calculations along with the introduction of strange frequencies by the windowing functions and negative aliasing, it would not be in anyone's interest to do so.

EXAMINATION OF THE FFT BATCH PROCESS WITH CALCULATION OF THE RMS ENERGY

Author: Jack D. Peters

VIATC 2022 VIBRATION INSTITUTE ANNUAL TRAINING CONFERENCE & EXPO | MFPT ANNUAL CONFERENCE

2022 CONFERENCE PROCEEDINGS

VIATC TRAINING CONFERENCE & EXPO
SAVANNAH, GEORGIA · 2022

EXAMINATION OF THE FFT BATCH PROCESS WITH CALCULATION OF THE RMS ENERGY

Jack D. Peters, Technical Director of Training
Vibration Institute
2625 Butterfield Road, Suite 128N, Oak Brook, IL 60523

Certified as a Category IV Vibration Analyst in accordance with ISO 18436-2, Mr. Peters is Technical Director of Training, an Instructor and a Member of the Board of Directors for the Vibration Institute. He is a Certified Maintenance & Reliability Professional (CMRP) in accordance with the Society for Maintenance & Reliability Professionals (SMRP). Prior to his retirement in 2017, Mr. Peters was Vice President of International Sales for Connection Technology Center. He worked for Eastman Kodak company from 1977 – 2004 completing his career at Kodak as the technical manager for world wide condition monitoring on photographic film and paper manufacturing machines. As the sole member of Accelent Technology LLC, Mr. Peters works with the Vibration Institute to oversee and improve training materials and training programs.

ABSTRACT:
The Fast Fourier Transform (FFT) is used in vibration analysis. It is well known that Window Filters (Hanning, Flat Top, etc.) can be applied to the time waveform during the batch processing of the FFT, and as a result, decrease the resolution of the FFT. The resolution is decreased by a numerical Window Factor based on the applied filter. It is not well known that the Window Factor (WF) can also be described as the Equivalent Noise Bandwidth (ENBW) or Noise Power Bandwidth. This paper will explain how to calculate the RMS Energy of the FFT when the WF or ENBW used to process the FFT has a numerical value greater than one.

KEY WORDS:
Time Waveform, Fast Fourier Transform, FFT, Window Factor, Window Filter, Equivalent Noise Bandwidth, Noise Power Bandwidth, RMS Energy, RMS Power, Lowest Resolvable Frequency, Vibration Analysis

INTRODUCTION:

The Fast Fourier Transform or FFT was developed by James Cooley (IBM) and John Tukey (Princeton) in 1965. The FFT included the introduction of computer processing of the Fourier Series originally developed by a French mathematician and physicist, Joseph Fourier. This highly improved the accuracy of vibration analysis for resolving closely spaced frequencies in the spectrum. [1]

In the form of an algorithm or calculation, the FFT is a batch process that is applied to individual blocks of time waveform data that typically represent the mechanical vibration of the machine as measured by an accelerometer, velocity sensor or displacement probe. Figure 1 illustrates a typical time waveform and FFT.

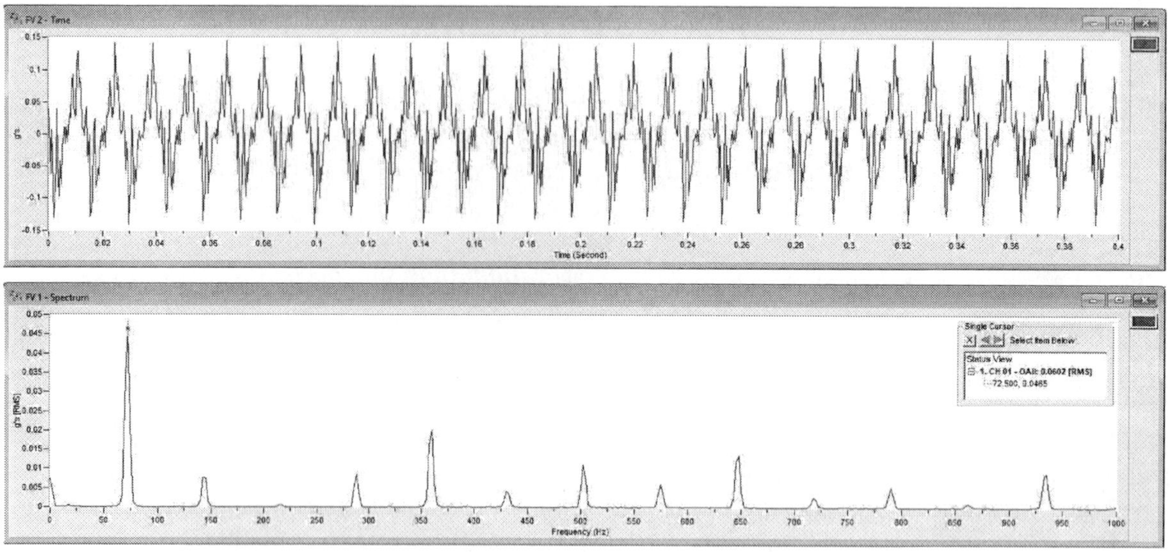

Figure 1 Time Waveform (upper trace) & FFT (lower trace)

The FFT shown in Figure #1, has a vertical axis in acceleration (g's RMS) and a horizontal axis in frequency (Hz). The individual frequencies identified in the FFT each have an amplitude value. The highest amplitude, 0.0465 g's RMS, is identified at 72.5 Hz. The amplitudes of the individual frequencies in the FFT are important in assessing the condition of the machine (i.e., Unbalance, Misalignment or Looseness) or the condition of individual components in the machine (i.e., Rolling Element Bearings or Gears). Although the amplitudes of the individual frequencies in the FFT could be trended for a change in amplitude indicating a change in condition, it would require a significant amount of processing and data storage. This may not provide an efficient method for general evaluation of the machine. The Overall RMS Value of the FFT describes the RMS energy contained in the FFT. In Figure 1, the Overall RMS Value is 0.0602 g's RMS and it describes the RMS Energy contained in the FFT from 0 – 1000 Hz. This overall value of the RMS Energy is easy to trend and alarm. If the energy increase or decreases, it indicates a change in condition. In some applications, it may be more appropriate to measure the RMS Band Energy to trend known changes in specific machine conditions or components. The RMS Band Energy is calculated for a range of frequencies instead of the total frequency span of the FFT. The RMS Band Energy from 50 – 100 Hz would specifically track the amplitude changes in a range that includes the 72.5 Hz, in Figure #1, but ignores all other amplitudes.

BINS AND LINES:
The horizontal axis on the FFT is defined as the frequency axis. Typical units will be Hz (Hertz), CPS (cycles/second) or CPM (cycles/minute). When a frequency is displayed on the horizontal axis, it is presented as a Line that represents all of the frequencies in the Bin. The number of Lines in the FFT is a user selectable feature of the data collector, vibration analyzer or dynamic signal analyzer being used in the measurement process. User selectable Lines typically include; 100, 200, 400, 800, 1600, 3200, 6400, 12800 and 25600. Examples of Bin width are shown:

Example 1: Frequency span is 0 – 500 Hz, Number of lines selected are 400

$$Bin\ Width = \frac{Frequency\ Span}{Number\ of\ Lines} = \frac{500\ Hz}{400\ Lines} = \frac{1.25\ Hz}{Line}$$

Example 2: Frequency span is 0 – 500 Hz, Number of lines selected are 1600

$$Bin\ Width = \frac{Frequency\ Span}{Number\ of\ Lines} = \frac{500\ Hz}{1600\ Lines} = \frac{0.3125\ Hz}{Line}$$

As indicated by Examples 1 and 2, if the number of lines is increased and the frequency span remains constant, the Bin width is decreased. It would indicate that a smaller Bin width would have better resolution. In fact, the Bin width describes the Lowest Resolvable Frequency (LRF) that can be identified in the FFT display. Given the 0 – 500 Hz frequency span and 400 Lines of Example #1, identifiable frequencies (Hz) in the FFT would be; 0, 1.25, 2.50, 3.75 - - - 496.00, 497.25, 498.5 and 500 Hz. The cursor in the FFT display could be successfully moved from Line to Line indicating the frequency and amplitude of each Line. As shown in Figures 2, 3 and 4, the Line is always in the center of the Bin width. A frequency of 35 Hz has been used to illustrate the Bin width, and reflects the calculations of Examples 1 and 2.

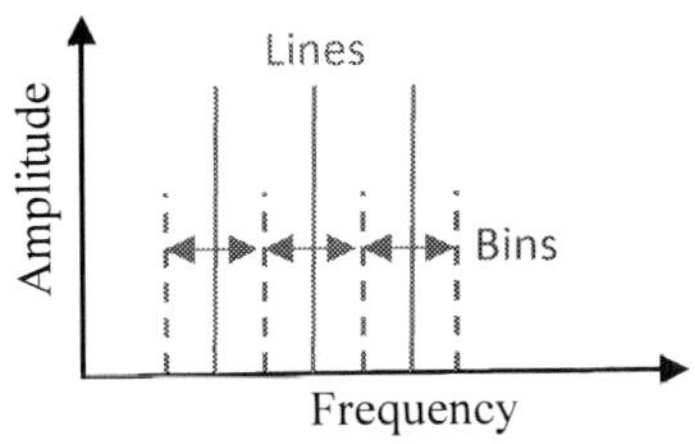

Figure 2 Bins & Lines

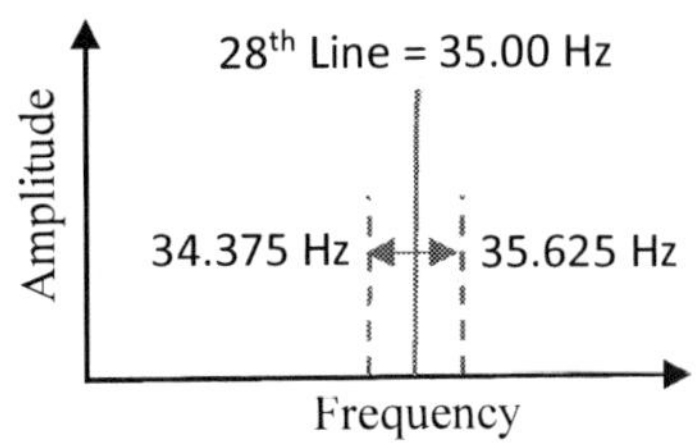

Figure 3 1.25 Hz/Line

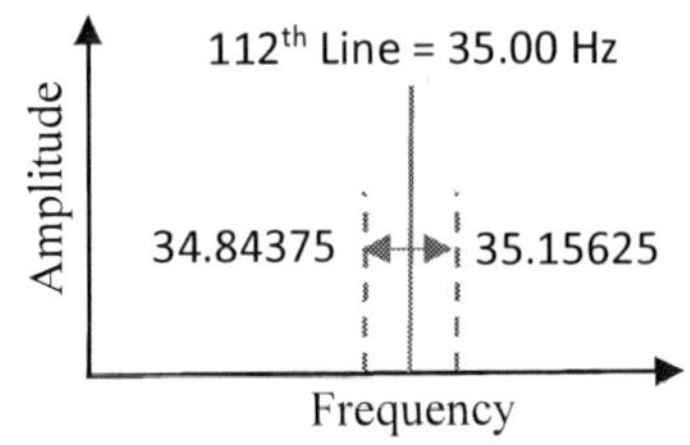

Figure 4 0.3125 Hz/Line

WINDOW FILTERS:

The FFT algorithm requires continuous data for calculation and processing. If the data in the time waveform is discontinuous, the frequencies displayed on the horizontal axis of the FFT will be poorly defined or as it is commonly described, they will have leakage. In vibration analysis, the FFT is typically a batch process using multiple samples of time waveform data, often referred to as averages or time blocks. Although each time block contains a sample of the time waveform data, the data may not be periodic and when the time blocks are averaged to calculate the FFT, the data is discontinuous. To correct the discontinuity between the individual time blocks a Window Filter can be applied. The Window Filter will modify the time waveform so that it starts and stops at a zero amplitude in each individual time block forcing the data to be periodic. In this manner, when the time blocks are averaged, they simulate a continuous time waveform and conform to the requirement of the FFT algorithm. There are multiple Window Filters available to the user. The most popular window filters for vibration analysis are Hanning and Flat Top. An illustration of the Window Filter effect on the time waveform is shown in Figures 5, 6 and 7.

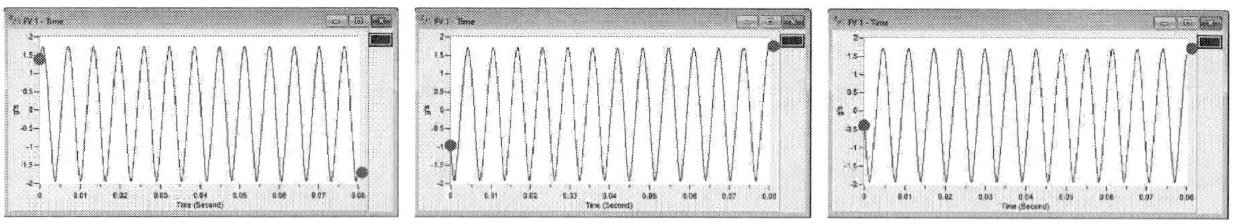

Figure 5 No Window Filter, Non-Periodic, Discontinuous Time Waveform Data

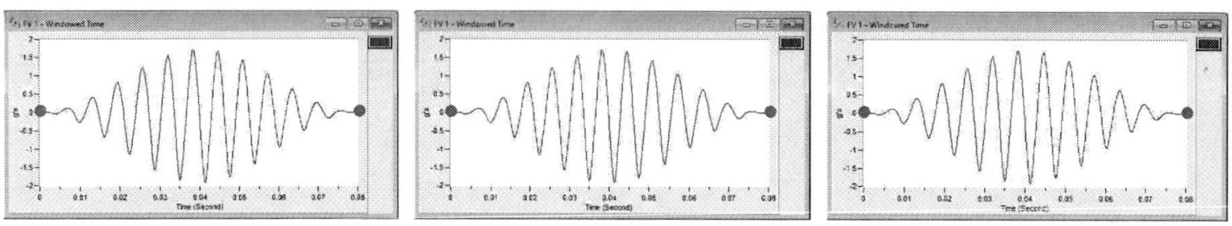

Figure 6 Hanning Window Filter, Periodic, Continuous Time Waveform Data

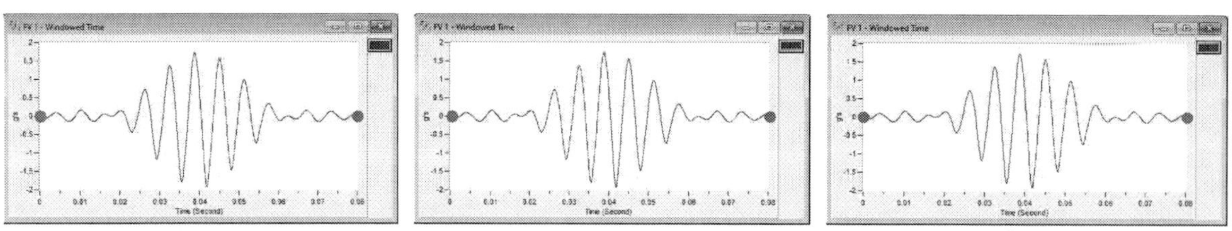

Figure 7 Flat Top Window Filter, Periodic, Continuous Time Waveform Data

WINDOW FILTERS & RESOLUTION:
Applying the Hanning or Flat Top Window Filter to the time waveform data will decrease the frequency resolution of the FFT. This means that closely spaced frequencies will not have enough resolution to separate them and they will appear as one frequency.

Window filters have an associated Window Factor (WF) with them. This Window Factor is a numerical value describing the Equivalent Noise Bandwidth [2] or Noise Power Bandwidth [3]. Window Factors (WF) or Equivalent Noise Bandwidths (ENBW) are shown in Table 1.

Window Filter	**WF or ENBW**
Uniform	1.00
Rectangular	1.00
Hanning	1.50
Flat Top	2.96
Flat Top	3.82

Table 1 Window Factors or Equivalent Noise Bandwidth

It should be noted in Table 1 that some Window Filters will have multiple values for the ENBW based on the calculated design of the Window Filter by the manufacturer of the data collector, vibration analyzer or dynamic signal analyzer. This author has two vibration analyzers. Both of them have an ENBW of 1.5 for the Hanning Window Filter, but for the Flat Top Window Filter, one of them has an ENBW of 2.96 and the other one has an ENBW of 3.82. The user should always refer to the technical specifications for their analyzer or ask the manufacturer to ascertain the ENBW of the Window Filter they are using.

In Examples 1 and 2, the Bin Width or Lowest Resolvable Frequency (LRF) was identified by dividing the Frequency Span by the Number of Lines. When a Window Filter is used, the Window Factor or ENBW must be combined with the LRF. This combination will identify the Bandwidth [4] of the FFT, as shown in Examples 3 and 4.

Example 3: Frequency span is 0 – 500 Hz, Number of lines are 400, Hanning Window

$$Bandwidth = \left(\frac{Frequency\ Span}{Number\ of\ Lines}\right) ENBW = \left(\frac{500\ Hz}{400\ Lines}\right) 1.5 = \frac{1.875\ Hz}{Line}$$

Example 4: Frequency span is 0 – 500 Hz, Number of lines are 1600, Hanning Window

$$Bandwidth = \left(\frac{Frequency\ Span}{Number\ of\ Lines}\right) ENBW = \left(\frac{500\ Hz}{1600\ Lines}\right) 1.5 = \frac{0.46875\ Hz}{Line}$$

At the Vibration Institute Annual Training Conference in 1997, Jack Frarey demonstrated that the Bandwidth was not sufficient to separate two closely spaced frequencies [5]. Mr. Frarey's work indicated that the true resolution of the FFT was calculated by multiplying the Bandwidth by a factor of two. The calculated Resolution of the FFT is shown in Examples 5 and 6.

Example 5: Frequency span is 0 – 500 Hz, Number of lines are 400, Hanning Window

$$Resolution = 2\left[\left(\frac{Frequency\ Span}{Number\ of\ Lines}\right) ENBW\right] = 2\left[\left(\frac{500\ Hz}{400\ Lines}\right) 1.5\right] = \frac{3.75\ Hz}{Line}$$

Example 6: Frequency span is 0 – 500 Hz, Number of lines are 1600, Hanning Window

$$Resolution = 2\left[\left(\frac{Frequency\ Span}{Number\ of\ Lines}\right) ENBW\right] = 2\left[\left(\frac{500\ Hz}{1600\ Lines}\right) 1.5\right] = \frac{0.9375\ Hz}{Line}$$

To better understand Mr. Frarey's findings, a portion of his experimental work has been repeated and presented in Figures 8, 9, and 10. A Frequency Span 0 – 400 Hz, 400 Lines, and a Hanning Window Filter have been used. This combination of frequency, lines and window filter provide a calculated bandwidth of 1.5 Hz. [Bandwidth = (400 Hz/400 Lines) x 1.5 = 1.5 Hz]

In Figures 8A & 8B, frequencies of 120 Hz and 121 Hz are presented. The delta difference between these frequencies is 1 Hz, and with a Bandwidth of 1.5 Hz, separation of frequencies is not expected in the FFT.

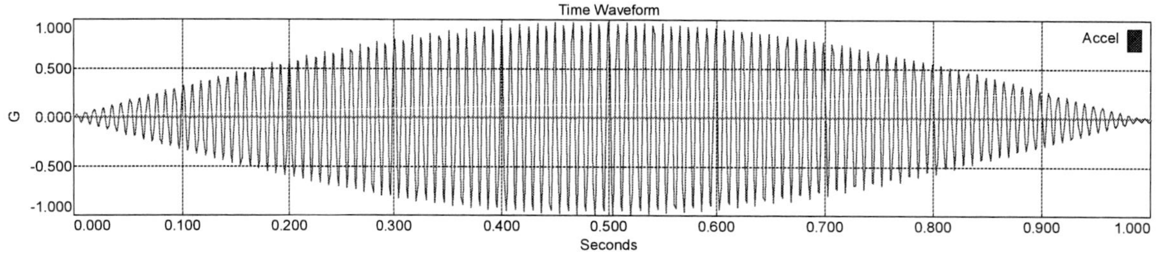

Figure 8A Sine Waveform 120 Hz + Sine Waveform 121 Hz = ($\Delta\ 1Hz$)

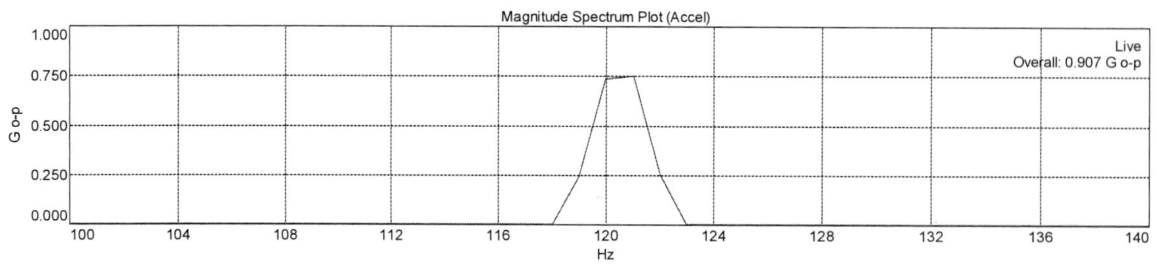

Figure 8B BW = (400 Hz/400 Lines)1.5 = 1.5 Hz = No Separation of Frequencies

In Figures 9A & 9B, frequencies of 120 Hz and 122 Hz are presented. The delta difference between these frequencies is 2 Hz, and with a Bandwidth of 1.5 Hz, separation of frequencies is expected in the FFT. As illustrated by the FFT, the two frequencies were not separated by the Bandwidth of 1.5 Hz, although it is less than the delta difference of 2 Hz between the frequencies.

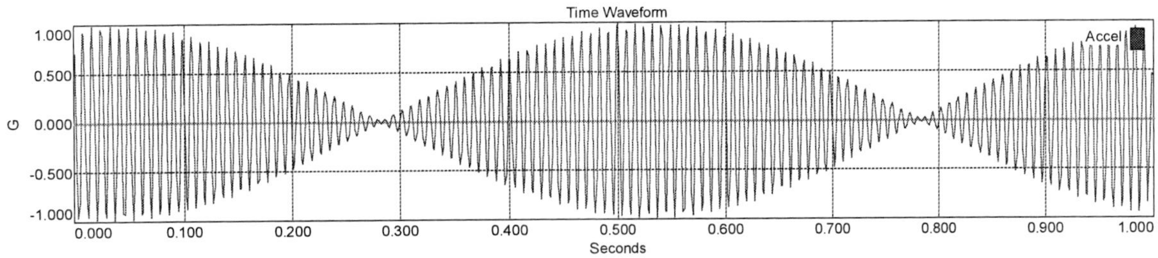

Figure 9A Sine Waveform 120 Hz + Sine Waveform 122 Hz = $(\Delta\ 2Hz)$

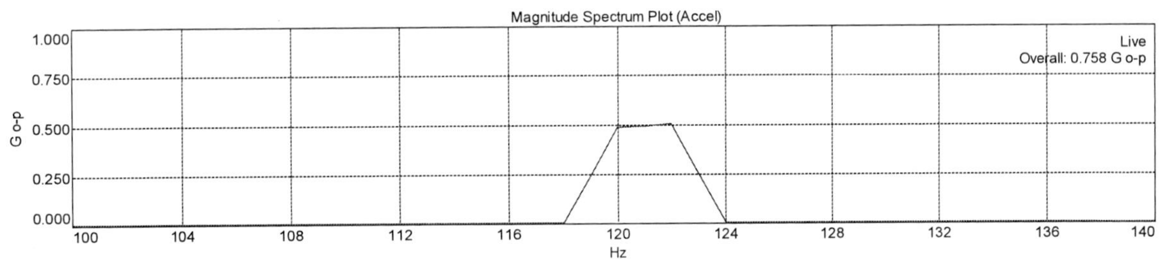

Figure 9B BW = (400 Hz/400 Lines)1.5 = 1.5 Hz = No Separation of Frequencies

In Figures 10A &10B, frequencies of 120 Hz and 123 Hz are presented. The delta difference between these frequencies is 3 Hz, and with a Bandwidth of 1.5 Hz, separation of frequencies is expected in the FFT. As illustrated by the FFT, the two frequencies were separated by the Bandwidth of 1.5 Hz. The delta difference between the frequencies is exactly twice the Bandwidth. The true resolution of the FFT requires the Bandwidth to be multiplied by a factor of two.

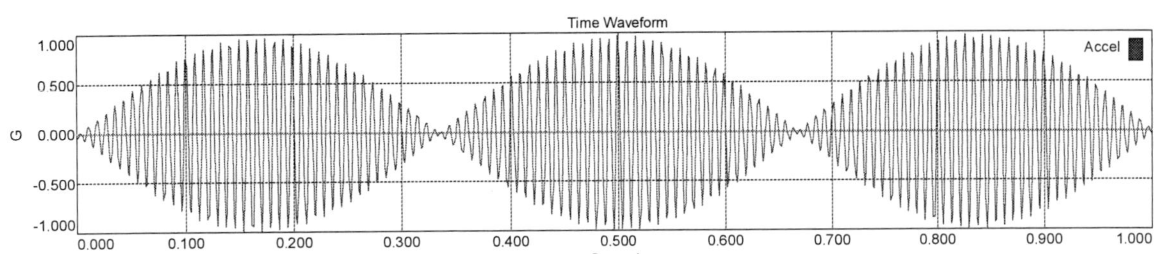

Figure 10A Sine Waveform 120 Hz + Sine Waveform 123 Hz = $(\Delta\ 3Hz)$

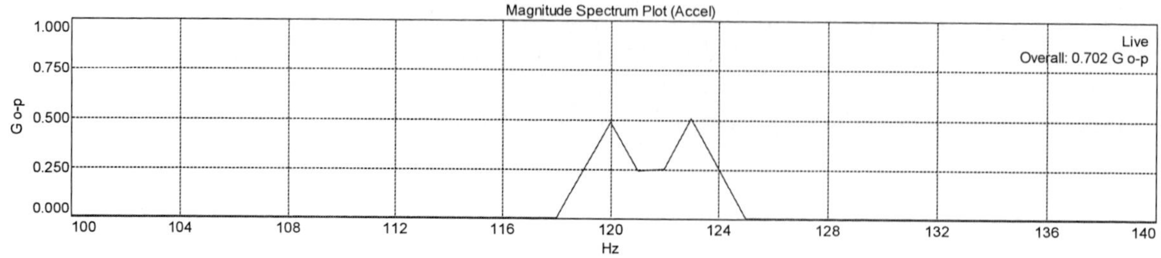

Figure 10B BW = (400 Hz/400 Lines)1.5 = 1.5 Hz = Separation of Frequencies

Resolution = 2(400 Hz/400 Lines)1.5 = 3.0 Hz

WINDOW FILTERS & AMPLITUDE:
Applying the Hanning or Flat Top window filter to the time waveform data can affect the amplitude of the frequency components in the FFT. Figures 11 and 12 illustrates these two Window Filters as they appear in the FFT [6].

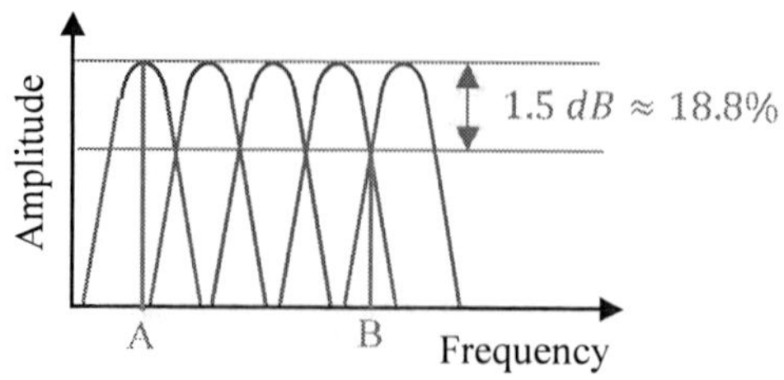

Figure 11 Hanning Window Filter

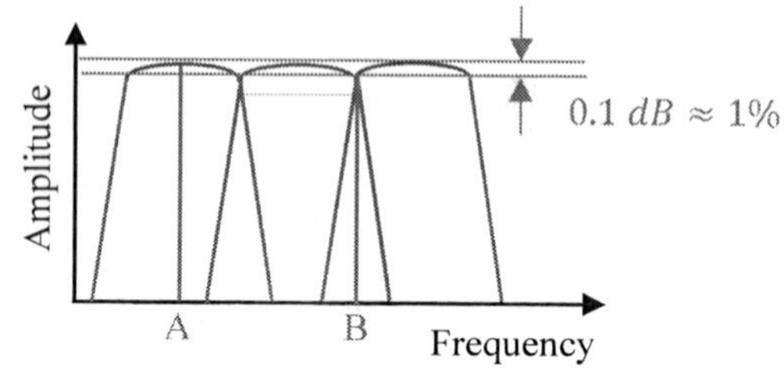

Figure 12 Flat Top Window Filter

The narrow Bandwidth of the Hanning Window Filter in Figure 11 provides good frequency resolution for general purpose vibration measurements but has a high amplitude uncertainty. Frequency A in Figure 11 is centered in the filter band and will have an accurate amplitude value. Frequency B in Figure 11 is at the intersection of two filter bands and will have an amplitude value that is attenuated by approximately 18.8%. If the frequency is not centered in the Hanning Window filter band, the amplitude error will be approximately 0 – 18.8%.

The wider Bandwidth of the Flat Top Window Filter in Figure 12 provides poor frequency resolution for general purpose vibration measurements, but has a low amplitude uncertainty. Frequency A in Figure 12 is centered in the filter band and will have an accurate amplitude value. Frequency B in Figure 12 is at the intersection of two filter bands and will have an amplitude value that is only attenuated by approximately 1%. The Flat Top Window Filter provides much better amplitude accuracy, approximately 0 – 1%, but the poor frequency resolution limits the use of this filter in vibration analysis.

WINDOW FILTERS AND RMS ENERGY:

Applying the Hanning or Flat Top Window Filter to the time waveform data will affect the RMS Energy of the FFT. In Figure 13, the fundamental frequency is 72.5 Hz and there are ten additional orders with visible amplitudes (g's RMS). The vibration analyzer has calculated the overall RMS Energy at 0.0602 g's RMS.

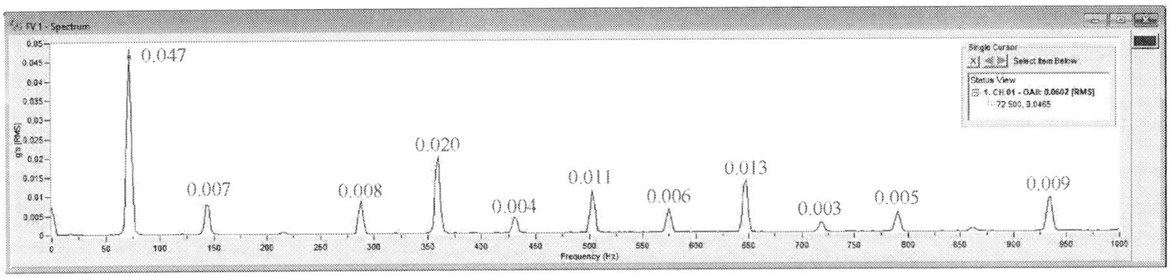

Figure 13 FFT with Multiple Orders of the Fundamental Frequency

An approximation of the overall RMS Energy could be made by using the individual amplitude of the fundamental frequency and amplitude of each order. This calculation could be made by; square root sum of squares, as shown in Example 7:

Example 7: Square Root Sum of Squares

$$RMS\ Energy = \sqrt{.047^2 + .007^2 + .008^2 + .020^2 + .004^2 + .011^2 + .006^2 + .013^2 + .003^2 + .005^2 + .009^2}$$

$RMS\ Energy$ =0.0564 g's RMS

The approximation in Example 7 has not used all of the energy that is contained in the FFT from 0 – 1000 Hz. It has only calculated an approximation of eleven amplitudes. To correctly calculate the overall RMS Energy of the FFT the equation in Example 8 should be used.

Example 8:

$$RMS\ Energy = \sqrt{\frac{A_1^2 + A_2^2 + A_3^2 + \cdots A_N^2}{ENBW}} \quad or \quad RMS\ Energy = \sqrt{\frac{A_1^2 + A_2^2 + A_3^2 + \cdots A_N^2}{WF}}$$

Where:

$A = The\ Amplitude\ of\ each\ Line\ in\ the\ FFT\ (g's\ RMS\ or\ {}^{in}/_{s}\ RMS\ or\ {}^{mm}/_{s}\ RMS)$

$N = The\ last\ Line\ in\ the\ FFT$

$ENBW = Equivalent\ Noise\ Bandwidth$

$WF = Window\ Factor$

The FFT shown in Figure 10 utilized a Hanning Window Filter and 400 Lines. The ENBW or WF will be 1.5. To calculate the equation in Example 8, it would be necessary to export the frequency and amplitude data from the FFT in Figure 10 and import it to Excel or some other type of software that can process a large volume of data. Calculations for 400 Lines with amplitudes from 0 – 1000 Hz will be required. Table 2 is an example on how this could be done in Excel and contains 400 amplitudes from 0 – 1000 Hz. Each amplitude (g's RMS) is in column 2 and has been squared in column 3 (g's^2 RMS). The sum of squares has been calculated as 0.00543711 g's^2 RMS. This sum will be divided by the ENBW or WF, and since we are using a Hanning Window Filter, the divisor will be 1.5 providing a result of 0.00362474 g's^2 RMS. The square root of this value is 0.06020583 g's RMS and represents the overall energy of the FFT as calculated by the vibration analyzer.

(To reduce the size of Table 2, values between 10 Hz and 990 Hz are not shown, but they have been included in the calculations.)

If the square root sum of squares had been used to calculate the overall RMS Energy from 0 – 1000 Hz in Figure 13, the value of the RMS Energy would have been too high as shown in Example 9.

Example 9:

$$Square\ Root\ Sum\ of\ Squares = \sqrt{0.00543711}$$

$$Square\ Root\ Sum\ of\ Squares = 0.07373676\ g's\ RMS$$

The value calculated by using the square root sum of squares is approximately 22% too high when compared to the actual value that includes the influence of the ENBW or WF on the RMS Energy. In order to gain a better understanding of how the ENBW or WF effects the RMS energy of the FFT, the analysis will be limited to two single harmonic frequencies, 35 Hz and 122 Hz, with no additional noise or influence from other sources.

Hz	g's RMS	g's^2 RMS
0.0	0.00724940	0.00005255
2.5	0.00362080	0.00001311
5.0	0.00012540	0.00000002
7.5	0.00024780	0.00000006
10.0	0.00021800	0.00000005
-	-	-
-	-	-
-	-	-
990.0	0.00007931	0.00000001
992.5	0.00011210	0.00000001
995.0	0.00016160	0.00000003
997.5	0.00019500	0.00000004
1000.0	0.00030960	0.00000010
Sum of Squares 0 – 1000 Hz		0.00543711
Divide by 1.5		0.00362474
Calculate Square Root		0.06020583

Table 2 RMS Energy Calculation

UNDERSTANDING RMS ENERGY WITH A HANNING WINDOW FILTER:

If an FFT with 400 Lines, Hanning Window and 500 Hz frequency span is used, the LRF or Bin width will be 500 Hz/400 Lines = 1.25 Hz/Line. The 35 Hz frequency will be Bin centered because 35 Hz/(1.25Hz/Line) = 28 Lines. This indicates that the amplitude of 35 Hz frequency should be measured correctly by the 28th Line. The 122 Hz frequency will not be Bin centered because 122 Hz/(1.25Hz/Line) = 97.6 Lines. This indicates that the amplitude of the 122 Hz frequency will be measured by the 97th and 98th Lines of the FFT. This means that there will be shared amplitudes at

121.25 Hz and 122.50 Hz since the frequency is not Bin centered. The FFT in Figure 14 represents the 35 Hz and 122 Hz frequencies with a Hanning Window Filter.

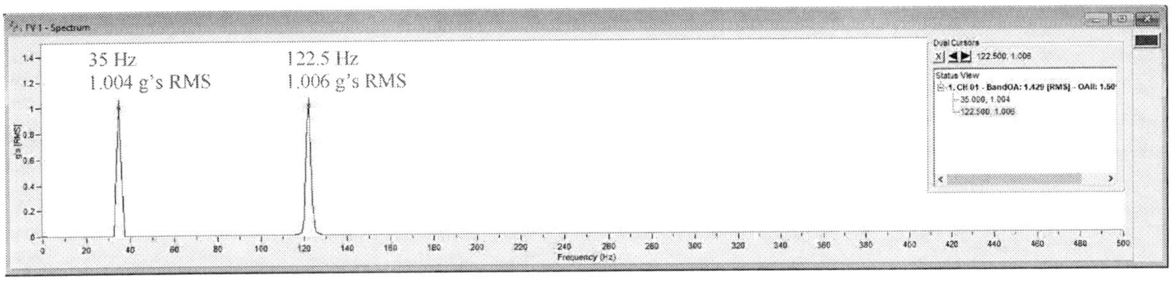

Figure 14 35 Hz and 122 Hz Representation in the FFT (Hanning Window)

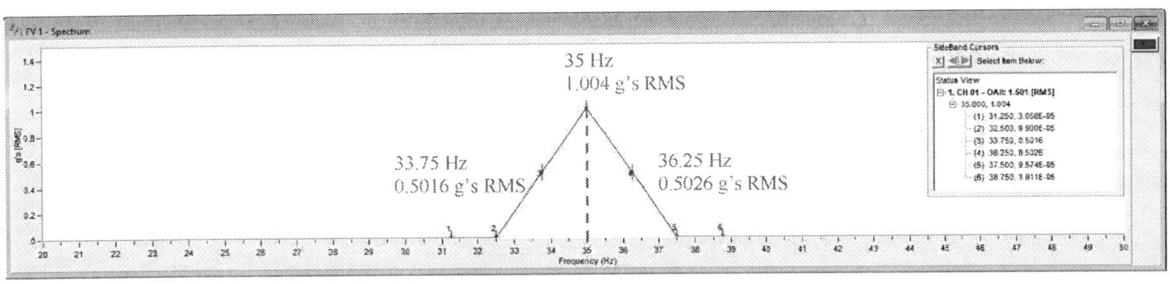

Figure 15 35 Hz Expanded View (Hanning Window)

In Figure 15, the FFT has been expanded to focus on the Bin centered 35 Hz frequency. The modulation like effect of the Window Filter has created what appear to be sideband amplitudes +/- one Bin of center. The ENBW or WF has created amplitudes above and below the 35 Hz frequency that must be accounted for. The RMS Energy for the 35 Hz frequency, using a Hanning Window Filter with an ENBW of 1.5 will be:

$$RMS\ Energy = \sqrt{\frac{0.5016^2 + 1.004^2 + 0.5026^2}{1.5}} = 1.004\ g's\ RMS$$

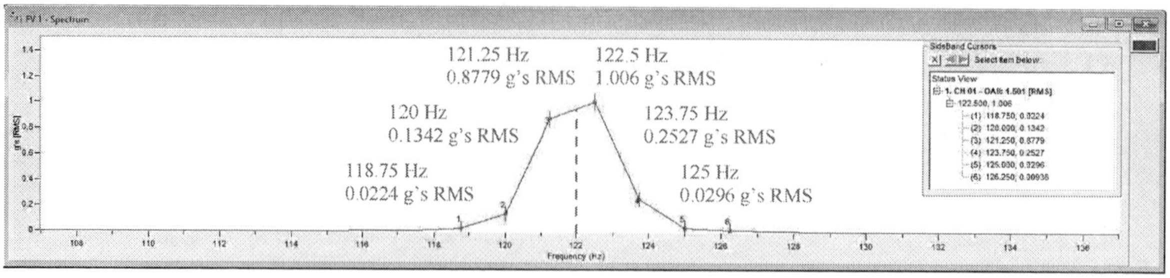

Figure 16 122 Hz Expanded View (Hanning Window)

In Figure 16, the FFT has been expanded to focus on the 122 Hz frequency that is not Bin centered. The ENBW or WF has created amplitudes above and below the 122 Hz frequency that must be accounted for. The RMS Energy for the 122 Hz frequency, using a Hanning Window Filter with an ENBW of 1.5 will be:

$$RMS\ Energy = \sqrt{\frac{0.0224^2 + 0.1342^2 + 0.8779^2 + 1.006^2 + 0.2527^2 + 0.0296^2}{1.5}} = 1.115\ g's\ RMS$$

The calculations for Figures 15 and 16 provide two very important pieces of information about calculating the RMS Energy in the FFT using a Hanning Window Filter.

1. If a single harmonic frequency is Bin centered (Figure 15) using a Hanning Window Filter, the RMS Energy will be equal to the RMS amplitude of that Bin centered frequency.

2. If a single harmonic frequency is not Bin centered (Figure 16) using a Hanning Window Filter, the RMS Energy must be calculated as shown because the frequency is distributed in more than one Bin.

UNDERSTANDING RMS ENERGY WITH A FLAT TOP WINDOW FILTER:
If an FFT with 400 Lines, Flat Top Window and 500 Hz frequency span is used, the LRF or Bin width will be 500 Hz/400 Lines = 1.25 Hz/Line. The 35 Hz frequency will be Bin centered because 35 Hz/(1.25Hz/Line) = 28 Lines. This indicates that the amplitude of 35 Hz frequency should be measured correctly by the 28th Line. The 122 Hz frequency will not be Bin centered because 122 Hz/(1.25Hz/Line) = 97.6 Lines. This indicates that the amplitude of the 122 Hz frequency will be measured by the 97th and 98th Lines of the FFT. This means that there will be shared amplitudes at 121.25 Hz and 122.50 Hz because the frequency is not Bin centered. The FFT in Figure 17 represents the 35 Hz and 122 Hz frequencies with a Flat Top Window Filter.

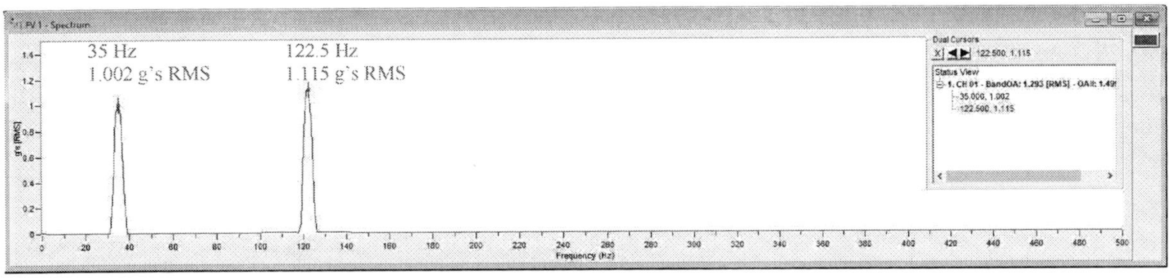

Figure 17 35 Hz and 122 Hz Representation in the FFT (Flat Top Window)

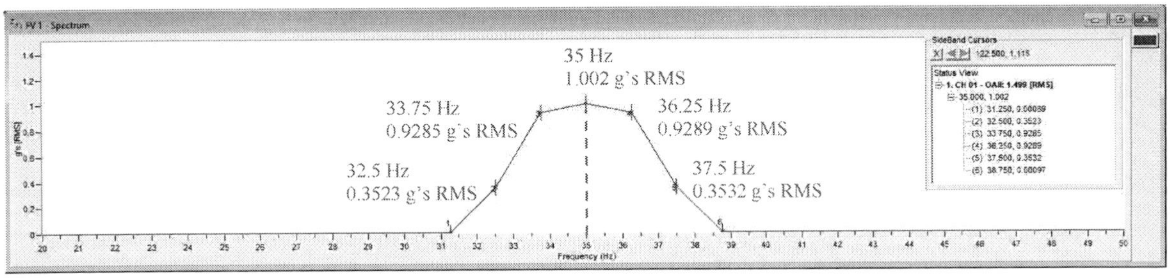

Figure 18 35 Hz Expanded View (Flat Top Window)

In Figure 18, the FFT has been expanded to focus on the Bin centered 35 Hz frequency. The modulation like effect of the Window Filter has created what appear to be sideband amplitudes +/- of the center Bin. The ENBW or WF has created amplitudes above and below the 35 Hz frequency that must be accounted for. The RMS Energy for the 35 Hz frequency, using a Flat Top Window Filter with an ENBW of 2.96 will be:

$$RMS\ Energy = \sqrt{\frac{0.3523^2 + 0.9285^2 + 1.002^2 + 0.9289^2 + 0.3532^2}{2.96}} = 1.002\ g's\ RMS$$

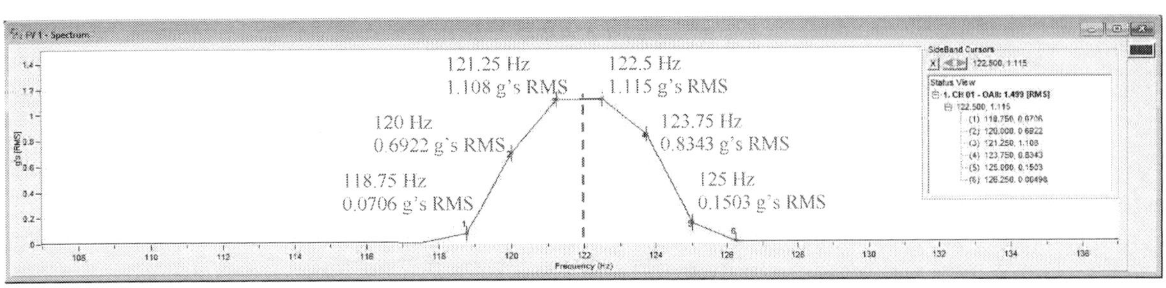

Figure 19 122 Hz Expanded View (Flat Top Window)

In Figure 19, the FFT has been expanded to focus on the 122 Hz frequency that is not Bin centered. The ENBW or WF has created amplitudes above and below the 122 Hz frequency that must be accounted for. The RMS Energy for the 122 Hz frequency, using a Flat Top Window Filter with an ENBW of 2.96 will be:

$$RMS\ Energy = \sqrt{\frac{0.0706^2 + 0.6922^2 + 1.108^2 + 1.115^2 + 0.8343^2 + 0.1503^2}{2.96}} = 1.114\ g's\ RMS$$

The calculations for Figures 18 and 19 provide two very important pieces of information about calculating the RMS Energy in the FFT using a Flat Top Window Filter.

1. If a single harmonic frequency is Bin centered (Figure 18) using a Flat Top Window Filter, the RMS Energy will be equal to the RMS amplitude of that Bin centered frequency.

2. If a single harmonic frequency is not Bin centered (Figure 19) using a Flat Top Window Filter, the RMS Energy must be calculated as shown because the frequency is distributed in more than one Bin, but since the Flat Top Window only has a 1% maximum amplitude error as previously discussed, the RMS Energy and amplitudes at 121.25 Hz and 122.5 Hz will be almost the same.

SUMMARY:
Window Filters are necessary for vibration analysis because they modify the time waveform data in each time block to simulate continuous data during the batch processing of the FFT. Modifying the original time waveform data creates cause and effect in the FFT that must be understood for measurement and analysis of the machine vibration.

The Window Factor or Equivalent Noise Bandwidth have been used interchangeably in this paper. The Window factor is known terminology that has been used with the calculation of frequency resolution for many years. The frequency resolution of the FFT can be calculated as shown:

$$Frequency\ Resolution = 2\left[\left(\frac{Frequency\ Span}{Number\ of\ Lines}\right) Window\ Factor\right]$$

The Window Factor is also known to affect the amplitude uncertainty of the individual frequencies being measured. If the frequency is not centered in the Window Filter there will be some amount of amplitude error. In a Hanning Window Filter the amplitude uncertainty can be 0 – 18.8%, but in a Flat Top Window Filter the amplitude uncertainty is only 0 – 1%. In general purpose vibration measurements, the Hanning Window Filter is preferred for better frequency resolution, but if the amplitude is critical to the measurement, then the Flat Top Window Filter can be used.

The Equivalent Noise Bandwidth is terminology better used to describe the calculation of the RMS Energy in the FFT. The amplitudes that are present in adjacent Bins of the FFT contribute to the noise bandwidth and must be accounted for when calculating the RMS Energy in the FFT as shown:

If FFT amplitudes (A) are in RMS units (i.e., g's RMS, in/s RMS or mm/s RMS) then:

$$RMS\ Energy = \sqrt{\frac{A_1^2 + A_2^2 + A_3^2 + \cdots A_N^2}{Equivalent\ Noise\ Bandwidth}}$$

If FFT amplitudes (A) are in Peak units (i.e., g's Peak, in/s Peak or mm/s Peak) then:

$$RMS\ Energy = 0.707\sqrt{\frac{A_1^2 + A_2^2 + A_3^2 + \cdots A_N^2}{Equivalent\ Noise\ Bandwidth}}$$

TERMINOLOGY USED IN "*CALCULATING THE RMS ENERGY OF THE FFT*":

Name	Abbreviation	Description
Window Factor	WF	A numerical value that describes the Equivalent Noise Bandwidth.
Equivalent Noise Bandwidth	ENBW	A numerical value that describes the Window Factor.
Lowest Resolvable Frequency	LRF	$\frac{Frequency\ Span}{Number\ of\ Lines}$
Bin Width	-	Lowest Resolvable Frequency.
Bandwidth	BW	$\left(\frac{Frequency\ Span}{Number\ of\ Lines}\right) Window\ Factor$
Resolution	Res	$2\left[\left(\frac{Frequency\ Span}{Number\ of\ Lines}\right) Window\ Factor\right]$
Line	N	A single frequency measure of the Bin Width.
Amplitude	A	A general term used to describe the amplitude value of a frequency or line measured in g's, in/s or mm/s.
Time Waveform	TW	A display of amplitude vs. time that represents the measured vibration of the machine.
Fast Fourier Transform	FFT	An algorithm that displays amplitude vs. frequency based on a calculation that was applied to the time waveform.
Window Filter	-	A special filter applied to the time waveform to make it periodic in the time block to simulate continuous data when processing the FFT.
RMS Energy (RMS units)	-	$\sqrt{\frac{A_1^2 + A_2^2 + A_3^2 + \cdots A_N^2}{Equivalent\ Noise\ Bandwidth}}$
RMS Energy (Peak units)	-	$0.707\sqrt{\frac{A_1^2 + A_2^2 + A_3^2 + \cdots A_N^2}{Equivalent\ Noise\ Bandwidth}}$

ACKNOWLEDGEMENTS:

1. Nelson L. Baxter, ABN Technical Services, Inc.
2. Dr. Ronald L. Eshleman, Vibration Institute
3. Michael Grace, Siemens Energy
4. Robert Mihata, Alta Solutions

REFERENCES:

1. Eshleman, Ronald L. and Peters, Jack D., *Category III Machinery Vibration Analysis*, Vibration Institute 2019.
2. Harris, Frederic J., *On the Use of Windows for Harmonic Analysis with the Discrete Fourier Transform,* Proceedings of the IEEE, Volume 66, Number 1, January 1978.
3. Cerna, Michael and Harvey, Audrey F., *The Fundamentals of FFT-Based Signal Analysis and Measurement,* Application Note 041, National Instruments Corporation, July 2000.
4. *Dynamic Signal Analyzer Applications,* Application Note 243-1, Hewlett-Packard Company, October 1983.
5. Frarey, Jack, *An Examination of Signal Resolution in an FFT Analyzer,* 21st Annual Meeting, Vibration Institute, July 17, 1997.
6. *The Fundamentals of Signal Analysis,* Application Note 243, Hewlett-Packard Company, 1985.

MEASURING ACCELERANCE AND DERIVING ALL OTHER FREQUENCY RESPONSE FUNCTIONS

Author: Eduardo Murphy Arteaga

VIATC 2022 VIBRATION INSTITUTE ANNUAL TRAINING CONFERENCE & EXPO | MFPT ANNUAL CONFERENCE

2022 CONFERENCE PROCEEDINGS

VIATC TRAINING CONFERENCE & EXPO
SAVANNAH, GEORGIA · 2022

MEASURING ACCELERANCE AND DERIVING ALL OTHER FREQUENCY RESPONSE FUNCTIONS

Eduardo Murphy Arteaga
Fundación M&P
Emiliano Zapata, Morelos, México
emurphya@aol.com

Eduardo Murphy Arteaga mechanical-electrical engineer began static balancing of fans in 1978. In 1980 he had the opportunity to work at the Institute for Electrical Research as a manager for the incipient Vibrations Laboratory directed by Dr J. Aguirre Romano. He attended his first Vibrations Institute Annual Meeting at Cherry Hill, New Jersey (1980), and knew Dr, Eshleman, Dr. Donald Bently, Charles Jackson, John Mitchell, and several other people doing important research and practical problem solutions in the vibration field. Since 1987, he, supported by his wife Teresa and other family members, works on vibration analysis, balancing and alignment. He has developed computer programs for frequency analysis and multichannel polar plot generation. He is currently a Category IV analyst.

ABSTRACT:
Nowadays, the most common sensor used to measure vibrations is an accelerometer, hence, the Frequency Response Function (FRF) "Accelerance" is naturally first calculated. However, if deeper analysis is required, perhaps on a foundation or a structure of a mechanical system, knowing the "Dynamic Stiffness" will give added insight to system properties.

In this paper, we transform Accelerance to other FRF's by using the appropriate magnitude scaling and phase shifts by transforming complex numbers in rectangular coordinates to polar format and vice versa for the "Real" and "Imaginary" graphs.

Finally, Coherence is calculated using similar procedures.

KEY WORDS: Accelerance; apparent mass; coherence; dynamic stiffness; mechanical impedance; mobility; polar form; rectangular form; resilience.

INTRODUCTION:
One of the frequency response functions used to characterize mechanical systems at low frequencies (1 – 500 Hz) is "Mobility", (Velocity / Force), however, sensors employed for measurements will probably be accelerometers due to their wide use and relative low cost. It is therefore necessary to convert Accelerance (Acceleration / Force) to Mobility by integration, that is, dividing accelerance magnitudes by the corresponding angular frequencies (radians/s) and subtracting 90° ($\pi/2$ radians) from the argument of the Cosine and Sine functions. Trigonometric functions arguments are calculated from the inverse tangent quotient of the Imaginary and Real parts.

The above procedure can be repeated to obtain "Resilience" (Displacement / Force) The inverse function, "Dynamic stiffness" (Force / Displacement) can be obtained as well as the "Mechanical Impedance" (Force / Velocity) and "Apparent Mass" (Force / Acceleration)

MEASURING ACCELERANCE:
Several methods used to apply a force to a system or structure include electrodynamic or hydraulic shakers, counter-rotating masses driven by a motor powered by a variable frequency drive, piezoelectric wafers, and an impact hammer with a transducer that outputs certain millivolts per unit of force (Newtons, lbf) Using an impact hammer and a group of accelerometers attached to the structure in different locations, requires several impacts applied in one or several positions. At least 4 or 5 impacts are necessary to evaluate coherence for each test.

One procedure is to capture a large amount of data (+20 s) with a medium sampling frequency (6400 samples / second) With this setup, 4 to 7 impacts can be applied with no data overlap. Once this file is stored, it can be "cut" at each impact, leaving 10 ms before the impact and generating files with 2048, 4096 or 8192 data points. This procedure has the advantage to view the waveform and discard those sections with low impact pulses and/or double impacts.

After obtaining the complex spectra coefficients (Real and Imaginary) of the signals (acceleration and force) a rectangular to polar conversion is done simplifying further calculations. Accelerance magnitude is the quotient of the magnitudes of acceleration and force. Phase is acceleration phase minus force phase. These results can be graphed in log – log display (magnitude) and log – linear for the phase. Real and Imaginary displays are simply obtained by multiplying magnitude by the Cosine and Sine of the angle. This procedure is done for all frequencies and for all the accelerometers used and repeated the number of valid impacts for the test, 4 or 5. If a different impact location is used in another test, the same procedure applies.

For a mechanical system or structure behaving close to, or as a linear system, relative angles between acceleration and force at any given frequency will not differ by large amounts on successive impacts. If relative angles do change by large amounts, the response is not due to the impact and/or the system is non-linear. Coherence is the quotient of the vectorial resultant magnitude to the algebraic sum of the magnitudes. Coherence equal to 1 at a given frequency, means that the relative angle was the same for all impacts. Good coherence is normally taken above 0.8.

Figure 1 shows the average of four tests for the Imaginary part of Accelerance, calculated at the middle of a demonstration rotor with three masses. Figure 2 shows the average for the Real part. Because displacement response of a mechanical system or structure lags 90° the applied force ***when at resonance,*** acceleration leads by 90°, so resonant peaks are displayed on the Imaginary part. Peaks between zero crossings in the Real part are used to calculate the quality factor "Q" or amplification factor "AF" and the ratio "δ" of damping coefficient "c" to the critical damping coefficient "Cc"

From Figure 1 we see that there are four natural frequencies between 10 and 500 Hz: 14, 25, 133 and 300 Hz. Figure 3 shows the Coherence function for Accelerance of the central mass. Low coherence is observed at 155 Hz (Coherence < 0.8) and above 480 Hz

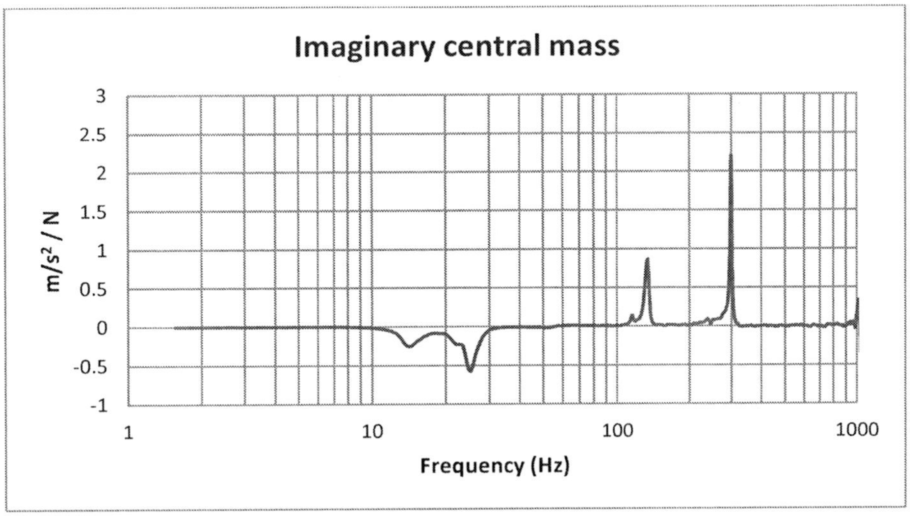

Figure 1

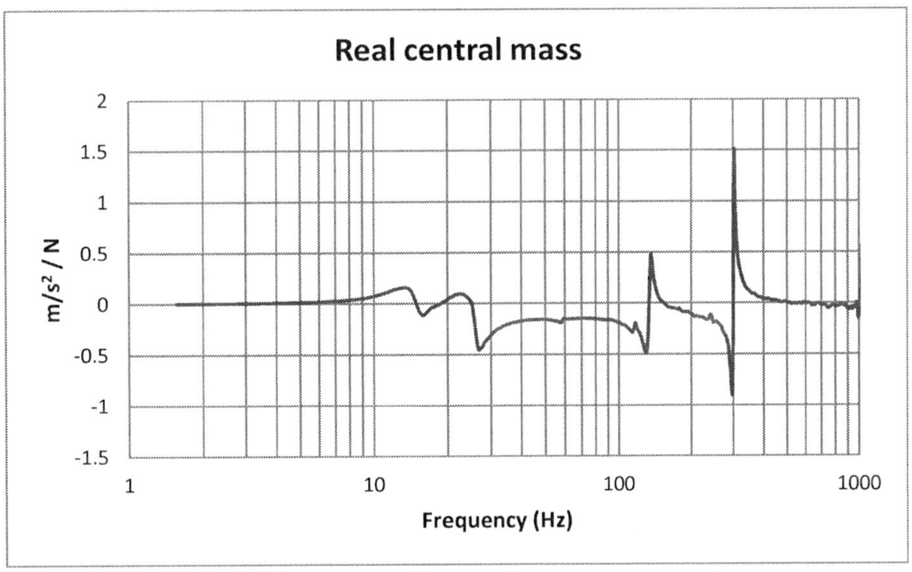

Figure 2

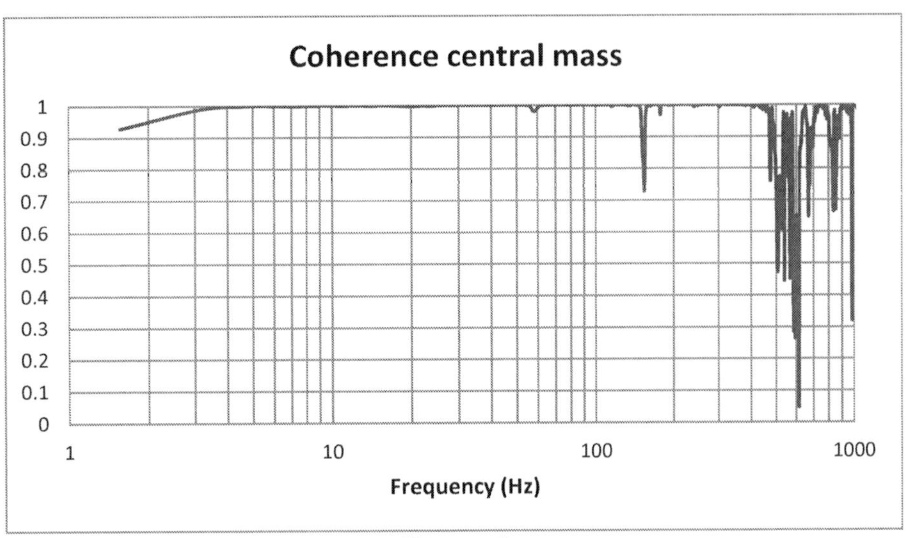

Figure 3

As an example of coherence calculation, Table I shows Accelerance data at 155 Hz. Four impacts were performed. Coherence is the Vector Resultant average divided by the algebraic addition average of magnitudes.

Table I
Coherence calculation at 155 Hz, four impacts

Impact	Magnitude (m/s^2 / N)	Phase, radians; (°)	Real (m/s^2 / N)	Imaginary (m/s^2 / N)	Vector resultant average (m/s^2 / N)	Coherence = Vector Resultant average / Algebraic magnitude average
1	0.0123788	1.26324 (72.3783°)	0.003747	0.0117979		
2	0.0092062	2.14365 (122.8222°)	-0.00499	0.0077365		
3	0.0118442	0.701231 (40.1776°)	0.009050	0.0076414		
4	0.0188413	-0.015779 (146.8728°)	-0.015778	0.0102968		
Σ / 4	0.0130676	(95.53°)	-0.001993	0.009368	0.0095778	0.73294

The above example shows operations to be performed for all frequencies in the bandwidth of interest.

QUALITY FACTOR "Q" OR AMPLIFICATION FACTOR "AF" CALCULATION:
Equation 1 (below) can be found in many texts for the determination of the quality factor "Q" by the half power method. "f_a" and "f_b" are the frequencies measured at the peaks above and below a resonant frequency in the Real part of accelerance or resilience, Imaginary part for mobility; δ is the damping coefficient divided by the critical damping coefficient ($\delta = c/Cc$)

$$Q = \frac{\left(\frac{fa}{fb}\right)^2 + 1}{\left(\frac{fa}{fb}\right)^2 - 1} = \frac{fa^2 + fb^2}{fa^2 - fb^2} \approx \frac{1}{2\delta} \qquad \text{Equation 1}$$

Equation 2 shows the exact relationship between Q and δ (c / Cc) for a system with a single degree of freedom and force proportional to ω^2

$$Q = \frac{1-2\delta^2}{2\delta\sqrt{1+\delta^2}} \qquad \text{Equation 2}$$

For the natural frequency at 300 Hz (Figure 1), Q is calculated from Figure 2 as follows (Equation 1):

$$Q = \frac{303.25^2 + 295.31^2}{303.25^2 - 295.31^2} = 37.7 \qquad \text{Equation 3}$$

And δ:

$$\delta \approx \frac{1}{2Q} = 0.013 \qquad \text{Equation 4}$$

We conclude that the natural frequency at 300 Hz has low damping (1.3% of Cc)

OBTAINING MOBILITY FROM ACCELERANCE:
Mobility is the FRF involving velocity divided by force. Acceleration has to be integrated by dividing its amplitudes by corresponding angular frequencies ω (2πf) and subtracting π/2 radians (90°) from the argument of the Cosine and Sine functions. Note that the arguments are decreased by a constant, so the relative angles between velocity and force will be the same as in Accelerance minus 90°. This means that Coherence will not change and will be valid for Mobility and Resilience. Equations 5 thru 8 show the above procedure:

$$|Mobility| = \left|\frac{500\ Accelerance}{\pi f}\right| mm/s$$ Equation 5

$$Mobility\ Phase = \theta = \left(Accelerance\ Phase - \frac{\pi}{2}\right) rad$$ Equation 6

Where[1]:

$$|Accelerance| = \sqrt{Real_{Accel}^2 + Imag_{Accel}^2}$$ Equation 7

$$Accelerance\ Phase = Tan^{-1}\left(\frac{Imag_{Accel}}{Real_{Accel}}\right)$$ Equation 8

And, from polar coordinates to rectangular form:

$$Real_{Mob} = |Mobility|\ Cos(\theta)$$ Equation 9

$$Imag_{Mob} = |Mobility|\ Sin(\theta)$$ Equation 10

Figure 4 shows the Real part of Mobility, figure 5 shows the Imaginary part.

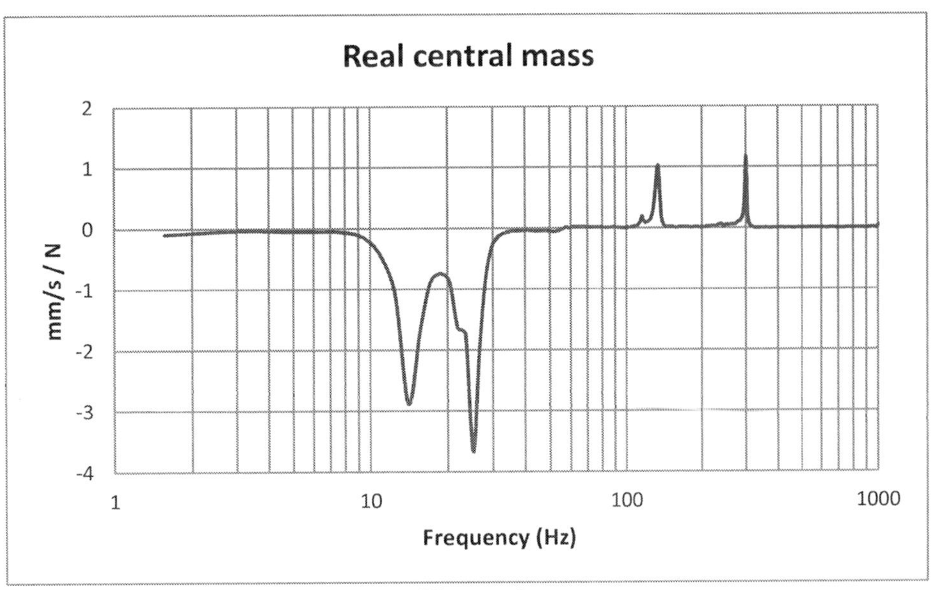

Figure 4

[1] MS Excel function "ATAN2(x, y)" gives the proper angle depending on the signs of "x" and "y"

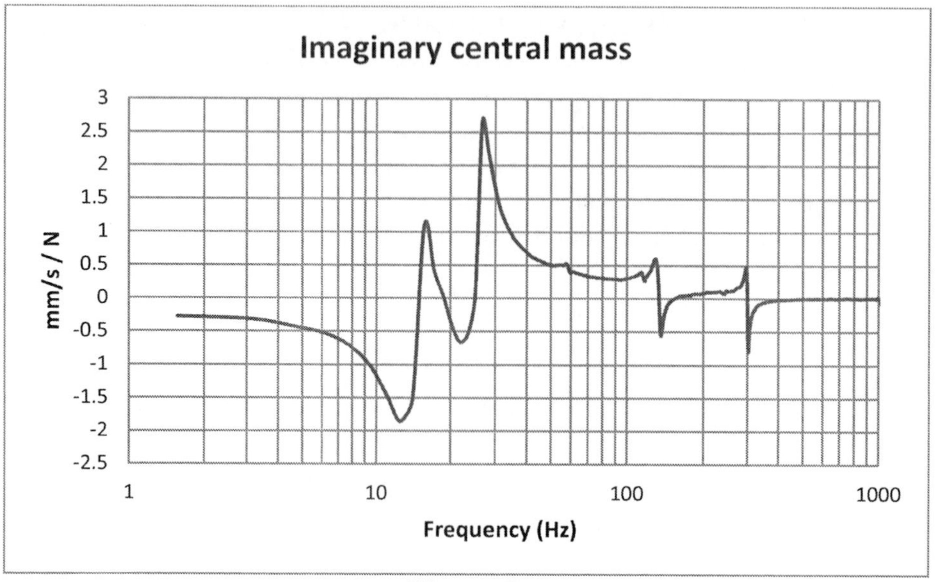

Figure 5

Because of the - 90° shift, the Imaginary part of Accelerance is now the Real part of Mobility (clockwise rotation) and the Real part of Accelerance is the minus Imaginary part of Mobility.

Resilience (Displacement / Force) is obtained by integrating Mobility:
Again, following similar procedures outlined in equations 5 to 10, the Real part of Mobility will become the minus Imaginary part of Resilience and the minus Imaginary part of Mobility will be the minus Real part of Resilience (clockwise rotations) Figures 6 and 7 show the Imaginary and the Real part of Resilience respectively.

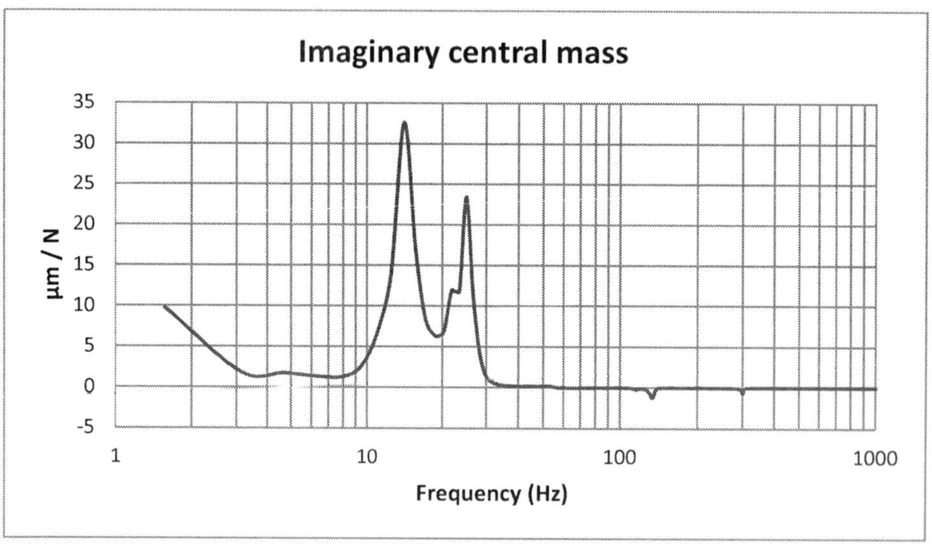

Figure 6

Note that for Resilience, low frequencies will be enhanced. For example, the peak at 14.1 Hz, 32.6 µm (figure 6) has an amplitude of 0.254 m/s^2 (See figure 1) One reason to use Mobility, is the capability to enhance medium frequencies as can be seen in figure 4.

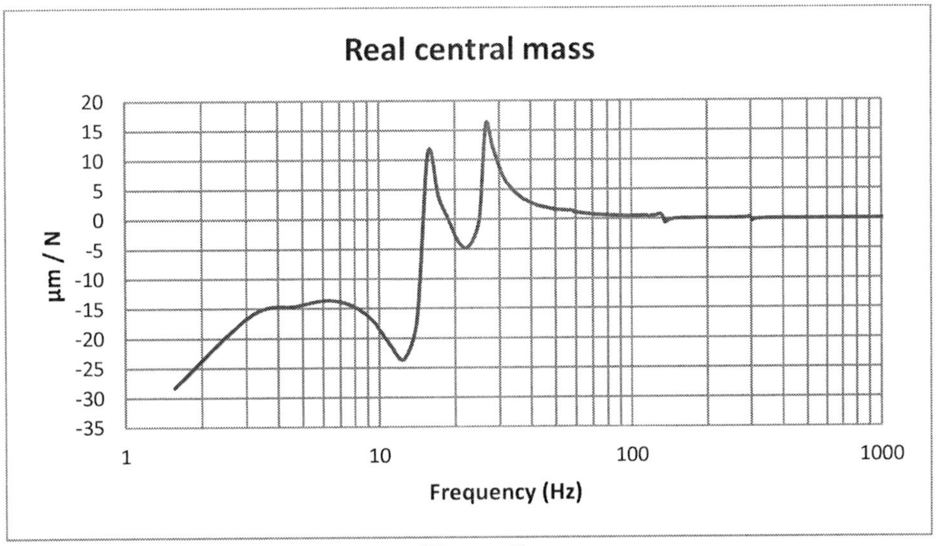

Figure 7

INVERSE FRF's:
Figure 8 shows the magnitude spectrum of one of the hammer impacts during the test. Amplitude is nearly constant (1 Newton) from 2 to 400 Hz. The Real and Imaginary parts of Accelerance, Mobility and Resilience have zero values at certain frequencies, however they will not occur simultaneously. Dividing Force by a zero value will result in infinite. To overcome this situation, Force divided by the ***magnitude*** of any FRF is used.

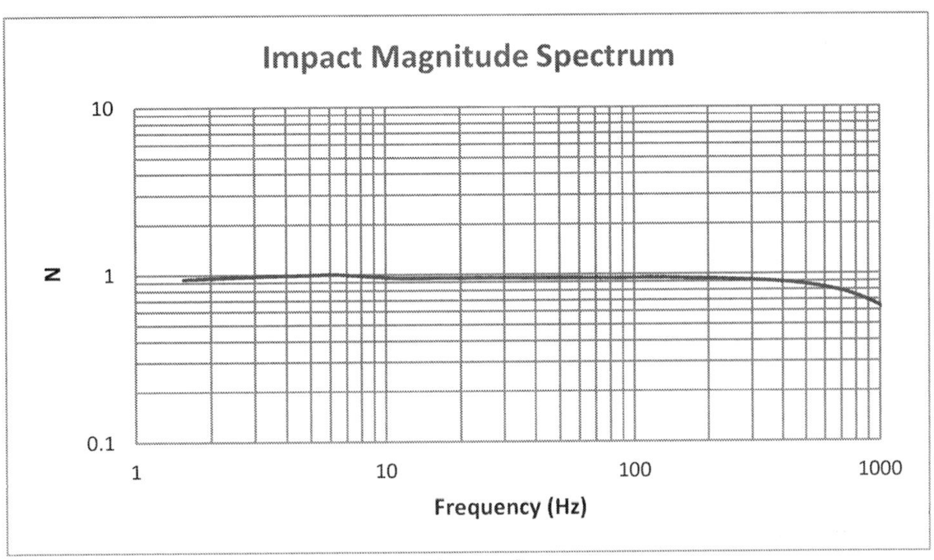

Figure 8

Because Accelerance is acceleration divided by force, it can be viewed as the inverse of the Mass involved in a particular mode of vibration. Rearranging terms, that is, Force divided by acceleration will result in "Apparent Mass". Figure 9 shows the inverse magnitude of Accelerance (1 / |Accelerance |) Apparent Mass at 300 Hz is 0.453 kg, its inverse will result in an accelerance of 2.21 / kg or 2.21 m/s^2 / N (See Figure 1)

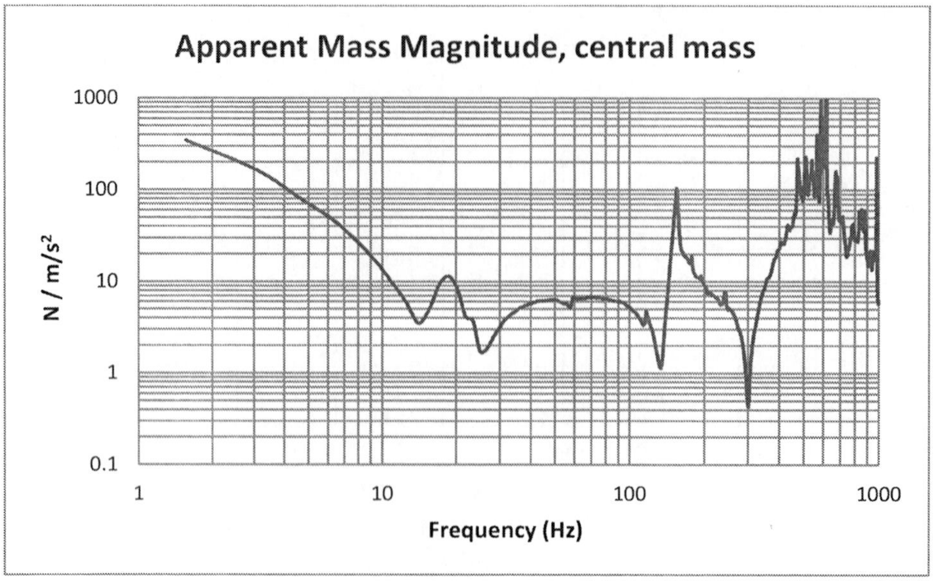

Figure 9

Back in the 60's or 70's, analogue computers were used to simulate mechanical systems. One simple circuit involved inductance, capacitance and resistance connected to a variable frequency voltage source. Inductance is analogue to mass, capacitance to the inverse of stiffness and resistance to the viscous coefficient of damping. Voltage is analogue to force and current to velocity. Mobility -velocity divided by force - (current by voltage) is the inverse of impedance, so Force divided by velocity is called "Mechanical Impedance"

Figure 10 shows the inverse of mobility magnitude, obtained from the Real and Imaginary parts, figures 4 and 5, respectively.

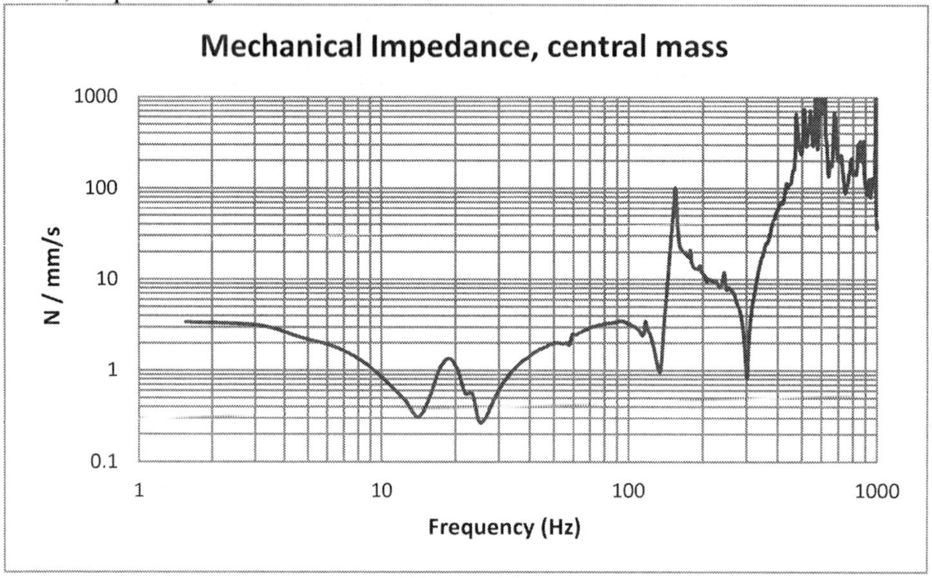

Figure 10

In an electrical circuit composed of an inductance, capacitance and resistance, resonance will exist when the inductive impedance equals minus the capacitive impedance, that is when only the resistance impedance remains. For the frequency at 300 Hz, figure 10 shows a mechanical impedance of 0.854 N / mm/s. Its inverse, mobility, is 1.171 millimeters / second per Newton as can be seen in figure 4.

Dynamic Stiffness can be obtained as the inverse of resilience. This inverse FRF is very useful when determining dynamic stiffness of foundations and mechanical systems. For example, a recommended value for a good concrete foundation is around 5000 MN / m (5000 N / µm) Mechanical structures have lower values, typically around 500 MN / m, so, to change a natural frequency (increase it) it is obvious to reinforce the mechanical structure. Figure 11 shows the Dynamic Stiffness obtained from Resilience, figures 6 and 7.

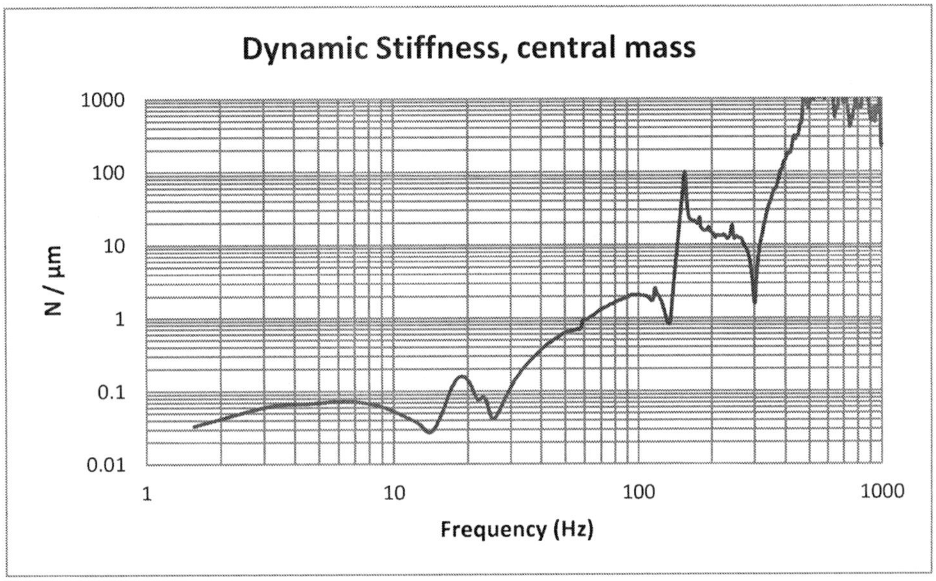

Figure 11

As can be seen, for this demonstration rotor, stiffness is low from 2 to 30 Hz. At 300 Hz, dynamic stiffness equals 1.564 N / µm. Resilience at 300 Hz is 0.64 µm / N (Figure 6)

NYQUIST PLOTS:

Figures 12 thru 14 show Nyquist plots for accelerance, mobility and resilience for the central mass. Plotting convention employed is:

- phase lag increases clockwise,
- frequency increases counterclockwise.

As can be seen, going from accelerance to resilience involves a 180° shift. Mobility involves a 90° shift. Both movements are clockwise. When plotting several points, it is possible to deduce the mode shapes of the structure at the natural frequencies involved.

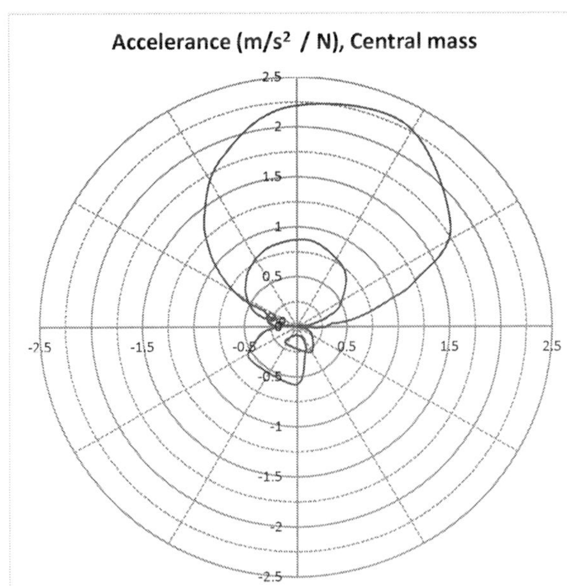

Figure 12

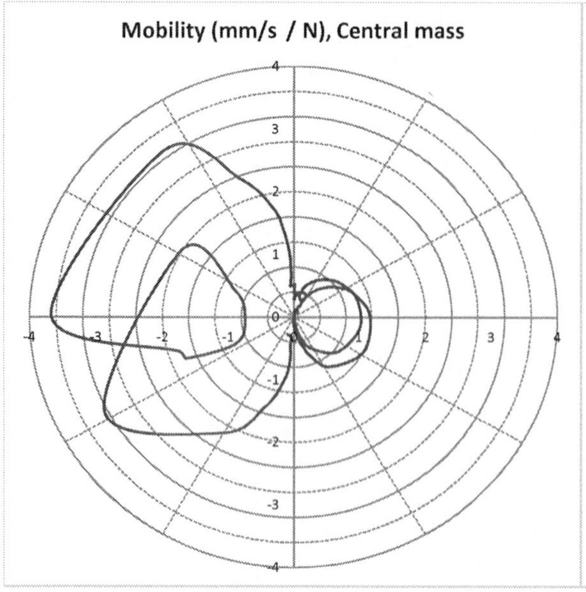

Figure 13

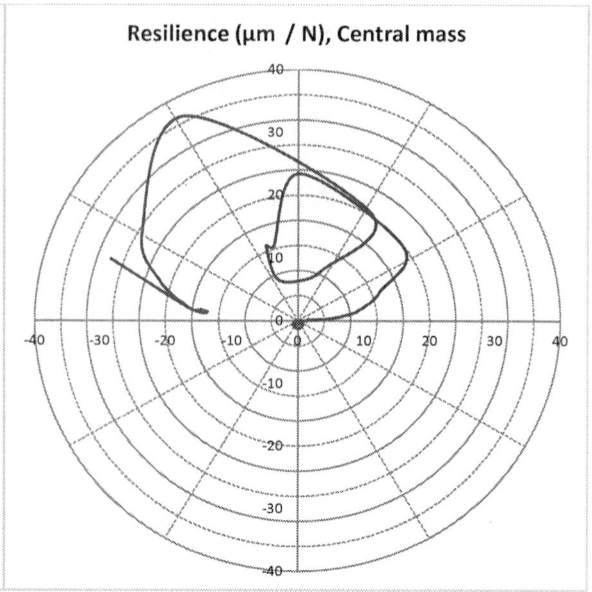

Figure 14

CONCLUSIONS

1. Accelerance can be used to obtain Mobility and Resilience. Inverse FRF's can also be obtained in terms of Magnitudes.
2. Rectangular to Polar conversion is convenient when dividing magnitudes by angular frequencies and integrating trigonometric functions ($\pi/2$ subtraction) Once in Polar coordinates, conversion to rectangular coordinates is straight forward.
3. Coherence can be calculated by dividing vectorial average magnitude resultants (4, 5 tests) by the algebraic average sum of magnitudes. Good Coherence is above 0.8
4. Using a large waveform file and "extracting" portions of it, allows users to select good data (single impacts, no overlap)
5. Nyquist plots, as well as Real and Imaginary parts are more suitable for identification of "modes" of vibration when several locations are examined.

BIBLIOGRAPHY:

Balakumar Balachandran and Edward B. Magrab, *Vibrations,* Thomson, 2004

Beer, Johnston and Cornwell, *Vector Mechanics for Engineers, Dynamics* (9th ed) McGraw Hill, 2010

Joseph A, Edminister, *Electric Circuits,* McGraw Hill Inc., 1969

Nelson L. Baxter, *Machinery Vibration Analysis III Course, Volume II*, Vibration Institute, 1995

Ray D. Kelm and Malcolm E. Leader, *Practical Modal Testing and Analysis Applied to Rotating Equipment Diagnostics,* Proceedings, 29th Annual Meeting, Vibration Institute, 2005

Robert Bond Randall, *Vibration-based Condition Monitoring,* John Wiley & Sons, 2011

Robert J. Sayer, *Modal Testing: Principles, Procedures and Case History Applications,* Proceedings, 32nd Annual Meeting, Vibration Institute, 2008

Robert E. Simpson, *Introductory Electronics For Scientists And Engineers,* Allyn and Bacon, Inc., 1975

Ronald Eshleman, *Machinery Vibration Analysis, Volume I: Diagnostic Techniques,* Vibration Institute, 2002

Singiresu S. Rao, *Mechanical Vibrations* (5th ed) Prentice Hall, 2011

Steve Goldman, *Vibration Spectrum Analysis* (2nd ed) Industrial Press, 1999

Spectral Dynamics Corporation, *Signal Analysis Using Digital Techniques,* 1976

William W. Seto, *Mechanical Vibrations,* McGraw Hill Book Co, Inc., 1970

PUMP CURVE AND PERFORMANCE

Author: Eugene Vogel

VIATC TRAINING CONFERENCE & EXPO
SAVANNAH, GEORGIA · 2022

PUMP CURVE AND PERFORMANCE

Eugene Vogel
EASA Pump and Vibration Specialist
St. Louis, MO
314-221-6121
gvogel@easa.com

Before joining EASA as a Pump and Vibration Specialist, Gene Vogel operated his own business, General Maintenance Equipment/Engineering, Inc. (GME), a marketing, service and training organization for industrial maintenance and related technologies. Among other courses, he taught a "Pumps and Pump Systems" seminar for American Trainco. Mr. Vogel has an extensive background in vibration and dynamic balancing, and chaired the St. Louis Chapter of the Vibration Institute from 1993 – 2000.

KEY WORDS: Affinity laws, best efficiency point (BEP), cavitation, net positive suction head (NPSH), pump performance, system curves

INTRODUCTION:

Many problems related to the performance of pumps, and the motors or other power sources that drive them, are related to the pump operating point. The pump operating point is measured in capacity and head. Understanding these parameters and knowing how to use them to trouble shoot pump performance is essential for technicians. This session will provide an introduction to reading the capacity, head and other critical parameters from the pump curve and how to use that information to troubleshoot a problem related to pump performance.

- Pump curve basics: Capacity, Head, Efficiency and Horsepower
- How the "system" affects pump performance
- Cavitation causes and solutions
- Common errors on pump installations
- A simple flowchart for troubleshooting pump performance

PUMP CHARTS & CURVES:

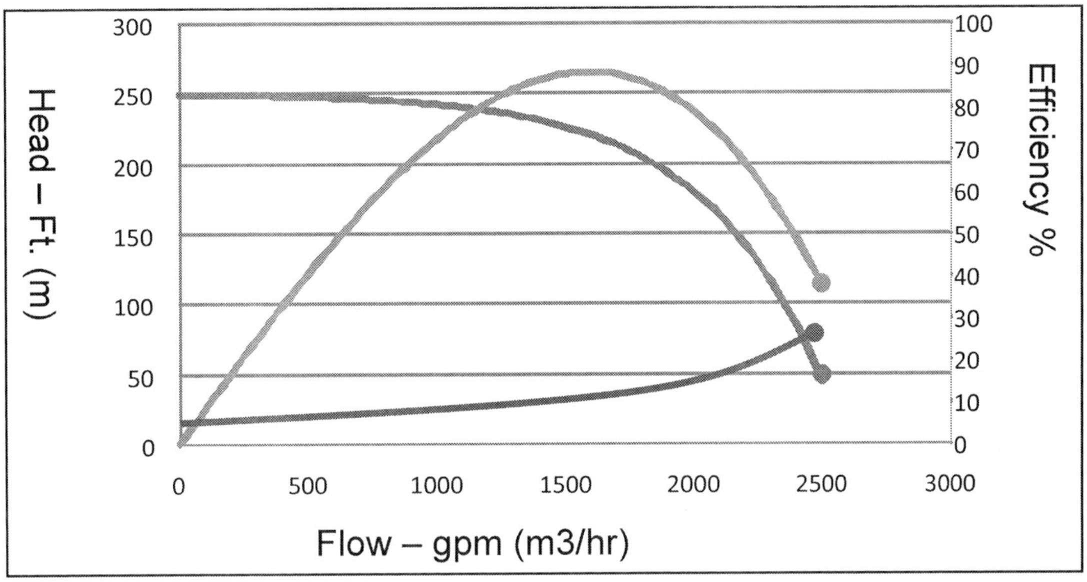

HEAD-FLOW CURVES:
The pump curve provides a graphic depiction of the design relationship between capacity and head for any operating point. Capacity is plotted on the horizontal axis and head is plotted on the left-hand vertical axis. When the pump discharge is closed, there is no flow and head is at its maximum. This is referred to as 'shut off' condition. As the discharge valve is opened, flow begins to increase and the head begins to drop. Finally, with the discharge fully open and flow unrestricted, maximum capacity and minimum flow occur. This is referred to as 'run out' condition.

The pump may operate at any point on the curve. Problems with the pump such as wear or blockage may cause operation at some point below the curve. The pump cannot achieve an operating point above the curve without an increase in speed or impeller diameter. The ideal operating point is near the middle of the curve at what is called the Best Efficiency Point, or BEP.

EFFICIENCY CURVES:
When the pump is operating against a closed discharge, at shut off, there is no flow and therefore the efficiency is zero. As the discharge is opened and flow increases, the efficiency also increases with the flow- but only to a point. As the discharge restriction is further reduced and flow increases, approaching run out, the efficiency again begins to drop due to turbulent flow through the impeller. An efficiency curve can then be plotted on the same horizontal capacity axis as the pump curve, and a vertical axis at the right-hand side. The high point on the efficiency curve is the BEP. A pump that operates very close to the BEP will run smoothest and use the least energy. As the operating point moves away from the BEP, to a lower or higher flow, rough operation and lost efficiency occur. A general recommendation is for pumps to operate with 15% of the BEP. In the case of operation below the BEP, in addition to rough operation the pump may run hot since more energy is lost to heat and flow is reduced. At or near shut off there will be insufficient flow and heat buildup and continued operation will cause imminent pump failure.

The efficiency curve will always start at the zero flow, zero efficiency point then rise to the BEP, and then fall toward run out condition. It is the shape of the efficiency curve that determines the shape of the pump curve. The head, capacity and efficiency in turn determine the horsepower/kW requirements at any operation point.

HORSEPOWER/kW RELATIONSHIP AND CURVE:

The horsepower required at any operating point can be calculated from the Capacity, Head and Efficiency by the following relationship:

$$\text{BHP} = \frac{\text{Q} \times \text{H}}{3690 \times \eta} \times s.g. \text{ where Q} = \text{Capacity, H} = \text{Head } \eta = \text{Efficiency and s.g.} = \text{specific gravity}$$

The factor 3960 adjusts for the units of gpm for Capacity and ft. for head.

For most centrifugal pumps, horsepower/kW increases as flow increases. For axial flow, propeller style pumps power may decrease with flow. When power increases with flow, (centrifugals), closing the discharge valve will cause the load on the motor to drop. While this may appear counter-intuitive, it is the result of less work being done by the pump. Since the horsepower/kW can be determined from the Head, Capacity and Efficiency curves, some older printed pump curves did not provide a power curve, but relied on technicians to calculate that value manually.

VISCOSITY FACTORS:

Viscosity is defined as the resistance of a fluid to shear forces, or flow. As such, viscosity represents a source of friction loss, reducing efficiency. More viscous fluids convert more of the pump's energy into heat, requiring more power to achieve the same flow as lower viscosity fluids. Generally, pump curves are provided for water and the effects of increased viscosity can be determined from correction charts. However, pump performance software that is discussed below allows fluid characteristics to be input for corrected pump curves.

MULTI-TRIM CURVES:

Often the desired operating point for a application is below the pump curve. To achieve the desired head or flow, a throttle valve or pressure regulator may be used. These devices 'waste' energy, converting it to additional heat. An alternate means of matching pump performance to system needs is to trim the pump impeller to a smaller OD. Pump charts may provide a series of pump curves for different impeller diameters. Pump manufacturers will provide pumps with impellers trimmed to customer's specifications, or end user's may trim the impellers within manufacturer's allowable range. The effect of trimming an impeller is to move the pump curve toward the origin of the chart, (down and to the left). Obviously, where some range of operation is needed, the impeller is only trimmed to the maximum operation point and other means are used to adjust for process demands. Impellers cannot be 'untrimmed', so use caution in specifying the trim amount. When an impeller is trimmed, some of the vane surface area is lost, and there is a corresponding minor loss of efficiency. This minor loss of efficiency is small compared to the loss of efficiency from throttling the discharge to achieve equivalent capacity or head conditions.

VARIABLE SPEED CURVES:

A popular alternative to trimming impellers is to provide a variable speed drive. Reducing the speed of the pump has an almost identical affect as trimming the impeller. However, the speed can be increased as well, providing greater flexibility and control. Another advantage to variable speed drives over impeller trimming the impeller is that there is no loss of vane surface area.

Efficiency may be slightly better with reduced speed vs. trimmed impeller. Finally, since increasing speed will raise the pump curve, it can be used to remedy a slightly deficient pump.

AFFINITY LAWS:

There are general proportions which describe the relationship between the pump chart parameters and changes to impeller diameter and speed. These are referred to as the Affinity Laws.

$$Q_2 = Q_1\left(\frac{D_2}{D_1}\times\frac{N_2}{N_1}\right) \qquad H_2 = H_1\left(\frac{D_2}{D_1}\times\frac{N_2}{N_1}\right)^2 \qquad BHP_2 = BHP_1\left(\frac{D_2}{D_1}\times\frac{N_2}{N_1}\right)^3$$

where D = Impeller Diameter and N = Pump Rotating speed.

These Affinity Laws show that either the impeller diameter or the pump speed will cause a linear change in Capacity, a squared change in Head and a cubic change in Power. So if the diameter of the impeller were reduced to half, the Capacity would be half, the Head would be one forth and the power required would be on eighth. If the speed of a pump were doubled, say from 1800 to 3600RPM, the Capacity would double the Head would be four times greater and it would require eight times the power. A 50 HP pump at 1800 would require 400 HP at 3600.

Two additional Affinity laws relate the effect of pump speed on the NPSHR and on stress and wear forces on the pump rotor.

$$NPSHR_2 = NPSHR_1\left(\frac{N_2}{N_1}\right)^2 \qquad Forces_2 = Forces_1\left(\frac{N_2}{N_1}\right)^2$$

The NPSHR is an indicator of the likelihood of cavitation. Both factors vary with the square of the speed, and are not affected significantly by impeller diameter. It should be noted therefore, that reducing the speed of a pump rather than trimming the impeller has the additional advantage of reducing the likelihood of cavitation, and reducing the wear on the impeller.

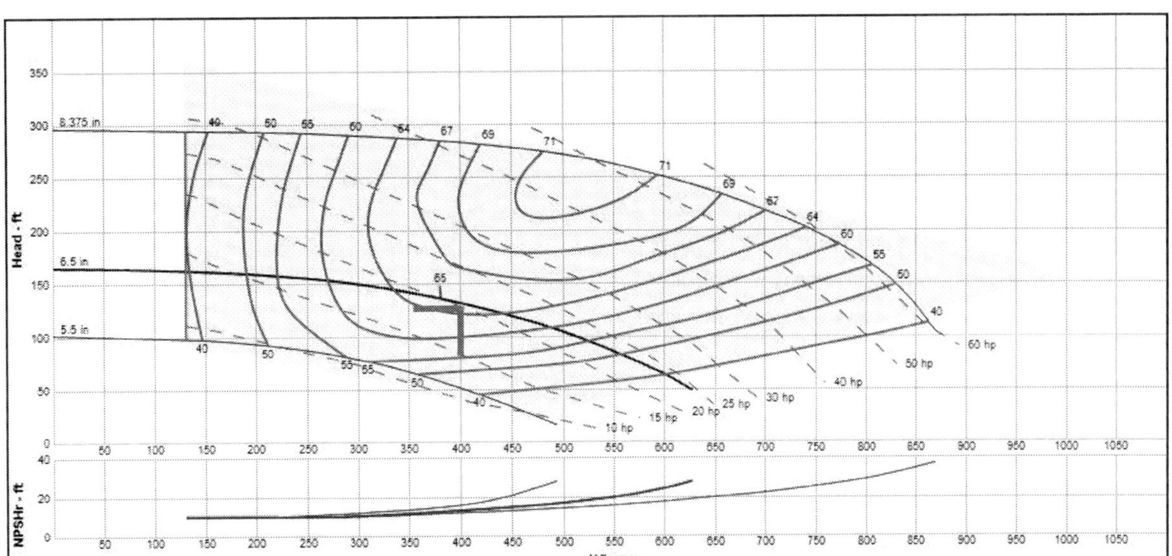

SYSTEM CURVES:
There are a variety of pipes, valves, fittings, tanks and other hydraulic equipment that contains the flow to and from the pump. These devices affect the flow through the pump in both positive and negative ways. Taken together, the sum of all such devices normally presents a resistance to flow from the pump, which is called 'system resistance'. The amount of resistance changes with flow rate. When the system resistance is plotted as a function of flow rate, or capacity, the result is a system curve. Since system resistance will increase as flow rate increases, the system curve will rise from left to right along the horizontal capacity axis. The system curve is then plotted with the pump curve to determine the maximum operating point.

- Static Head
  - Suction Level
  - Discharge Level

$$h_s = h_{Discharge} - h_{Suction}$$

\+ Velocity Head
  - Calculated $h_v = \dfrac{v^2}{2g}$ or $h_v = \dfrac{0.00259 \times gpm^2}{PipeDiameter^4}$

\+ Friction Head
  - Pipe Losses: From Table
  - Minor Losses: $h_f = k \times \dfrac{v^2}{2g}$

= **Total Dynamic Head, (TDH)**

The amount of system resistance at any flow rate is composed of three dynamic components: Static head, velocity head and friction head. Each is discussed below.

STATIC HEAD:
It is easy to imagine how a column of liquid of some height relates directly to the pressure at the bottom of the column. This pressure, when represented as an equivalent column of liquid is called head. The height of the liquid level at the suction or discharge of a pump is called static head or elevation head. Static head is measured from the centerline of the impeller to the surface of the liquid, which may be positive or negative if the surface is above or below the pump.

The total static head presented to the pump by the system is the difference between the static discharge head and static suction head. That is, the static suction head subtracted from the static discharge head. It is important to emphasize that static head can be positive or negative; depending on whether the surface level of the liquid is above or below the pump.

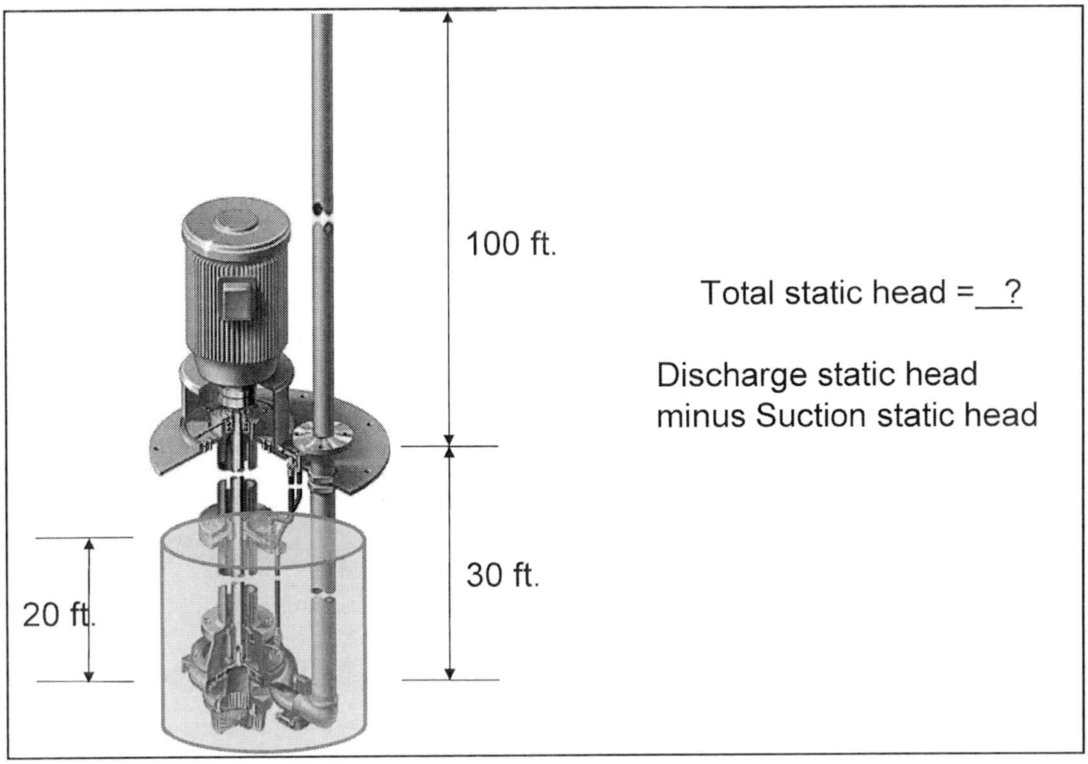

For example, a sump pump may have a suction level 10 ft. below the pump impeller centerline and be pumping to a tank 25 ft. above the centerline. Since the suction level is below the pump the static suction head is $^{-}10$ ft. (negative 10 ft.). So the total static head to the pump is 25 ft. - ($^{-}$10 ft.) = 25 ft. + 10 ft. = 35 ft., since subtracting a negative is the same as adding a positive.

It is interesting that some pumps may have a negative static discharge head. For example: A petroleum pipeline pump moves crude from a well in the state of Wyoming to a refinery on the Gulf Coast of Texas. The well head in Wyoming is more than 5000 ft. above sea level and the refinery on the Gulf is virtually at sea level. For an explanation of why a pump is needed to move liquid downhill, see the section on friction head below.

VELOCITY HEAD:
Liquid has mass, (similar to weight), and it takes energy to make mass move faster; that is, to increase the velocity of the fluid. Therefore, if a pump has to take fluid which may be stopped, (or moving slow), and speed it up to flow through a system, that takes energy. How much pressure the pump has to exert against the fluid to increase the velocity is a measure of the amount of that energy. Since pressure can be measured in units of ft. of head, we calculate this energy as 'velocity head' in ft.

The common relationship for calculating velocity head is $H_v = \frac{V^2}{2g}$, where V is the velocity of the fluid through the pipe and g is the standard gravity: 32ft./sec^2. However, a more useful relationship is: $H_v = \frac{0.00259 \times (gpm)^2}{(pipeDiameter)^4}$, since it does not require calculating the velocity of the fluid through the pipe. From this second relationship it is apparent that the velocity head increases with the square of the flow rate and decreases with the fourth power of the pipe diameter.

Consider several systems configurations and the impact of velocity head. First, liquid is in a sump below the pump and is pumped into a tank above the pump. Since the liquid is basically stopped in the sump, the pump must supply the velocity head to get it moving through the smallest diameter in the suction line. Now consider a system with the suction tank above the pump. Gravity will now provide a force to get the fluid moving and the pump benefits from that additional force. The resistance from velocity head will be lower in this system. Now consider a closed loop system with suction and discharge lines of the same diameter. When the pump is running the fluid moves through the loop at a constant velocity, so the total velocity head is zero. However, if the suction line is larger than the discharge, (which is common for most pumps), the pump must now supply the velocity head to move the fluid faster out the discharge.

For most systems, the velocity head is not a major contributor to the total system resistance. However, you will find it is included in any accurate analysis of pump performance.

FRICTION HEAD:

Fluid flowing through a pipe or other hydraulic device creates friction as it is in contact with the walls of the device. The area of contact, the roughness of the surface and the velocity of the fluid all affect the amount of resistance. The actual mechanism of fluid friction is more complex, incorporating the concepts of laminar vs. turbulent flow. However, the simple model of fluid in contact with the roughness of the surface is sufficient for calculating basic system resistance. The calculated resistance due to friction is called the friction head, following the manner of static head and velocity head.

With the wide variety of devices that conduct fluid to and from a pump there is the need for some standard method of calculating the total friction head. However, there are actually several methods commonly in use. Generally, the friction head for pipe is determined from a table and added the friction that is calculated for all other system components.

PIPE LOSSES:

The friction losses for pipe are usually obtained from "pipe loss tables" which have been calculated for common pipe materials and sizes. Pipe loss tables are printed in a variety of pump and engineering handbooks. The table data is arranged by pipe type, diameter and flow rate. For each selection by these criteria, the tables report the "feet of head loss per 100 ft. of pipe". To apply the data from the table, determine the length of that pipe, divide by 100 and multiply by the value from the table. The result is the friction head for that piece of pipe. Systems with several sizes of pipe will require calculations for each size pipe.

3 in.			
Flow Rate	Friction Loss (ft/100 ft)		
(gal/min)	Steel Pipe Schedule 40	Copper Pipe Type L	PVC Pipe Schedule 40
50	0.7	0.8	0.6
100	2.5	2.7	2.2
150	5.3	5.7	4.8
200	9.1	9.5	8.1
300	19.8	19.8	17.1
400	34.2	33.4	29

4 in.			
Flow Rate	Friction Loss (ft/100 ft)		
(gal/min)	Steel Pipe Schedule 40	Copper Pipe Type L	PVC Pipe Schedule 40
100	0.7	0.7	0.6
150	1.4	1.5	1.3
200	2.4	2.5	2.2
250	3.6	3.7	3.3
300	5	5.1	4.6
400	8.7	8.6	7.8

Pipe Loss Tables

MINOR LOSSES:
For traditional manual calculations, the friction losses for anything other than pipe are lumped together under 'minor losses'. Most often, the total minor losses will exceed the pipe losses. Thus, they are not "minor" and cannot be omitted. The traditional method for calculating minor losses is based on K values provided by the device manufacturer. This K factor is multiplied by the velocity head for that diameter fitting. K factors may be available in pump reference and engineering books, or from the component manufacturer.

Many hydraulic device manufacturers provide friction head data in the form of "equivalent pipe length". Suppose a heat exchanger has 2" pipe flanges and reports a friction head as equivalent to 15 ft. of pipe. The 15 ft. is simply added to the length of the 2" pipe and friction head is determined from the pipe loss table value. Alternately, a manufacturer may report that the device produces a certain pressure drop at a specified flow rate; as 10 psi drop @ 150 gpm. The flow rate is the maximum flow rated for that device, and the 10 psi can be converted to head with the 2.31 conversion factor. In this case, the friction head at lower flow rates is not known.

MANUAL CALCULATION:
The individual results from static head, velocity head, pipe loss and minor loss are simply added up to determine the total system head at some flow rate. The traditional manner for performing this calculation is to total all values for the suction side of the pump, and total all values for the discharge side, then add the suction and discharge values. Be certain to observe proper positive/negative sign values.

EXAMPLE:

360 gpm

Suction:
- ✓6 ft. Elevation
- ✓20 ft., 4" Sched40 Pipe
- ✓90° Elbow
- ✓2, 4" Gate Valves

Discharge:
- ✓30 ft. Elevation
- ✓35 ft. 3" Sched40 Pipe
- ✓1 - 90° Elbow
- ✓3" Globe Valve

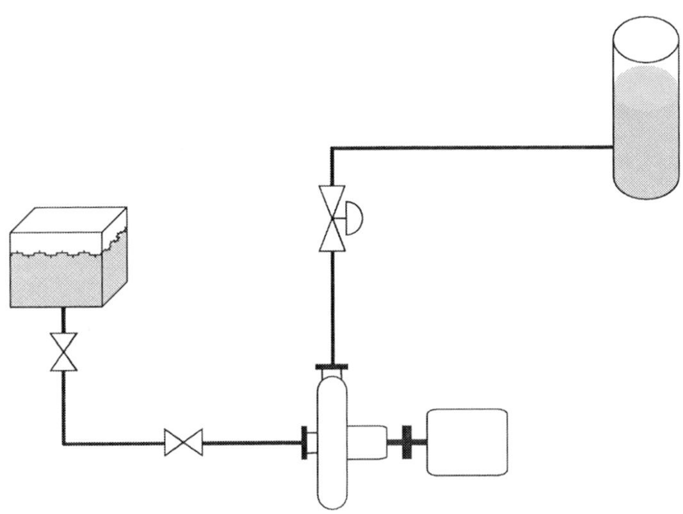

Example System

Suction Head									
flow - gpm	100	200	300	360	400	500	600	700	800
pipe diam - in.	4	4	4	4	4	4	4	4	4
pipe Length - ft	20	20	20	20	20	20	20	20	20
Velocity Head	0.1	0.4	0.91	1.31	1.61	2.52	3.64	4.95	6.47
Friction Loss ft/100 ft.	0.7	2.7	6.0	8.6	10.8	16.9	24.3	33.1	43.3
Pipe Loss	0.14	0.54	1.2	1.72	2.16	3.38	4.86	6.62	8.66
Minor Loss									
- Gate Valves 'K'	0.14	0.14	0.14	0.14	0.14	0.14	0.14	0.14	0.14
- Elbow 'K'	0.51	0.51	0.51	0.51	0.51	0.51	0.51	0.51	0.51
Total Minor Loss	0.07	0.31	0.71	1.03	1.27	1.99	2.87	3.91	5.11
Elevation	6	6	6	6	6	6	6	6	6
Total Suction Head	5.89	5.55	5	4.56	4.18	3.15	1.91	0.42	-1.3
Discharge Head									
pipe diam - in.	3	3	3	3	3	3	3	3	3
pipe Length - ft	35	35	35	35	35	35	35	35	35
Velocity Head	0.31	1.27	2.87	4.14	5.11	7.99	11.51	15.66	20.46
Friction Loss ft/100 ft.	3.1	12.1	26.8	38.4	49.9	77.9	112.2	152.7	199.5
Pipe Loss	1.09	4.24	9.38	13.4	17.5	27.3	39.27	53.45	69.83
Minor Loss									
- Globe Valve 'K'	6.1	6.1	6.1	6.1	6.1	6.1	6.1	6.1	6.1
- Elbow 'K'	0.54	0.54	0.54	0.54	0.54	0.54	0.54	0.54	0.54
Total Minor Loss	2.05	8.43	19.05	27.5	33.9	53.1	76.42	104	135.9
Elevation	30	30	30	30	30	30	30	30	30
Total Discharge Head	**33.4**	**43.9**	**61.3**	**75.1**	**86.5**	**118**	**157.2**	**203.1**	**256.1**
Total Dynamic Head	**27.6**	**38.4**	**56.3**	**70.5**	**82.3**	**115**	**155.3**	**202.7**	**257.4**

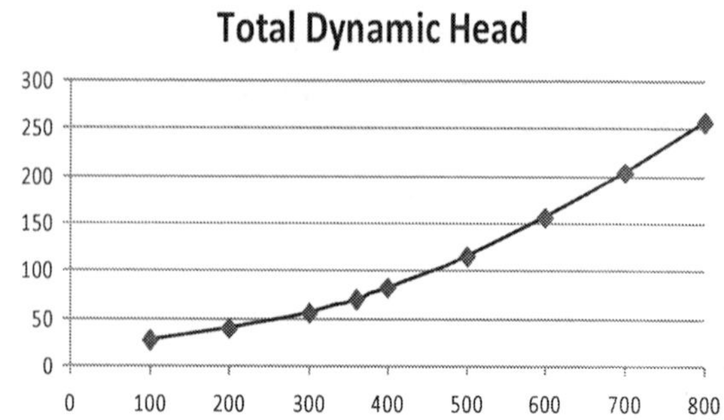

Determining Operating Point

Observe from the above discussion of pump curves and system curves that both curves have capacity on the horizontal axis and head on the vertical axis. Further, the pump curve will slope down from left to right, and the system curve slopes up from left to right. When both curves are plotted on the same chart there will be a point of intersection. That point of intersection will be the maximum operating point. That is to say, the capacity at that point is the maximum capacity available without changing either the pump or the system resistance.

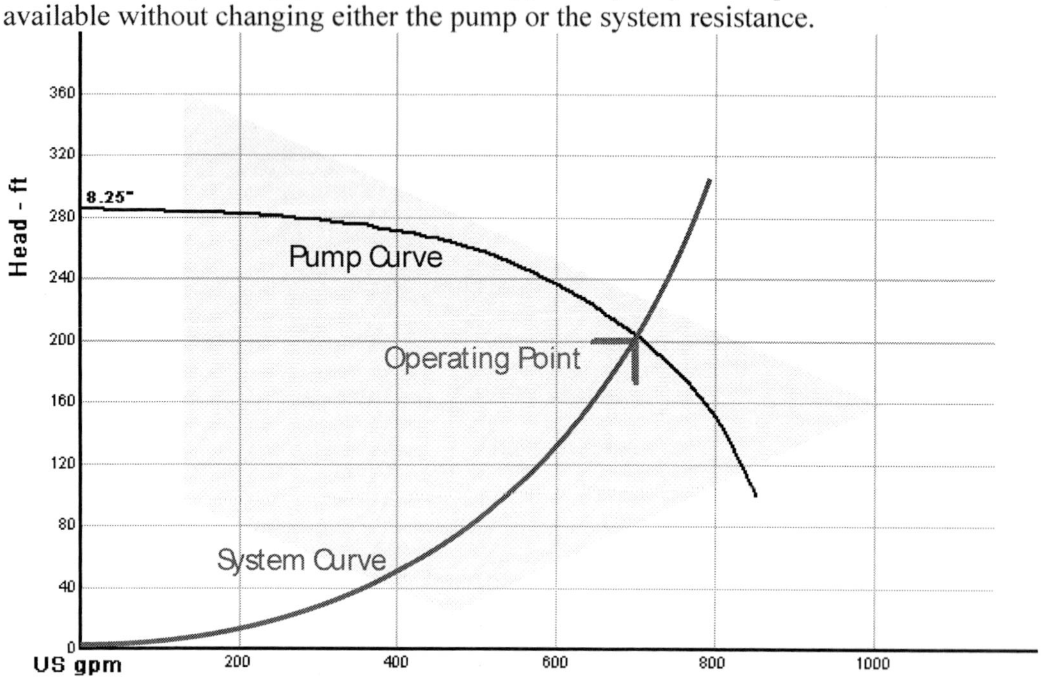

Of course, any flow rate less than the maximum can be achieved by throttling or speed reduction. The effect of throttling the discharge is to tilt the system curve up and the left, so that the operating point move back on the pump curve. The effect of reducing pump speed is to shift the pump curve down and to the left, moving the operating point back on the system curves.

CAVITATION:

Phenomena

Cavitation is a common occurrence when liquid is subjected to pressure changes, as through a pump, through changes in pipe diameter, or whenever there is turbulent flow. At atmospheric pressure, many fluids remain liquid because their vapor pressure is below atmospheric pressure. (Vapor and liquid are considered fluids.) When pressure is sufficiently reduced the liquid will vaporize. When a liquid is warmed, it takes more pressure for it to remain liquid; that is to say the vapor pressure of a fluid increases with temperature.

Using room temperature water as an example, when the pressure on the water is reduced enough, the water vaporizes. Suppose there were water in an air conditioning line. A vacuum pump is connected to the system reducing the pressure, so the water vaporizes and is pump out. Warming the line would make it easier for the water to vaporize and be pumped out.

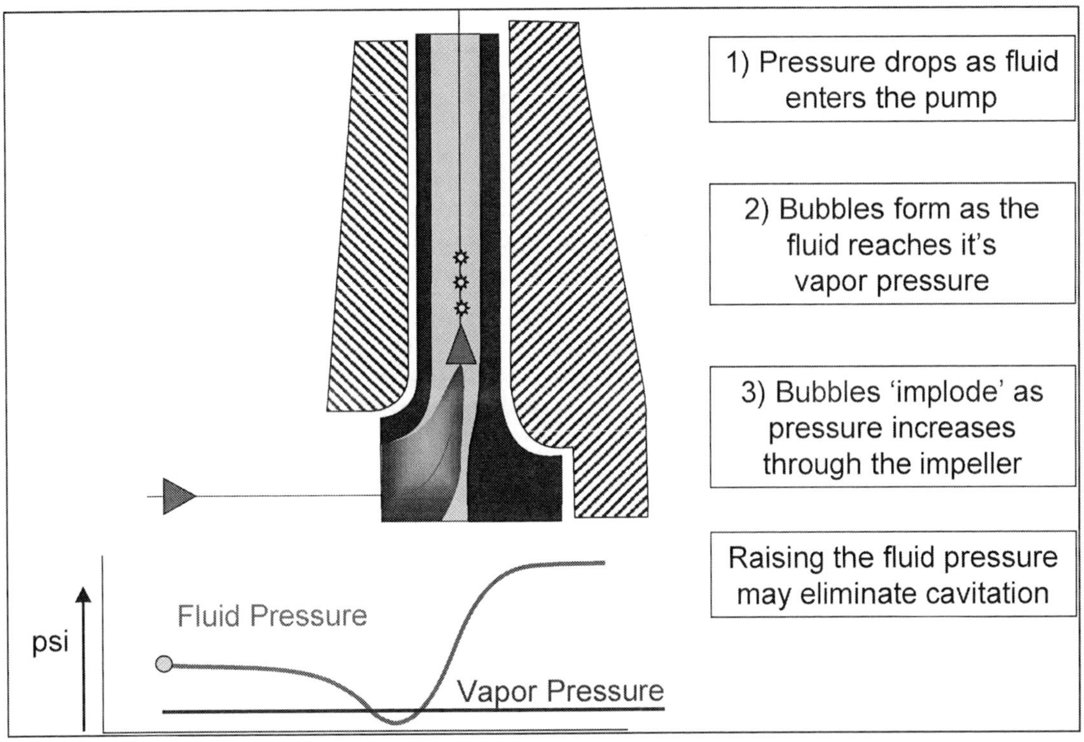

As liquid enters the suction eye of the impeller, the pressure drops as the liquid picks up speed through the impeller, (remember Bernoulli's Principal). If the pressure on the liquid drops below the vapor pressure for that liquid at that temperature, some of the liquid will vaporize forming bubbles. Now as the fluid containing bubbles passes through the impeller the pressure then increases and the bubbles implode releasing large quantities of energy. This phenomenon of bubbles forming and imploding due to pressure changes is called cavitation. The energy released when the bubbles implode causes surface damage to the impeller and imparts vibratory energy to the impeller, shaft and bearings. The sound of cavitation has been described as similar to sand, gravel, rocks or marbles passing through the pump. Cavitation is a very damaging to pumps. Another factor contributing to the formation of bubbles is agitation. Shake a bottle of water and see the bubbles that are formed. Agitation occurs as the leading edge of the impeller vanes contact the incoming fluid. The rotating speed of the pump is directly related to the amount of agitation, so higher speed pumps are more likely to cavitate. The shape and attack angle of the impeller vanes also plays an important role in the amount of agitation and resulting cavitation.

Impeller Damage

During cavitation, when the bubbles implode, the surface of the impeller which is in contact with the bubble is damaged. Small pieces of material are broken away from the surface. Over time, the impeller develops significant pits and may develop deep erosion marks that may grow into holes completely through the impeller. With classic cavitation this damage is along the "working side" of the vanes between the suction eye where the bubbles form, and the discharge edge of the impeller. In fact, the impeller may have the appearance that sand or gravel have been pumped through it. However, the damage will be limited to the working surfaces of the impeller, with no damage to the outer surfaces, (as there would be from debris going through the pump).
Impeller made of stainless steel will show cavitation damage as a mottled surface, due to the 'tough' nature of stainless steel. Impellers may be hard coated to reduce damage from cavitation, however, the damaging affects on the bearings and seal will remain.

NPSHA – NPSHR

From the explanation above it should be apparent that if the pressure on the liquid going through the pump never falls below the vapor pressure, no bubbles will form and there will be no cavitation. So a comparison must be made between the pressure required to prevent cavitation and the pressure available at the suction of the pump. But atmospheric pressure is part of that comparison, and atmospheric pressure changes with elevation above sea level, and with weather conditions. Therefore absolute pressure is used when making this comparison. Remember that atmospheric pressure is 14.7 psi or 39.96 ft. head. When measured in absolute pressure/head units, all values will be positive. (Negative pressure, vacuum, can only exist to the extent that there is atmospheric pressure to pull vacuum against.) Also, when making this comparison of pressure required to pressure available, the units will be ft. of head, (or meters of head). Thus the terms become: Net Positive Suction Head or NPSH. There is NPSHR, the Required, and NPSHA the Available.

The NPSHR is provided by the pump manufacturer via the pump chart. It is most often plotted as a function of flow, similar to horsepower. However, on some style pump charts the NPSHR data is provided in isometric form. The NPSHA is calculated from four terms: Atmospheric Head (33.96 ft.) + Suction Static Head - Suction Friction Head – Vapor Pressure. Remember that Suction Static Head will be negative if the fluid level is below the pump, (suction lift).
In order to prevent cavitation, the NPSHR must be greater than the NPSHA. Of course, the greater the difference the better. API 610 provides a guide to recommended ratios of NPSHR to NPSHA. This ratio is called the NPSH Margin and should be 1.15 or greater in all cases.

Vapor Pressure

Most liquids will vaporize, or boil, when the pressure on them is reduced, or when the their temperature is raised, sufficiently. The point at which they vaporize, based on pressure and temperature, is call their vapor pressure. The vapor pressure for a fluid is normally presented in a table of related values or a graph. For instance, water vaporizes at 212°F at 14.7 psi, (atmospheric pressure at sea level). However, in Denver at about 5000 ft., the atmospheric pressure is only 12.2 psi, and water vaporizes at 201°F.

Each fluid that is pumped will have such a temperature/pressure relationship. When determining the vapor pressure of a fluid, be certain of the temperature of the fluid as it enters the pump; it may be quite different than ambient temperature. Many common fluids have vapor pressures far below atmospheric pressure at normal room temperatures. Hydrogen, nitrogen, propane, natural gas and similar fluids are considered cryogenic liquids, since they must be brought to extremely low temperature, or very high pressure, to remain liquid and not vaporize.

Calculation Examples

The NPSHA is calculated from four terms: Atmospheric Head (33.96 ft.) + Suction Static Head - Suction Friction Head – Vapor Pressure:

hAtmosphere	+ 33.96
hsuction static	+ 6
hsuction friction	– 2.76
hvapor pressure	– 0.78
NPSHA	36.4 ft.

The NPSHR is read from the pump chart; in this example it is 13.2 ft.

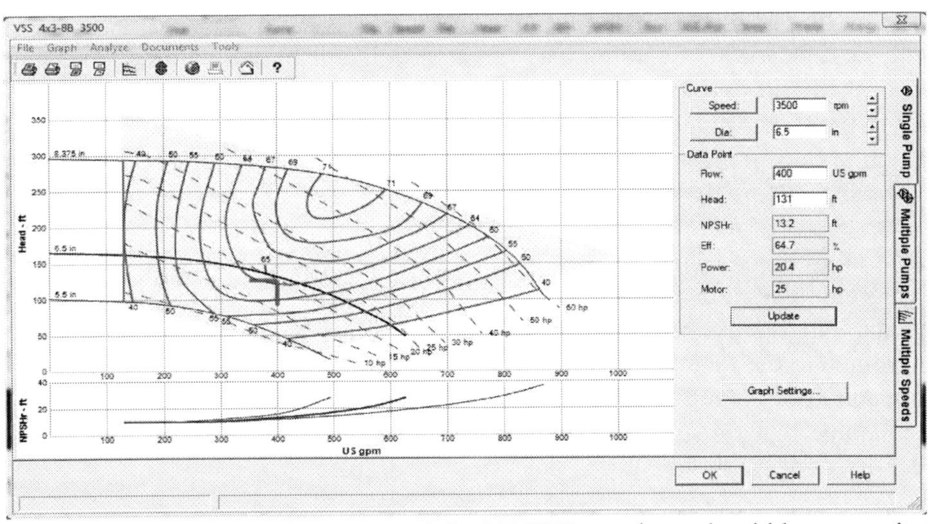

In this example the NPSHA exceed the NPSHR, so there should be no cavitation. It is always beneficial for the NPSHA to be well above the NPSHR. The ratio of NPSHA to NPSHR is called the NPSH Margin. For most pumps the NPSH Margin should be 1.2. The API sets recommended NPSH Margin numbers for many pump applications.

COMMON ERRPORS ON PUMP INSTALLATIONS

Over-sizing the pump

Installing a 100 hp motor on a 50 hp load makes for a very reliable, if somewhat expensive, installation. However, installing a 100 gpm pump on 50 gpm system results in a very inefficient and unreliable system. The pump operates with higher vibration and higher temperature than a smaller pump that is matched to the 50 gpm system. It is common is mechanical design to include a "safety margin" when sizing components; and iterative safety margins can add up to a much larger component than is required. So it is very common to find too large of a pump for the application, which results in turbulent recirculation, vibration and heat.

Mismatched pumps in series and parallel
The rule is: pumps in parallel – match head, pumps in series – match flow. When this rule is ignored, one of the pumps will overpower the other resulting in low flow and the associated vibration and heat.

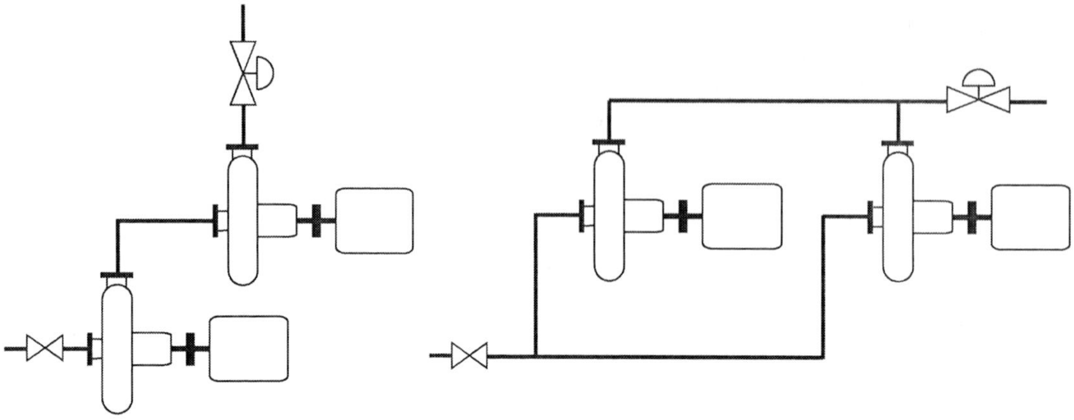

Restrictions on the suction line or insufficient suction head
The most convenient physical location for a pump may not an acceptable hydraulic location for the pump. If the source of the liquid is remote from the pump and too small of diameter suction line is used, the pump will have insufficient NPSHA and will cavitate. Isolation valves, strainers and extra elbows in a suction line all create friction losses that reduce NPSHA. High resistance valves such as globe valves should not be used in suction lines.

Poor base, grouting and alignment
For pumps with non-integral motors, the base must be rigid enough to resist torque and piping strains yet have enough damping to dissipate inherent pump vibration. This normally requires that the pump base be grouted to the foundation. Lack of grouting and insufficient pump base stiffness allow the pump and driver to move creating misalignment and also allow vibration to propagate through the machine. The result is a rough running pump that subjects the sensitive bearings and seals to excessive dynamic forces.

Even when the base structure is adequate, lack of proper alignment will subject the pump to unnecessary vibratory forces. And since pumps and drivers often undergo different thermal changes during operation, the alignment often must be offset to account for those differences. A thorough evaluation of pump alignment includes inspection of the pump base and foundation, and measurement of movement from thermal and torque strains.

Unsupported piping and strained piping connections
Any excessive stress that is applied to the pump nozzles (flanges and threaded fittings) will cause distortion of the internal alignment of pump bearing and pump seals. In addition, stress from piping will make resonant frequencies more likely in the pump and base structure. All piping connected to the pump should be self supported. Proper adjustment of piping hangers and supports should also account for thermal expansion or contraction of those components.

Pump materials lack protection from erosion and corrosion
Pump impellers and volutes are often constructed of cast materials that are susceptible to erosion from abrasive debris in the pumpage. Evidence of erosion will be clearly visible on impeller

vanes and at the cutwater of the volute. Erosion can be differentiated from cavitation damage by the location and appearance of the damage. Pumps that will be subjected to abrasives in the pumpage should have impellers, volutes and wear rings that are either hard coated or constructed from abrasion resistive materials.

Likewise, pumpage which is highly acidic or caustic will damage pump internal parts. In all cases the damaging characteristics of the pumpage must be considered when specifying a pump for any application.

TROUBLE SHOOTING PUMP OPERATION

When a problem occurs, the pump is often the first suspect. However, there are three possible sources, the pump, the fluid (at operating temperature and vapor pressure), and the system. Using the process of elimination, the primary problem source can usually be identified. Since multiple problem sources often exist, it is important to be thorough.

Is it the Pump?

When a low Flow or low Head condition exists, the first step is to determine if the problem is the pump, the fluid or the system. Measure TDH and Flow at the pump and compare them to the pump curve. If the operating point is on the pump curve, the pump is operating properly; the problem is the system, or possibly the fluid. If the operating point is below the pump curve, the problem may be either the pump or the system or possibly the fluid. Important! If, on startup, there is little or no Head or Flow: Shut off the pump immediately and verify there is sufficient fluid in the pump, (pump primed).

<u>Running the pump dry will damage the seal.</u>

Is it the Fluid?

Verifying the fluid properties is good place to start. Tests for viscosity, specific gravity and temperature of the fluid are readily available and inexpensive. Ambient conditions, (weather), can affect viscosity. When pumpage is a strong acid or base, dilution can change the specific gravity.

Is it the System?

With fluid properties verified, the problem must now be either the pump or the system. Again, if the pump operating point is on the pump curve, then the pump is working properly and the problem must be the system to which the pump is connected. If the pump is not operating on its curve, there still are some system problems that need to be eliminated before the pump is identified as the culprit. If vapor is getting into the pump, either by air entrainment or by cavitation, the pump will not operate on its curve, but there may be nothing wrong with the pump. So the next step is to test for cavitation, and for air entrainment.

If cavitation and air entrainment can be eliminated, and the pump is not operating on its curve, then there is very likely a problem with the pump.

What could be wrong with the pump?

When a pump does not operate on its curve and cavitation and air entrainment have been eliminated, the most likely causes are damage to the impeller, blockage in the pump impeller, blockage in the pump volute, or excessive wear ring or impeller clearance. Other causes would be related to the speed of the pump, such as the shaft spinning in the impeller or an incorrect drive speed. While driver speed can be verified externally, investigating the other causes will involve opening the pump.

What could be wrong with the system?

When a pump is operating on its curve and fluid properties have been verified, the problem must be with the system. There are two possible cases: Either the flow is too low, (and therefore the head is too high), or the head is too low, (indicating the flow is too high). When considering head and flow, remember, the pump is operating on its curve, so if one is too low the other must be too high. Of course one other possibility is that it is the wrong pump, either by improper design or by mistaken installation; including installing the wrong impeller.

Low flow, (head too high)

A low flow condition most often indicates a restriction in a line. If the restriction is in the suction line there will likely be cavitation, otherwise the restriction is likely in the discharge line. A second possibility is that the suction static head is too low, or the discharge static head is too high. A suction tank may have a float switch to shut off the pump at some level, below which the pump should not operate. Similarly, a discharge tank may have a high level switch which has malfunctioned.

Low head, (flow too high)

A low head condition indicates there is too much flow, and likely, the flow is not going where it should. System leaks can be internal or external. A diverter valve that is allowing too much flow to bypass, or a failed check valve that lets flow circulate backwards through a parallel pump would result in too much flow and low head. On a municipal water system with buried water mains, a major leak or line break will allow too much flow and result in low head, (low line pressure).

Searching for system problems

Looking for blockage or leaks in a hydraulic system is very similar to looking for opens and shorts in an electrical system. Except you can't just touch test leads to the pipe to measure pressure as you would to measure voltage. Just as with electrical circuits, however, it is easier to measure pressure than flow, (like voltage being easier to check than current).

Simplified Pump Troubleshooting Chart

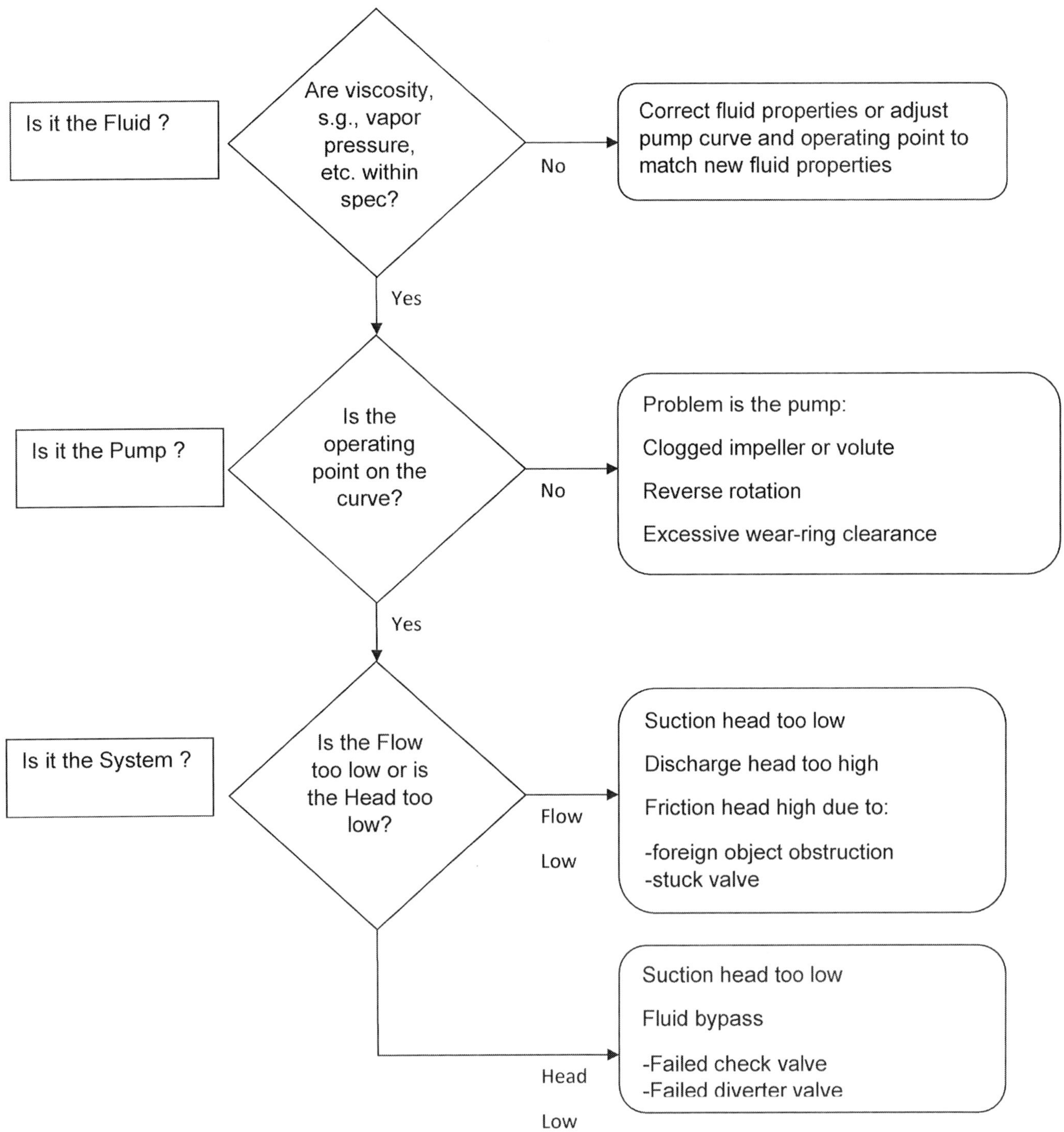

VERTICAL PUMP RESONANCE & PRACTICAL FINITE ELEMENT MODELING

Author: Monroe Voyles III, PE

VIATC TRAINING CONFERENCE & EXPO
SAVANNAH, GEORGIA · 2022

Vertical Pump Resonance & Practical Finite Element Modeling

Monroe Voyles III, PE
ITT Goulds Pumps

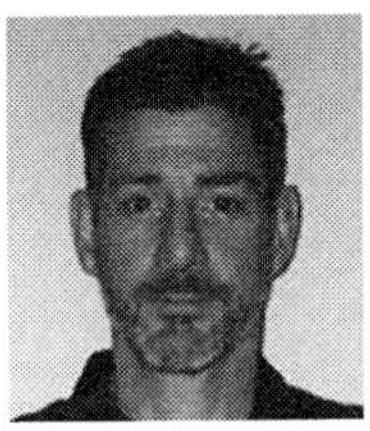

Mechanical engineer with 27 years of experience; 7 petrochemical, 3 refining, and 17 pump OEMs. Experience with large electric motors, gas turbines, centrifugal compressors, reciprocating compressors, steam turbines, large API pumps, ANSI pumps, high pressure boiler feed water pumps, non-contacting gas seals, high temperature bellows seals, mag-drive pumps, liquid ring compressors, cooling tower fans/gear boxes, and large compressor gear boxes. Work included RCFA, vibration analysis, performance testing, and upgrading systems/equipment to resolve problems or increase performance. Proficient with finite element modeling (CREO/Simulate/DyRoBes).

B.S.M.E Louisiana State University 1994.
Professional Engineer Louisiana 2002/Texas 2014.
CAT IV Vibration Analyst (Vibration Institute 2010 & Mobius 2014).
Instructor for the Vibration Institute.

ABSTRACT:
Vertical pumps can experience large amplitude vibration when a system natural frequency is at or near a machine generated forcing frequency like 1X shaft (running) speed. This resonant condition is often unexpected because it was either not considered in the purchase specification, or finite element modeling may have been conducted before purchase and an acceptable frequency separation margin was predicted. The author has analyzed many cases where resonance was not predicted by industry standard finite element modeling, but the actual machine exhibited high amplitude vibration due to resonance when installed in the physical system. Experimental modal data was used to evaluate these applications and the corresponding finite element models; significant issues were discovered with regard to boundary conditions and simplified motor models. This paper describes potential modeling problems and techniques to improve the finite element model so the predicted natural frequencies and mode shapes will better correlate with the physical machine responses (experimental modal data). Industry standard finite element modeling is discussed and areas of improvement are presented. Boundary conditions can be affected by improper installation and/or degradation; resonance may begin on a "good" machine after many years of operation with low vibration amplitude. A case history example is provided to demonstrate some of the various issues with finite element modeling and machine/system degradation over time.

KEY WORDS: Broadband noise; bump test; experimental modal test; finite element method; finite element model; force hammer; machine frequencies; natural frequency; resonance; mode shape; boundary condition

INTRODUCTION:

As a vibration analyst I often found experimental modal data did not correlate well with finite element analysis (FEA) results. The FEA would predict large frequency separation margin, but the machine would exhibit large amplitude vibration due to resonance. I started to do FEA so I could better understand the issues and hopefully improve the modeling for problem applications. After a few years of comparing experimental modal data with FEA predictions I discovered a few areas of potential problems. This paper discusses the phenomenon of resonance and how vertical pump natural frequencies are affected by the motor, foundation, piping, and fluid inside the pump. Finite element models need to be adjusted with field experimental modal data when possible or known boundary condition issues should be considered. Two of these problems (motor modeling and foundation assumptions) are shown in a case history. Practical modeling issues are discussed regarding boundary condition assumptions. Details of the finite element modeling process and software performance are not covered. Vertical pump designs, and common mode shapes will be discussed to aid in better field data collection and modeling of problem machines. Items to consider when reviewing pre-purchase FEA reports are presented to reduce the potential of resonance on future installations.

RESONANCE:

Resonance occurs when a forcing frequency is at or near a system natural frequency. Resonance amplifies the vibration amplitude response to a given force input; the dynamic stiffness approaches zero at resonance. A natural frequency is a frequency that readily absorbs energy. Any vibration energy applied at a natural frequency will have a larger response or amplitude as compared to other frequencies. Broad band energy can excite a natural frequency. An impact or impulse can also excite natural frequencies (example: impact test or "bump" test), but the largest responses and typically when excessive vibration is a problem occur when a forcing frequency is at or near a natural frequency. In this case an optimal timing relationship between the applied periodic force and the machine response cause each cycle of vibration to increase amplitude until damping and or non-linear forces create an upper limit in vibration amplitude. Without damping or non-linear stiffness a system at resonance will create an infinite amplitude response. The amplification factor at resonance is shown in Figure 1 [1]. Amplification factor is a ratio of the response amplitude to the static deflection of a given force. Damping in the system limits the amplitude response at resonance to a finite amplitude rather than infinity. Increasing levels of damping are shown with different colored traces; the red trace has the least damping and the green trace has the largest damping. The ratio f/f_n is the ratio of forcing frequency to the natural frequency; resonance occurs when $f/f_n = 1$. Note the amplitude axis in Figure 1 is log scale so the amplification with low damping at resonance is 50 times higher than without resonance. Resonance can be thought of as an amplifier so that it will increase the response of a given input forcing frequency amplitude by the amplification factor; if the force input amplitude is reduced by 50% the response will be reduced by 50%. If the forcing frequency is reduced or increased the amplification factor will change according to the curves in Figure 1; a slight change in frequency above or below the natural frequency will have a significant effect on the vibration response amplitude.

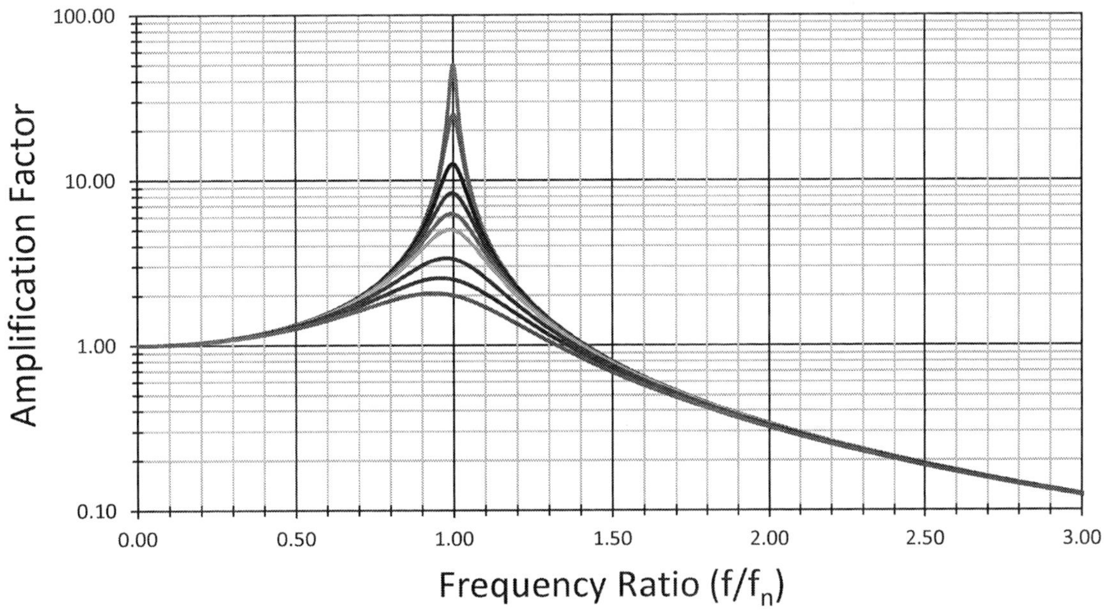

Figure 1. Amplification v.s. natural frequency response for constant amplitude force; multiple damping curves. Lowest damping (red trace). Highest damping (green trace).

A real machine system has many natural frequencies. These natural frequencies and their associated mode shapes are primarily controlled by the mass and stiffness distribution. Usually when resonance is a problem one or more of the lower natural frequencies may be near a machine forcing frequency like 1X shaft speed frequency. In some cases two or more modes may be near the same frequency and the response is affected to some degree by these multiple modes. For the 1X shaft speed resonance case a "normal" amplitude force from residual rotor mass imbalance, hydraulic imbalance, bent shaft, etc. can excite a natural frequency and create a much higher amplitude at 1X than would be present without resonance. There can be many forcing frequencies depending on the machine design and use; for vertical pumps 1X is the most common issue to cause high vibration amplitudes.

Typically the solution to resonance vibration involves moving the natural frequency well away from the forcing frequency or the forcing frequency will be changed to create sufficient frequency separation margin. Damping can be increased to reduce the amplitude response at resonance, but this is often less practical. Reducing the applied force magnitude from the machine generated frequencies will also reduce the vibration response amplitude. Note reducing the force magnitude will have a corresponding effect on the response amplitude, but the resonance response amplification ratio is typically much higher than 1. Often the amplification from resonance can be 5-20 times higher than without resonance. So performing a very precise field balance on a machine will have some effect to reduce vibration amplitude at 1X resonance, but it is usually far less effective than increasing frequency separation margin and it is not easy to reduce forces and keep them low throughout the life of the machine.

To increase or decrease a natural frequency a full understanding of the problem mode shape and the machine mass/stiffness distribution is needed. Real machines are complicated and often it is difficult to understand or predict the mode shapes and natural frequencies without finite element modeling. After an accurate finite element model is created a trial and error process can be used to find the optimum modification of the system mass and stiffness distribution to provide a sufficient frequency separation margin. Typically a frequency margin 20% above and 15% below the forcing frequency is used as a design criteria. Most machines don't need margins this large to have acceptable vibration amplitude, but the larger margin mitigates the effects of unknown variables and errors. It is also critical that a modification does not move another natural frequency near the same or another machine forcing frequency.

Often a single degree of freedom (SDOF) system is used to describe natural frequencies; see figure 2. The simple formula often cited is $\omega = \sqrt{K/M}$. Where ω is the natural frequency in radians/second of a simple system consisting of one discrete mass (M) and one discrete spring or stiffness (K). This system with one degree of freedom can only have one natural frequency. A real machine does not have one mass and one stiffness; it has a continuous distribution of mass and stiffness. Therefore, a real machine will have an infinite number of natural frequencies. Usually a real machine will have one or several relevant natural frequencies that are near a machine forcing frequency so all the other natural frequencies and modes not near a forcing frequency are of no concern. Higher frequency modes will have relatively lower motion than the lower frequency modes.

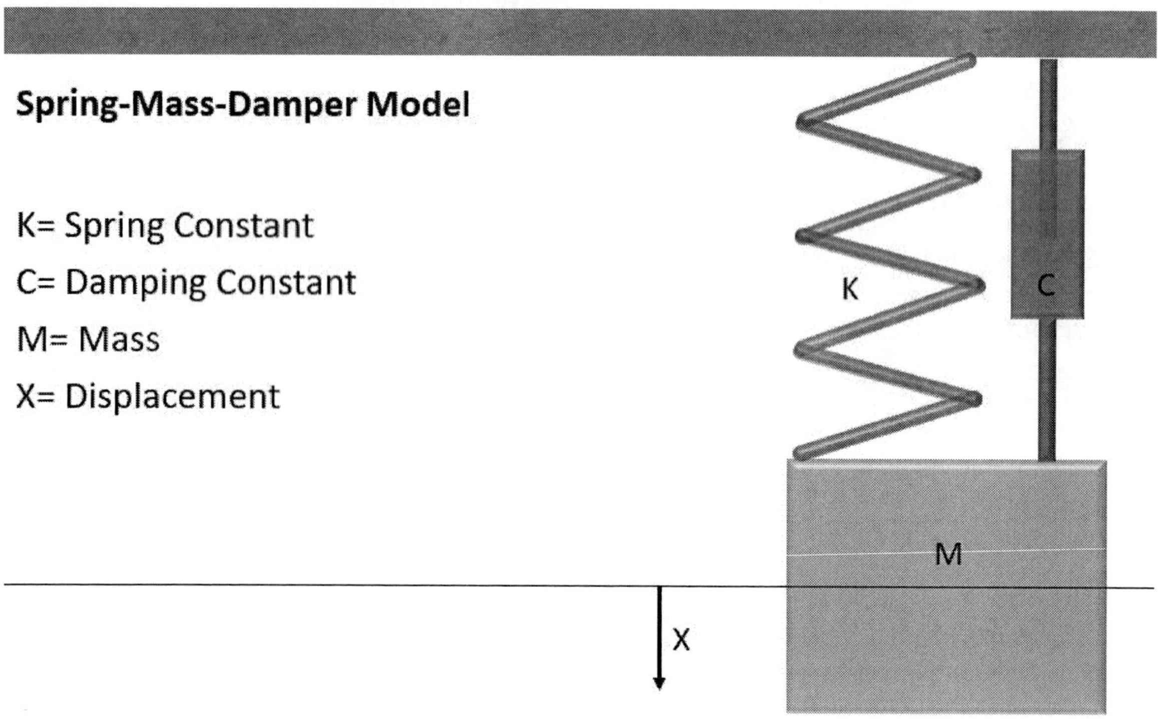

Figure 2. Single degree of freedom model.

MODELING:

Finite Element Modeling, later referred to as modeling, is a mathematical process of approximating the real machine system that has a continuous mass and stiffness distribution with a simplified model of discrete cells that have an associated mass and stiffness. The finite element method is then used to simultaneously solve the partial differential equations related to each cell or finite element. These solutions provide the Eigen values or natural frequencies and mode shapes for the simplified model system over a frequency range specified; natural frequencies not near forcing frequencies are typically not a concern. Note the simplified model likely has a frequency range that it will be valid or provide reasonable output; higher frequencies may not be accurate if the model does not have sufficient details and geometrical accuracy. The finite element model is an approximation to the real system. The approximate model may also generate modes that are not observed on the real machine and are sometimes ignored if the modes of concern a valid. The analyst must understand the machine, the experimental modal data (if available), and the model assumptions to ensure the predicted modes and natural frequencies are reasonable.

Often details regarding components in the system are unknown or are too complex to model with high fidelity. In these cases an even more simplified models of these components may be used and the geometry and material properties will be adjusted to provide a reasonable response in the model at the frequencies of interest. Poor assumptions in this process can create significant errors in the model output, but the modeling time and cost will be significantly reduced. The goal is to provide enough model detail to have sufficiently accurate model output while minimizing the modeling time and cost. Inaccurate but simple assumptions may be chosen to reduce modeling time and cost, but the accuracy of the results can be significantly affected.

Components that frequently create errors in modeling are the motor, foundation, piping, and fluid added mass effects.

MOTOR MODELING:
Unlike the pump model the motor model is typically a very crude approximation of the actual machine. A motor has complex internal structures and supports that are not provided by the original equipment manufacturer (OEM) to protect their intellectual property. In addition there are flexible components like stator windings that are very complicated to model and again the OEM will not provide details even if it was desired to model the motor with full fidelity and details. The motor rotor is also a complication for a static structural analysis; its influence depends on the design and the bearing system. Motor base designs may be sensitive to bolt torque; the reed critical frequency can vary with bolt stress/torque. This reed critical frequency variation is more likely with small base contact areas and fewer bolts; poor surface contact and bolt forces can create variable stiffness. So the designer is left to create a crude and simplified approximation of the motor for the F.E. system model. Changing a motor size or type on an existing machine may create resonance. Even if the horsepower rating is the same a different OEM or model will likely have a different mass and stiffness. An FEA should be conducted when a motor size or type is changed.

FOUNDATION MODELING:
A concrete foundation is common on vertical pumps. The concrete foundation may at first consideration seem very rigid. Many designers make this rigid assumption in the modeling process and significant errors can occur depending on the system design. In fact concrete is typically not very rigid compared to the stiffness of pumps, motors, etc. This is especially true when slabs are relatively thin and cover large spans like in a cooling tower application. The concrete stiffness is likely difficult to calculate and varies with grade of concrete and even age. Further, there is probably a significant amount of steel rebar in the foundation that provides significant stiffness. You can see that a high fidelity model could include a very large foundation structure with many complex structures. Even with this level of detail there are many unknowns regarding the effective stiffness of the foundation where the pump is attached. The pump contact to the foundation is another variable that can be affected by bolting (bolted joint effects) and imperfect contact. A degraded foundation can also have cracks, loose anchor bolts, loose soleplates, and other damage that can significantly reduce the effective stiffness.

Unfortunately designers often remove all these concerns and significantly simplify the modeling process by assuming the pump discharge head is fully constrained at the foundation. This allows a much smaller model, faster processing, and fewer difficult questions that must be answered. The only problem is this short cut may create significant errors in the predicted natural frequencies and mode shapes. Structural steel foundations probably have more negative impacts on the modeling process because the stiffness is usually even lower than a concrete foundation and the steel structure will probably combine with the pump, motor, and piping to create even more system natural frequencies and mode shapes. A detailed system model including the foundation should always be used when a vertical pump is mounted on a structural steel foundation.

PIPING MODELING:

The effect of piping on a vertical pump with regard to natural frequencies can range from insignificant to extremely important. The piping location on the pump structure, the mode shape of concern, and the type of pipe connection to the pump are the primary issues to consider. Flexible joints often limit the piping effect on the pump system. Relatively small diameter piping that connects to the pump near the foundation may have limited effect for most modes of concern. Large and rigid piping that connects near anti-nodes (locations of large motion) for the modes of concern may add significant modal mass and stiffness. Pipe supports, tie-rods, and anchors can have a significant effect and should be fully understood. The designer conducting the modeling must evaluate how far to extend the system model with regard to piping. There are a couple techniques to assist with evaluating piping models. The first is to conduct a sensitivity study. Add increasing levels of piping detail and calculate the model output. If adding more piping detail (further away from the pump) has little effect (for the modes and frequencies of concern) then you can probably stop the piping model at that point. The second method is to use experimental modal data to calibrate the system model. If the system model does not provide accurate mode shapes and frequencies compared to the actual physical machine then more or more accurate details need to be added to the model.

FLUID ADDED MASS EFFECTS:

Fluid mass contained inside the pump and piping directly affects the effective modal mass and must be added to the model. In addition fluid mass surrounding the pump also interacts with the pump structure and increases the effective mass of the pump. This interaction is complicated, but essentially as the pump moves through the surrounding fluid the fluid must accelerate and move around the pump. This acceleration of the surrounding fluid creates forces on the pump that collectively provide an equivalent mass equal to the displaced volume or the volume of the pump. The added mass calculations are inviscid and assume high Reynolds numbers. There are empirical factors to adjust the effective mass based on the aspect ratio (L/D) of the submerged parts of the pump and based on a surrounding can or pipe structure around the pump. A surrounding can or cylinder around the submerged pump structure can amplify the added mass effect and a low aspect ratio can reduce the added mass effect. Fluid added mass is not modeled directly in the analysis; an equivalent mass is applied to the structure by increasing the effective density of the parts that have fluid mass effects.

CASE HISTORY:

A case history on a large VS1 cooling water pump (Figure 3) is presented next to demonstrate the possible effects of the motor and foundation modeling on system model results. The motor is rated for 3350 H.P. and turns at 506rpm. This case history goes through the modeling process, compares system model output to actual field measured vibration data, and describes the mode shapes produced by this common VS1 pump design. Mode shapes are important when collecting vibration data, evaluating vibration data, and when adjusting finite element models.

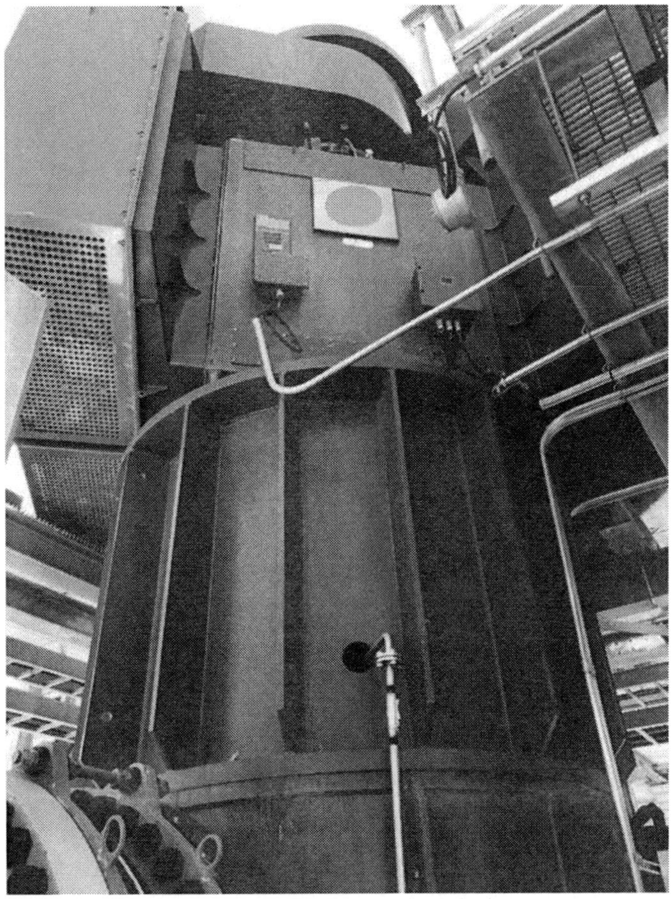

Figure 3. Above grade section of VS1 cooling water circulation pump.

HISTORY:

There are three identical cooling water circulation pumps and all three had a history of high motor top vibration amplitudes since plant commissioning. The pump OEM installed a more rigid motor stand on all three pumps but vibration amplitudes were still excessive. The owner even tried adding threaded jacks under the motor corners to increase stiffness; this did not improve the vibration amplitudes significantly. Peak vibration amplitudes were near .7ips and the motor motion was elliptical. The major axis of the vibration was approximately 30 degrees left of the pump discharge piping (Figure 4). The orbit shown is the motor top structural orbit created with two accelerometers oriented orthogonally. Resonance was suspected due to the elliptical vibration shape. Impact test results were not acceptable due to the background noise of other operating pumps and the low frequency of 1X operating speed. A high average FFT spectrum was able to identify natural frequencies near 1X running speed frequency on both operating pumps (Figure 5). ODSFRF data was used to obtain modal data (Figure 6). ODSFRF is a feature of MeScope that uses operating vibration to excite natural frequencies; these FRFs can then be used to animate a 3D model. This animation of field operating data is an operating deflection shape (ODS) but natural frequencies and mode shapes can be identified and presented. The ODSFRF shape for the problem resonance frequency of 8.48Hz or 1X is shown in Figure 7. Note the axis of motion for this mode is indicated with the yellow dashed line; it corresponds to the major axis of the orbit ellipse.

MeScope is a 3D animation software that allows animations of a model using operating vibration data (ODS) or modal vibration data (Modal). ODS and modal animations present relative motions of a machine model for a given frequency. Modal animations use the same or similar model to present natural frequency relative motions (mode shape). The amplitude scales are selectable, but relative motion at each point is maintained. A time based animation is possible but requires a sensor for every measurement point and is less common.

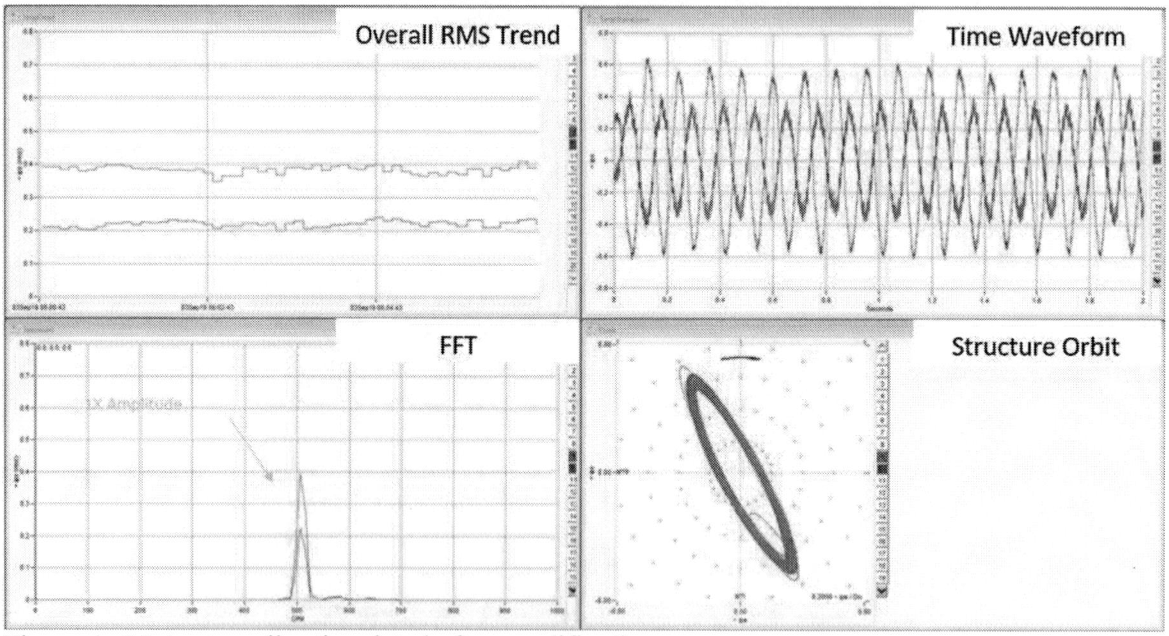

Figure 4. Motor top vibration data before modifications.

Figure 5. High average FFT from motor top accelerometers (zoomed axes near 1X noise floor).

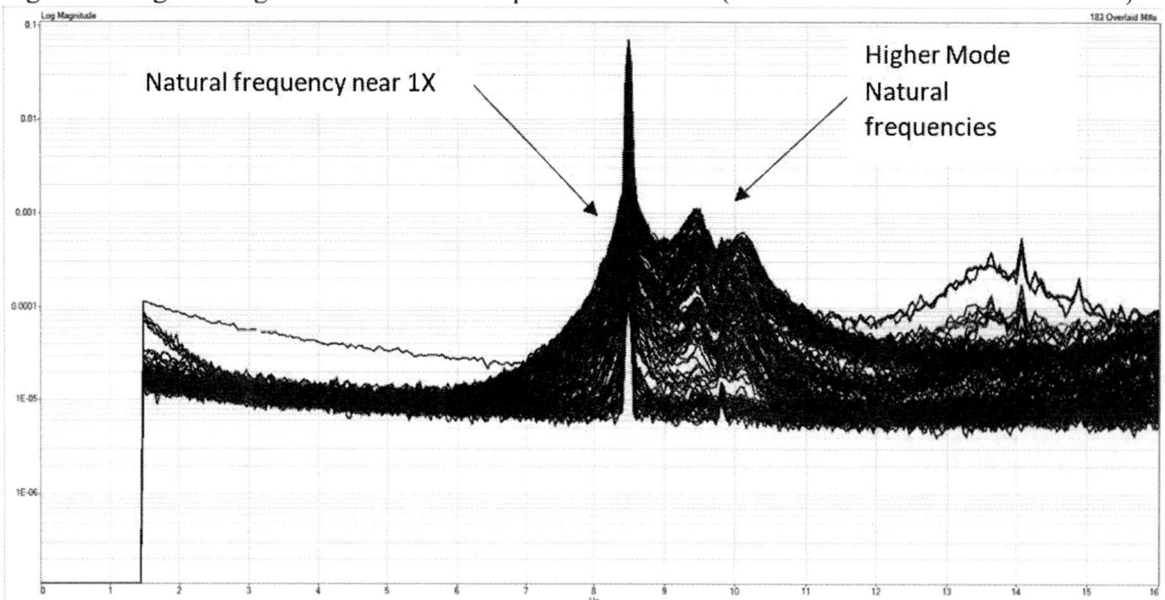

Figure 6. ODSFRFs for ODS test (all measurements overlaid).

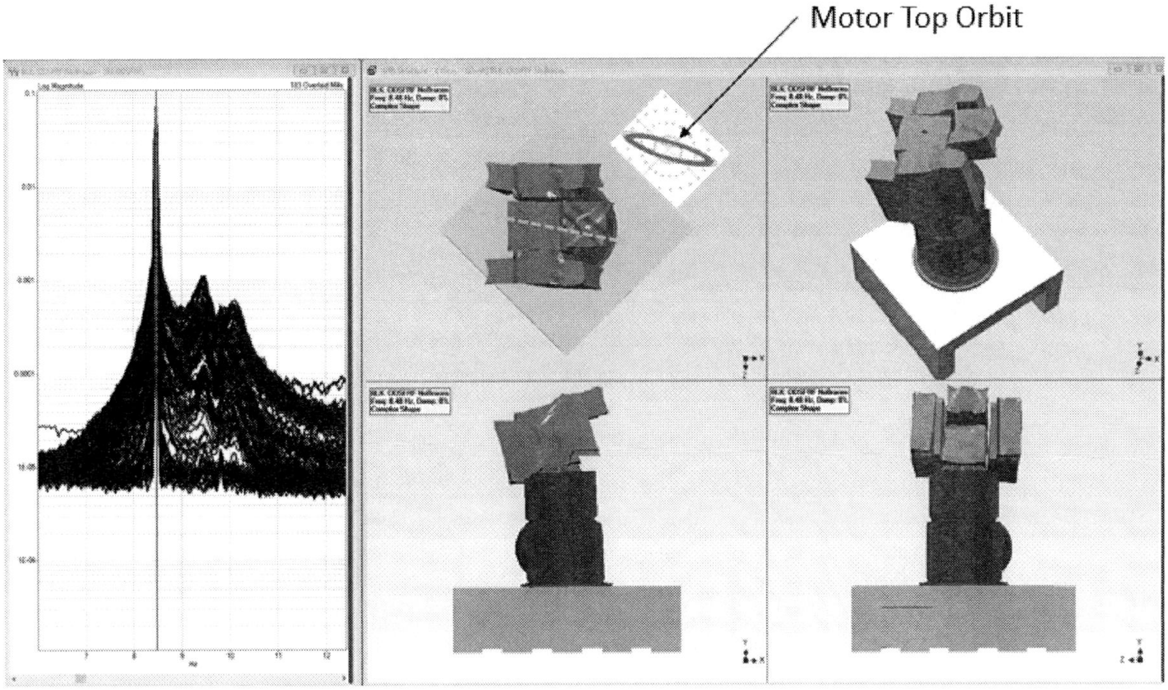

Figure 7. Mode shape for 8.48Hz frequency; resonant frequency at 1X. Measured actual field modal data.

SYSTEM MODEL:

Finite element modeling was needed to properly design a solution for the high vibration amplitudes and to fully understand the system characteristics. Usually most of the system on a vertical pump is below grade or inside a can. Without access to the entire structure an accurate modal or ODSFRF visual animation cannot be made. The ODSFRF data will only allow visualization of the above grade part of mode shapes. The lack of measurement points complicates the model tuning process. Most natural frequencies can be identified with ODSFRF data, but the finite element model is needed to understand the mode shapes. The final system model mesh is shown in Figure 8. More than 75000 cells were required and this does not include the foundation used in the full system model. Piping was excluded because of the soft expansion joint and the location on the pump near the foundation; this is an approximation that will certainly have some small effect. It seemed the effect of piping would be small based on experience and the unknowns in the piping and expansion joint were likely greater than any effect on the system model. The problem mode near 1X shaft speed frequency is shown in Figure 9.

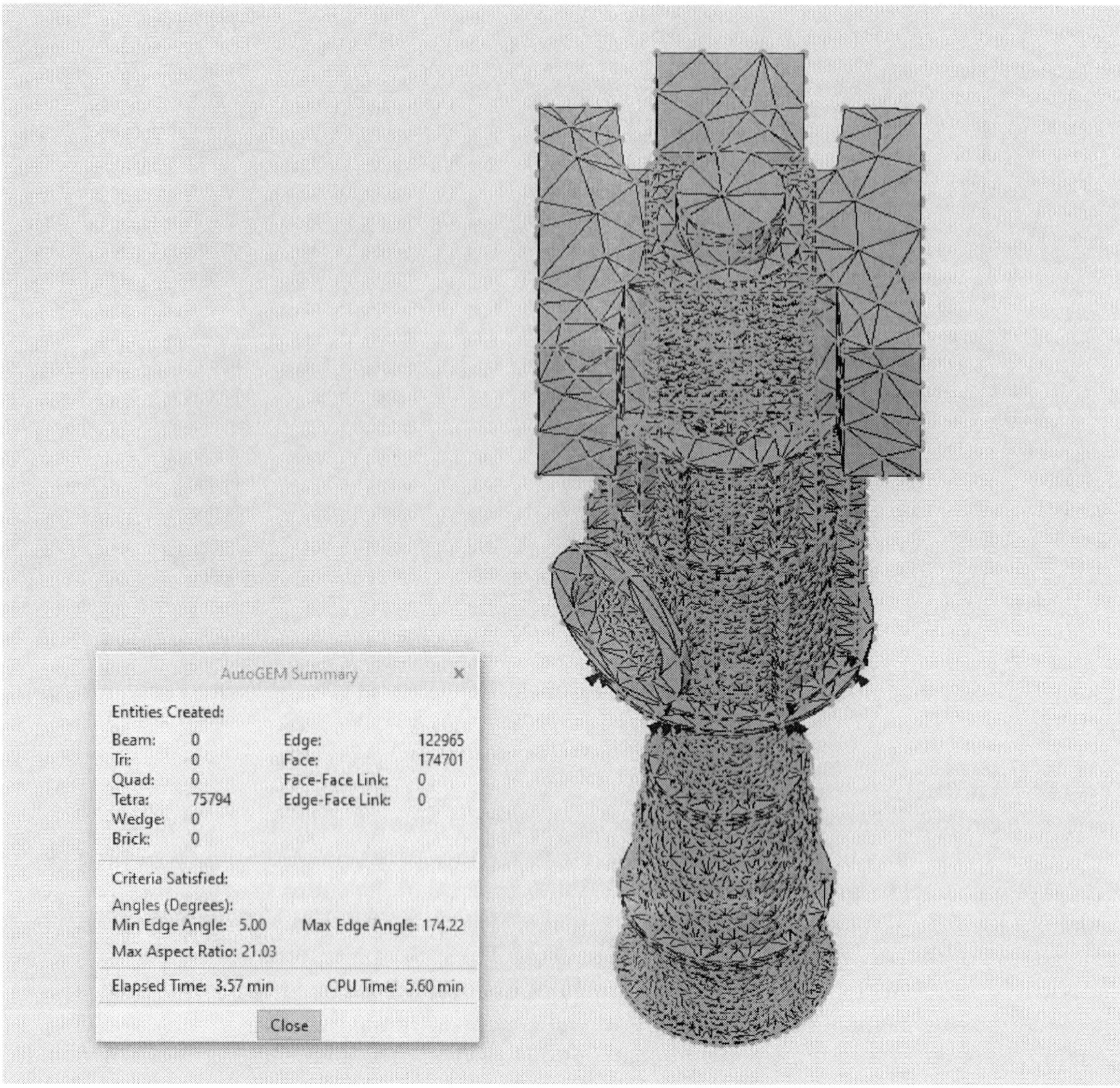

Figure 8. CREO SIMULATE mesh for pump and motor structure.

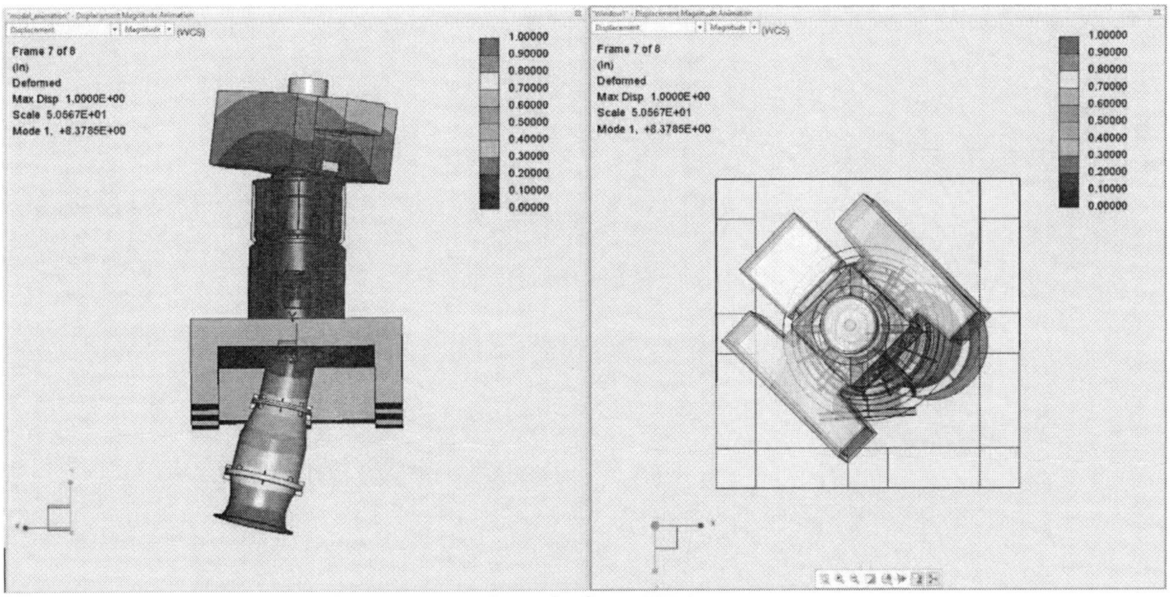

Figure 9. Problem natural frequency and mode from FEA.

Foundation details were available so that an approximate foundation model could be made. Pump details were limited; therefore, a site survey was made to measure the above grade component dimensions. The bowl and end bell were approximated with available drawings and by using similar components from known machines. The component masses were provided to ensure the approximate below grade components had the appropriate mass. It was assumed there would be some error in the model due to limited information. A significant difficulty was the motor model. Motor structural details were not available; mass, center of gravity, and reed critical frequency (RCF) were provided by the OEM. However, only a single RCF was provided with a +/- 10% accuracy. The motor is not axis symmetric so it was assumed there would be two natural frequencies corresponding to the major and minor stiffness axes.

MOTOR MODEL:

A combination of site measurements and outline drawings were used to estimate a motor structural model as close as possible to the actual motor. After a realistic motor structure model was created the RCF of the model was adjusted to match the OEM provided RCF of 13Hz. Later after the full system model was completed further adjustments of the motor model were required to properly adjust the system model. The result was a motor with high asymmetry and lower RCFs than provided by the OEM. The final motor model in-line with junction box RCF was 11.08Hz and the perpendicular RCF was 12.3Hz. (Figure 10) The OEM range for a RCF was 11.7Hz to 14.3Hz (+/-10%). The model required a high asymmetry and 5% lower RCF than the minimum provided by the OEM.

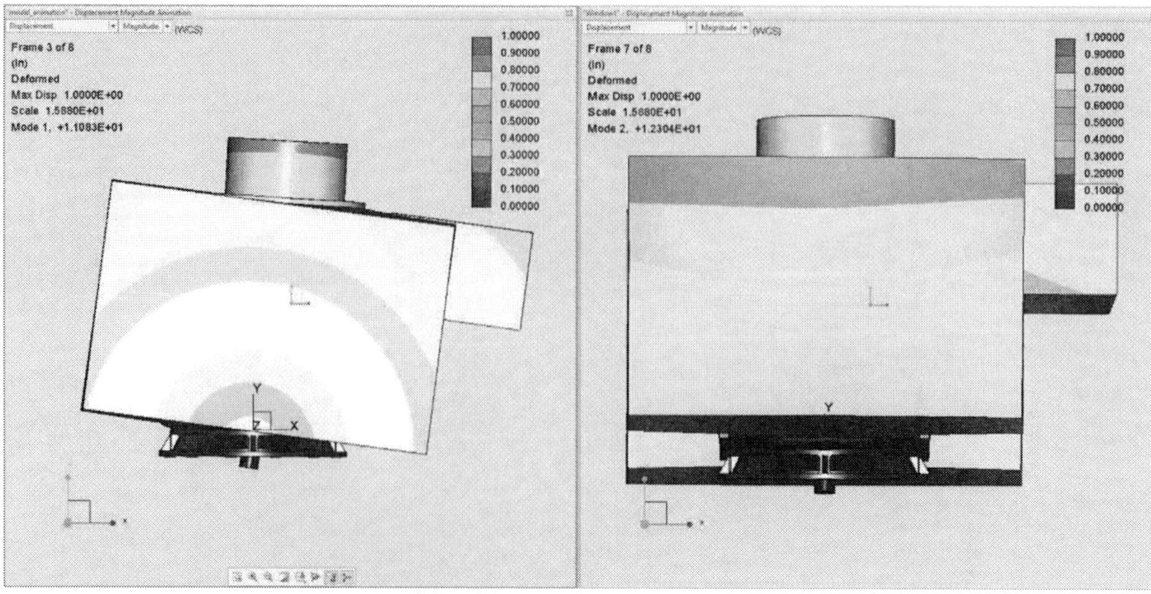

Figure 10. Detailed motor model RCF final adjusted properties.

SYSTEM MODEL ADJUSTMENT (TUNING):
Seven different natural frequencies and 4 different modes from the actual machine operating vibration data were used to adjust the system finite element model. The MeScope animations of these 7 actual natural frequencies and mode shapes (measured with operating vibration data) correspond extremely well with the system model (Figure 11). Model adjustment was a time consuming trial and error process.

Very low frequency separation margin (resonance)

mode	Description	Direction	Actual (operating)	Adjusted Model		% 1X
1	Column / Pivot	Perpendicular	8.40	8.38	-0.2	-0.8
		Parallel	8.60	8.55	-0.6	1.2
2	Motor Reed / Bending	45L	9.45	9.46	0.1	12.0
		45R	10.10	10.12	0.2	19.8
3	Torsional	Torsional	19.20	18.93	-1.4	124.0
4	2nd Bending	45R	24.25	24.94	2.8	195.1
		45L	28.00	28.22	0.8	234.0
		Direction	Hz	Hz	%	%

% difference between FE model and actual natural frequencies

Figure 11. System model frequency adjustment (tuning); un-modified system. The pump OEM previously installed a stiffer motor stand and this is included in the un-modified system.

Higher modes and natural frequencies were not involved with the high vibration amplitude problem, but they were important to ensure the model was accurate. An example of one such higher mode is the 45 degrees right motor reed at 10.1Hz (Figure 12). Note the large amplitude vibration at 1X is in the 45 degrees left direction, but the 10.1Hz natural frequency and mode detected with the ODSFRF is in the 45 degrees right direction. The operating amplitude of this higher mode is very small and is excited by impulse forces and broad band flow related energy. A high number of averages is needed to effectively process these natural frequencies with operating vibration data. If the model is only adjusted with the problem natural frequency it is more likely incorrect model changes can provide a correct natural frequency prediction. However, the model will not be accurate and any design changes or frequency separation margins for other modes will be incorrect. Model adjustment often requires several areas of uncertainty and boundary condition assumptions. An incorrect boundary assumption is unlikely to provide an accurate prediction for several modes. If a boundary assumption does not work for all modes then it is probably incorrect and another area of the model or boundary assumption needs to be evaluated. Adjusting or tuning with many modes is the best way to ensure an accurate model. The resonance issue involved the first natural frequency which is a pivot mode. For the pivot modes the pump/motor structure is essentially rigid and the deflection occurs at the foundation or the discharge head flange area. Figure 13 shows the pivot modes on top and the next higher frequency bending modes on the bottom. For the bending modes the pump/motor structure bends in a "C" shape around the foundation area with little deflection of the foundation and discharge head flange area. The axis for the motor/bending motion is shown in blue and the column motion is shown in red (Figure 14).

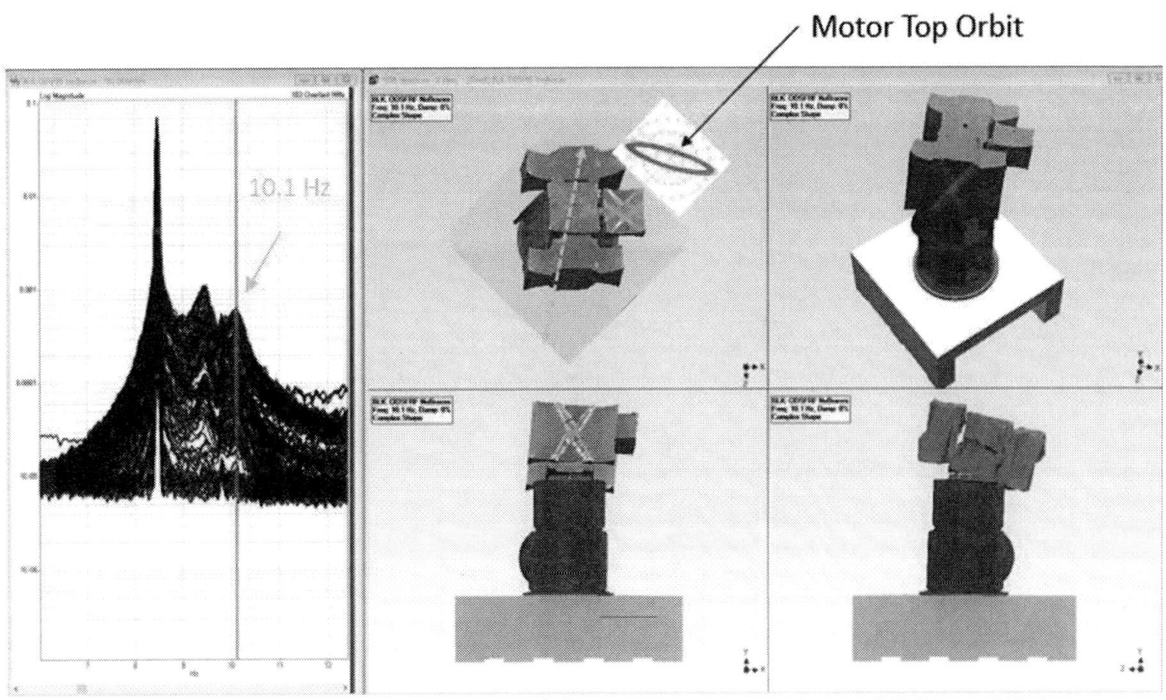

Figure12. ODSFRF shape for 10.1Hz natural frequency; measured actual field modal data.

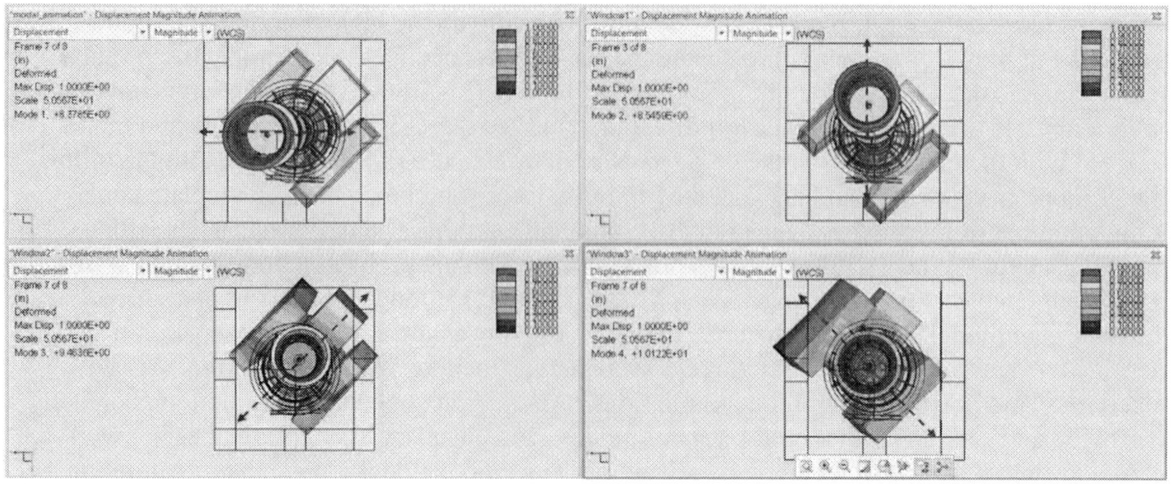

Figure 13. FEA results for adjusted system model. Transparent view from below grade; compare first 4 mode shapes.

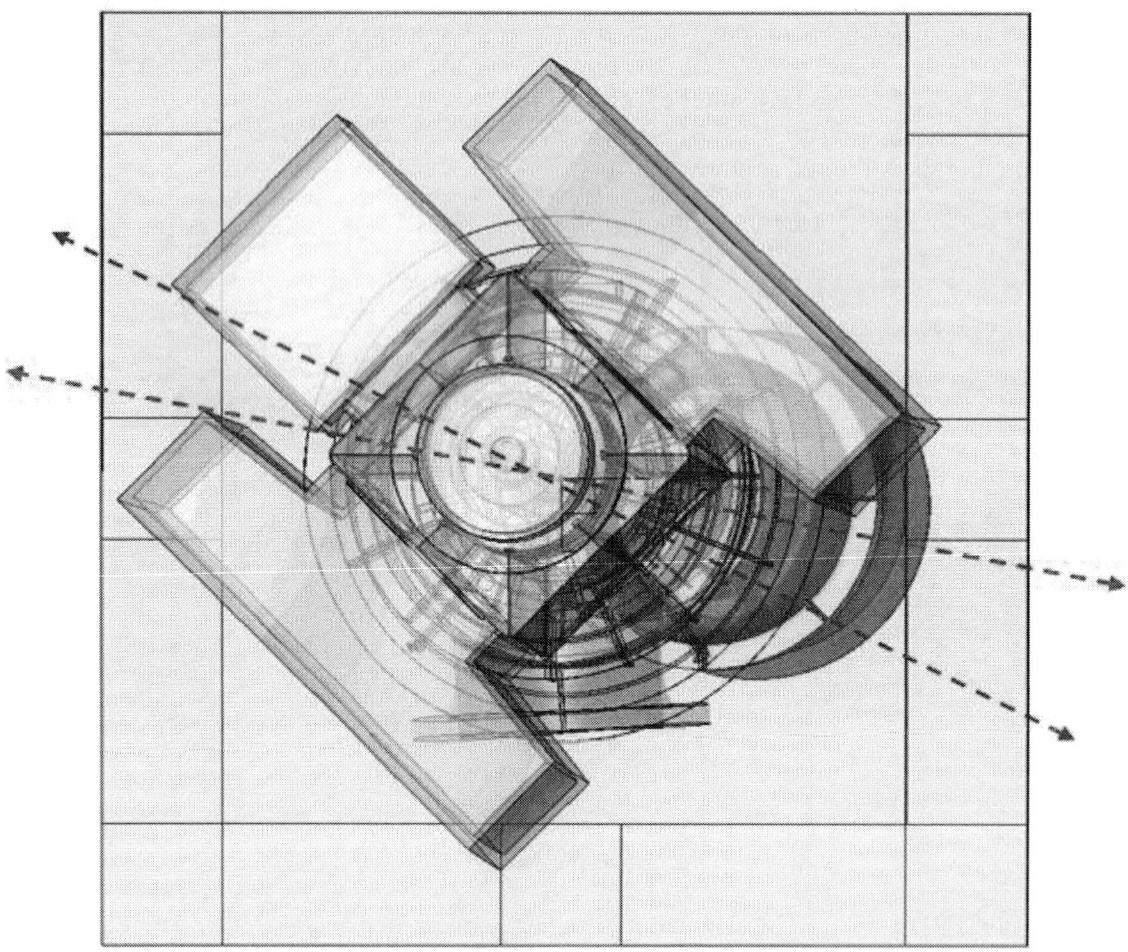

Figure 14. System model 8.37Hz mode shape; transparent to see above and below grade shapes.

A torsional mode was also used in the adjustment process. This torsional mode is a good indicator of the mass moments of inertia accuracy. It is a good mode for adjusting or tuning this complicated motor structure.

Note the below grade pump/column structure (pivot modes) moves in-line with and perpendicular to the discharge plane. For the (bending modes) the motor moves on planes closer to 45 degrees left and 45 degrees right of the discharge plane. Later in this paper a motor model with symmetric stiffness will be evaluated and all modes will align with the in-line and perpendicular to discharge planes. The bending modes mostly involve motor motion and are affected by motor asymmetric stiffness. The pivot modes mostly involve pump/column motion and are not significantly affected by the motor stiffness asymmetry. For all modes there is some transfer of motion or energy across the foundation plane. The foundation tends to isolate the above grade and below grade motions to varying degrees depending on the design of the pump and the foundation mass/stiffness.

MODIFICATIONS:

After the system model was adjusted to provide accurate modal predictions for 7 natural frequencies it considered accurate. A trial and error process was conducted in the model to evaluate possible changes to the structure. Simple and least costly changes were first attempted. Increasing levels of change were applied so that sufficient frequency separation margin could be achieved with the least amount of changes and cost. The final required changes involved adding a few ribs to the inside of the discharge head, adding ribs to the outside of the column, and adding a brace from each motor corner to the motor stand (Figure 15). The orange components in figure 14 required modifications to achieve a predicted 20% frequency separation margin from 1X. Vane-pass frequency is well above the modes measured and considered in the model so 1X margin was the only concern.

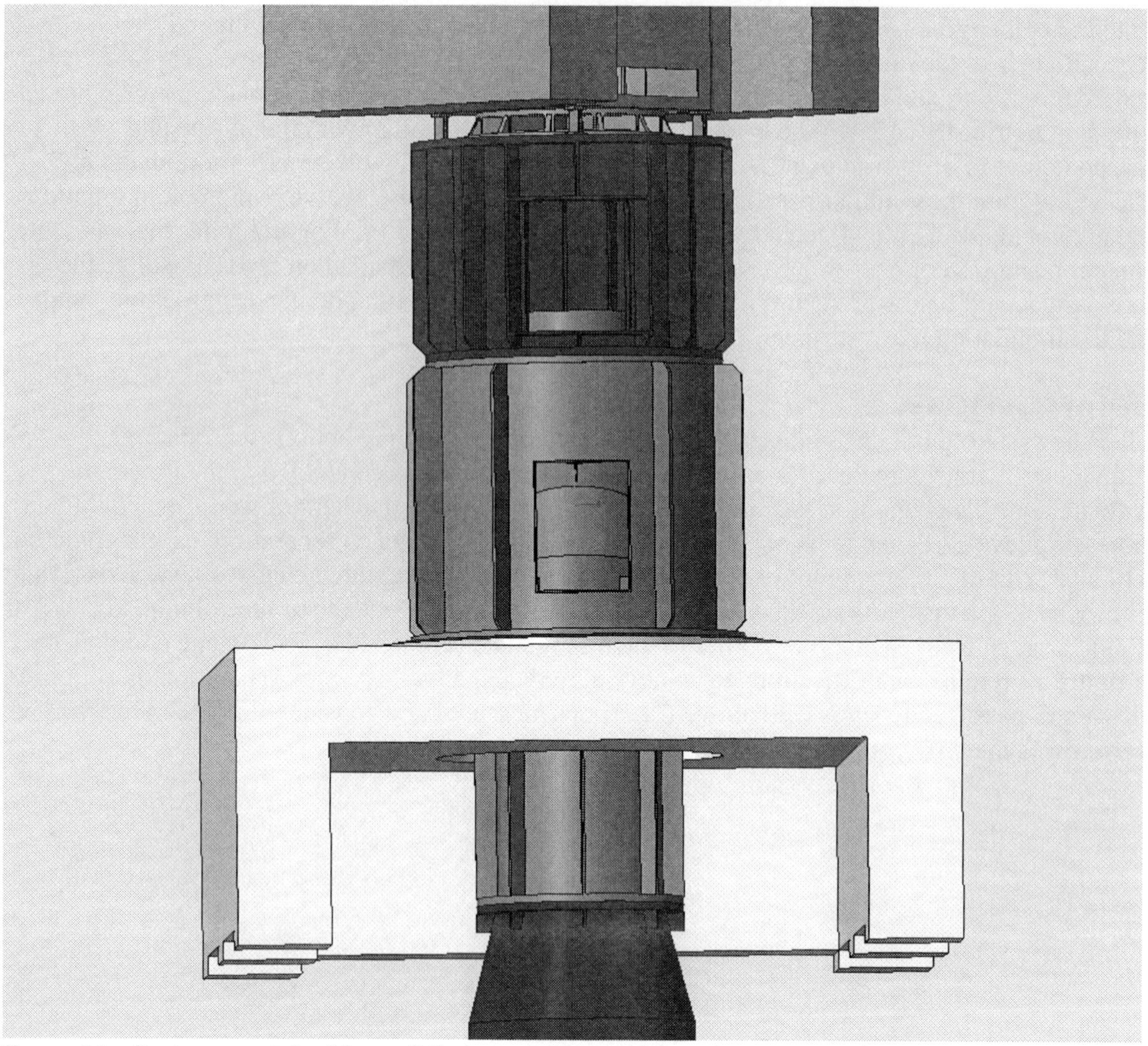

Figure 15. System model with modified components shown in orange color.

MODIFICATION RESULTS:

Site test data after the modification indicated the resonance was completely eliminated and operating vibration was extremely low (Figure 16). The vibration shape at 1X is no longer elliptical. The structure orbit is now circular with no amplification due to resonance. Vibration amplitude reduced from .7ips peak to less than .1ips peak. The vibration before modification can be compared to the after modification vibration; reference Figure 4 before and Figure 14 after modification. Note the amplitude scales on the plots are different because the amplitudes are significantly smaller after modification. Data from the owner's DCS system was also plotted before and after modification in Figure 17.

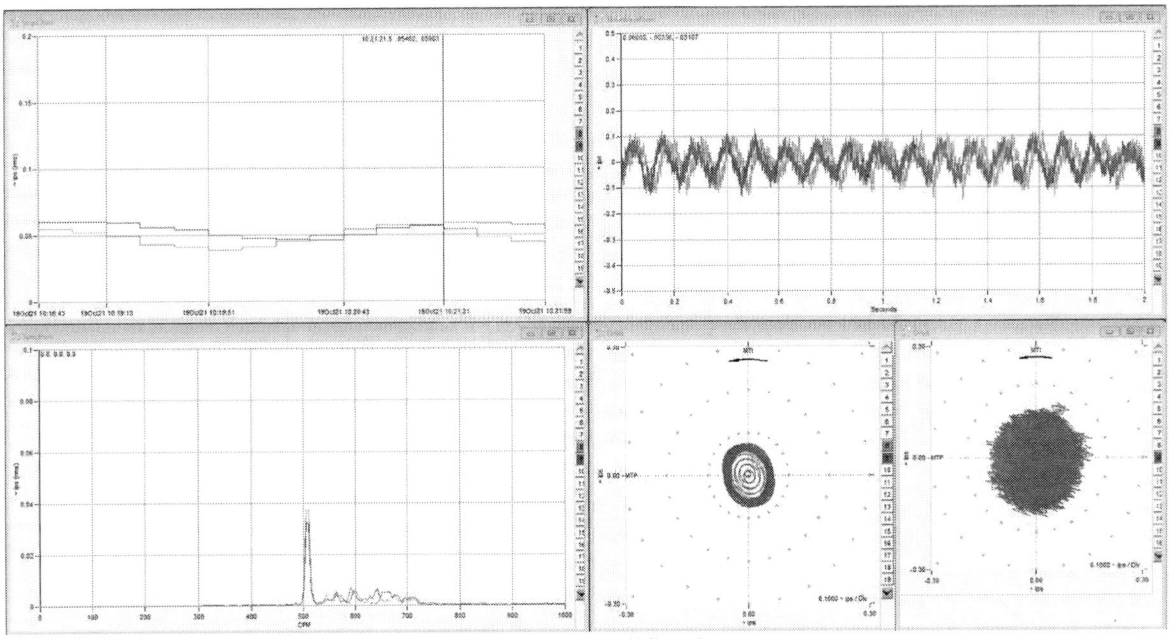

Figure 16. Motor top vibration data taken after modification.

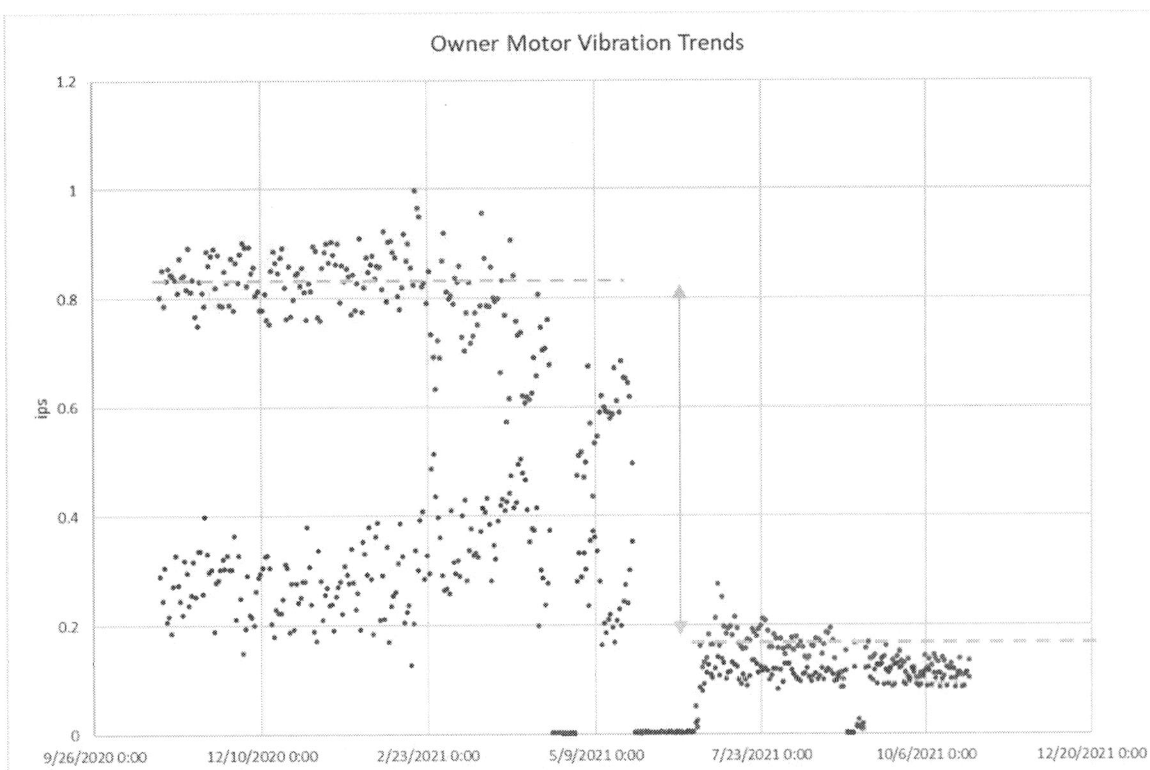

Figure 17. Owner DCS vibration data before and after modification.

The 1X frequency separation margin after modification was significantly lower than predicted; 20% was predicted and only 10% was realized. The modified model predicted natural frequencies are shown in Figure 18. A high average FFT was used after modification to measure the new natural frequencies (Figure 19). Fortunately the predicted margin was high enough to achieve the desired resolution of the resonance vibration problem. The most likely reason for the lower margin is the motor brace design and modeling assumptions.

Very good minimum frequency separation margin

mode	Description	Direction	Actual (operating)	Modified Model		% 1X
1	Column / Pivot	Perpendicular	8.40	10.40	23.8	23.1
		Parallel	8.60	10.15	18.0	20.1
2	Motor Reed / Bending	45L	9.45	11.55	22.2	36.7
		45R	10.10	12.23	21.1	44.7
		Direction	Hz	Hz	%	%

Figure 18. Modified system model FEA results. Modified pump and motor braces.

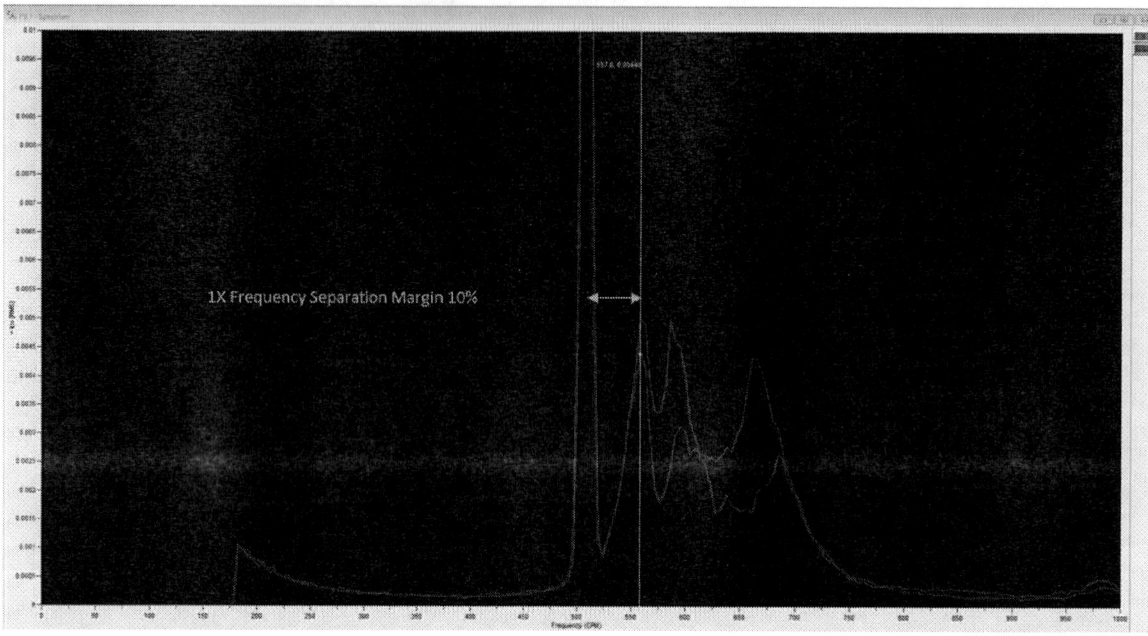

Figure 19. Motor top high average FFT after modification (zoom near 1X and noise floor)

Motor Brace Design:
The motor brace was originally designed to bolt horizontally to the motor structure and vertically to the motor stand. After the motor was removed during the outage existing bolt holes were found on the bottom of the motor. It was decided to modify the motor braces to fit the existing motor holes (Figure20). The new bolt locations caused a shimmed contact and offset contact points that reduced the effective stiffness. An assumed motor brace FEA modal prediction is shown in Figure 21. This modal deflection image shows the potential loss of stiffness due to poor contact at the bottom shim, a gap at the motor casing on the side, and off set centerlines of the bolts.

Figure 20. Final motor brace installation. Planned horizontal bolt design was modified to use a vertical bolt located in an existing motor threaded hole location. This required shims, gaps, and a bolt/contact centerline offset.

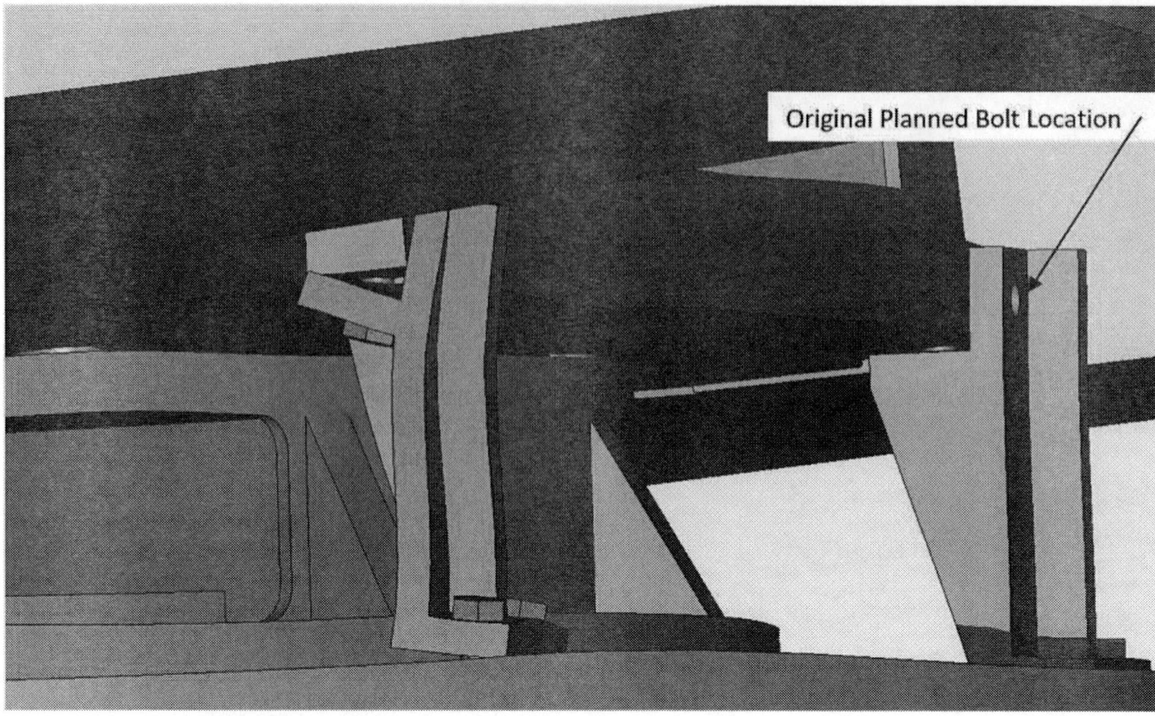

Figure 21. FEA Simulation indicating possible source of effective stiffness reduction due to contacts and offsets.

To evaluate the sensitivity of the modified system to the motor braces a simulation was made with all of the actual modifications except the motor braces. This simulation without motor braces gives a lower natural frequency limit if the motor braces are completely ineffective. The modified pump simulation without motor braces predicts 7.9% separation margin (Figure 22). This margin is less than the 10% realized but much less than the 20% for the ideal modified system simulation.

Pivot Mode Margin Decreased 12% Without Motor Braces

mode	Description	Direction	Actual (operating)	Pump Mod Only		% 1X
1	Column / Pivot	Perpendicular	8.40	9.12	8.6	7.9
		Parallel	8.60	9.79	13.8	15.9
2	Motor Reed / Bending	45L	9.45	10.78	14.1	27.6
		45R	10.10	10.97	8.6	29.8
		Direction	Hz	Hz	%	%

Figure 22. System model results; modified pump with added ribs; no motor braces.

It is interesting that the system simulation with motor braces and no pump modifications (Figure 23) provided less than a 1% increase in natural frequency. This shows how the relative stiffness of all the components in series must be raised or the lower stiffness components will control the system effective stiffness. Recall the effective stiffness K_{eq} for springs in series is given by the equation $1/K_{eq} = 1/K_1 + 1/K_2 + 1/K_3$…… This shows that lower stiffness components in a system will have a significant or controlling effect on the system stiffness when in series. A vertical pump can be visualized as an approximate system of springs in series. The motor is in series with the motor stand. The motor stand is in series with the discharge head. The discharge head is in series with the foundation, etc. This is why only changing stiffness on one component has little effect, but moderate changes to all or most components has a significant effect. Further this is the best explanation why the lower than expected realized margin is likely related to the motor braces not being well connected or having reduced effective stiffness. The motor brace connections added another lower stiffness spring in series that reduced the motor brace effective stiffness in the system. A second pump was recently modified with improved motor braces/contacts and is expected to raise the frequency separation margin on this second modified pump near the simulation predictions for the ideal modified system.

Very Small Improvement In Pivot Modes(<1%); Motor Modes Increased Significantly(20%)

mode	Description	Direction	Actual (operating)	Braces Only		% 1X
1	Column / Pivot	Perpendicular	**8.40**	8.42	0.2	-0.4
		Parallel	**8.60**	8.64	0.5	2.2
2	Motor Reed / Bending	45L	**9.45**	11.08	17.2	31.1
		45R	**10.10**	11.86	17.4	40.4
		Direction	Hz	Hz	%	%

Figure 23. System model results; no pump modifications; motor braces added. Simulation with motor braces as the only modification.

This springs in series approximation is also a good way to visualize why the OEM's initial upgraded motor stands had little effect in resolving the natural frequency issue. The original and less stiff motor stand was simulated to understand the original system before the OEM attempted to raise the problem natural frequency by installing a more rigid motor stand (Figure 24). The stiffer motor stand could not effectively stiffen the system because components with much lower stiffness were in series. The motor stand was not changed during the modification described in this paper because it was already stiff enough to be effective with the new modifications. The original OEM motor stand was simulated in the un-modified system model and this demonstrated the original motor stand moved a higher mode frequency (column parallel at 8.60Hz) lower and created a different resonance for this different natural frequency (Figure 25). It is interesting that the stiffer motor stand installed previously by the pump OEM moved the problem resonance from the column/pivot parallel mode to the column/pivot perpendicular mode. These two modes/natural frequencies are very close so the minor effective stiffness change in the system moved one slightly higher than 1X and moved the one slightly below 1X closer to 1X with no net vibration amplitude improvement. In the FRF plot you can see the 1X frequency is actually affected by three natural frequencies (Figure 25). The two column/pivot natural frequencies (perpendicular/parallel) and the motor reed/bending 45L natural frequency combine their effects on the system response at 1X because their bands of amplification overlap at 1X. The three natural frequency responses are estimated with the yellow dashed lines in Figure 26. The system response is a summation of the contributions from each mode/frequency interacting.

Original (Previous) Motor Stand **Existing Stiffer OEM Motor Stand**

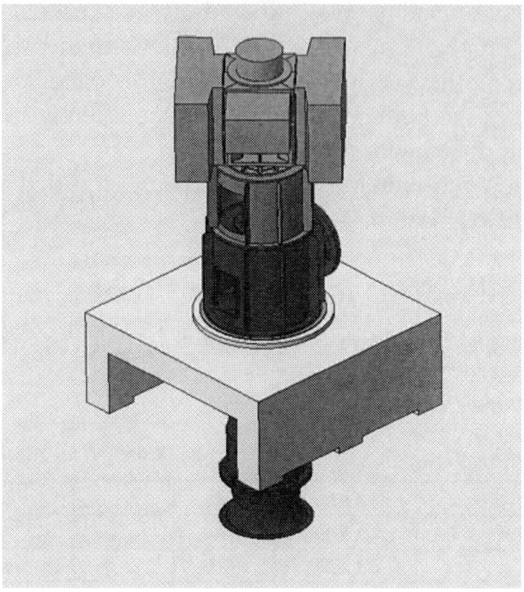

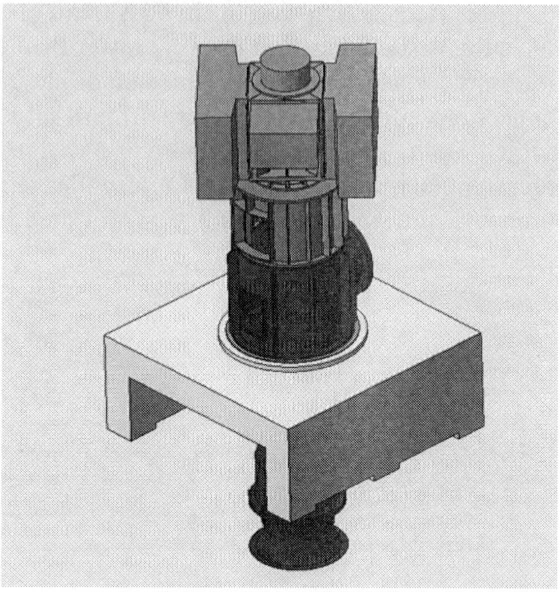

Figure 24. Adjusted models with existing motor stand (as found before the upgrade/modification presented in this paper) and the original OEM motor stand (previously the pump OEM installed a stiffer motor stand).

2% Reduction In Pivot Mode With Original Motor Stand; Note Parallel Pivot Mode Lowered Closer To 1X

mode	Description	Direction	Actual (operating)	Orig Mtr Stand		% 1X
1	Column / Pivot	Perpendicular	8.40	8.26	-1.7	-2.2
		Parallel	8.60	8.40	-2.3	-0.6
2	Motor Reed / Bending	45L	9.45	8.98	-5.0	6.3
		45R	10.10	9.35	-7.4	10.7
		Direction	Hz	Hz	%	%

Figure 25. System model results; detailed motor, original motor stand, and no pump modifications. Simulates the original system before the pump OEM installed a stiffer motor stand.

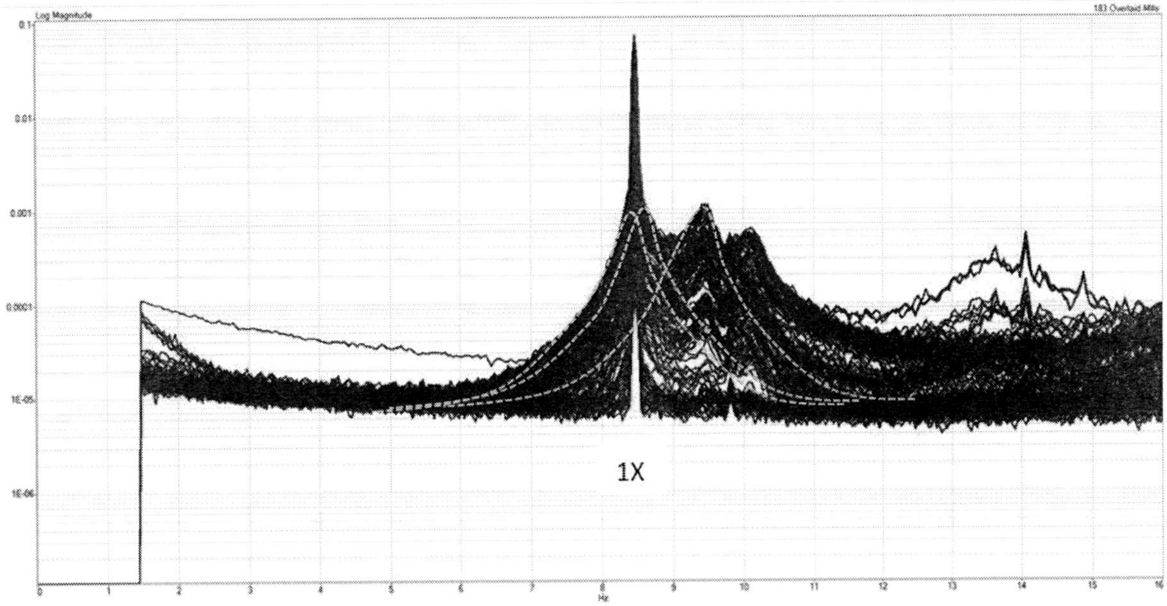

Figure 26. ODSFRFs. Operating data used to measure natural frequencies and animate 3D modal models. Three modes near 1X overlap and affect the 1X forced response; yellow traces are estimates of each mode natural frequency response.

STANDARD MOTOR MODEL:

An industry standard motor model was simulated to demonstrate the possible negative effects of using an inaccurate motor model on system with a complex motor structure. A typical method is to use a homogenous cylinder to model motors (Figure 27). This simple model was adjusted to the motor OEM published mass, center of gravity, and reed critical frequency. This standard motor model is desirable because it requires little time to create a model, is easy to adjust to the motor OEM provided properties, and reduces simulation run time. The simple standard motor model saves time, but in some cases it can provide incorrect system model predictions. It is the author's opinion that the standard motor model can be invalid with regard to mass distribution (mass moments of inertia) and in stiffness distribution. The simple cylinder model will have a curved deflection based on a uniform vertical stiffness distribution; a real motor usually has areas of reduced stiffness near the base so that the motion approximates a rigid body frame deflecting more near the base. This stiffness variation and the different mass moments of inertia can provide invalid responses; this is especially true on taller motors with high aspect ratios and when operating at higher speeds. Higher speeds can excite higher frequency modes where the motor mass and stiffness distribution can be more critical. The detailed motor model RCF (Figure 10) can be compared to the simple standard motor model (Figure 27). The detailed motor had a more realistic rigid body type motion of the motor frame relative to the motor base flange. Note the field modal animations correlate very well to the detailed motor model and they correlate poorly with the standard motor model.

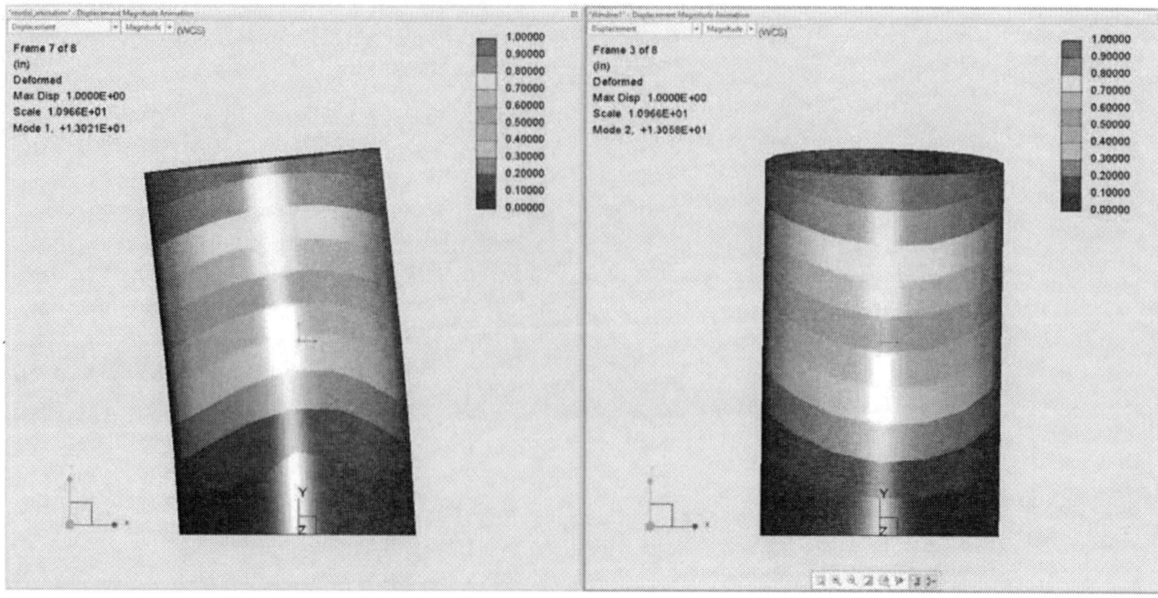

Figure 27. Standard motor model; RCF adjustment results. Homogenous cylinder fixed at the base is adjusted to motor OEM provided mass, center of gravity, and RCF.

The standard motor model was used on the unmodified system to compare the predicted natural frequencies and mode shapes on the original system. The system with a standard motor model predicted similar frequency separation margins for the column modes (Figure 28), but the motor reed/bending modes increased natural frequencies by 7% and 3% (Figure 29). The primary mode causing resonance at 1X is a column/pivot mode; the motor reed/bending mode also contributes to some degree. The standard motor model seems to provide a somewhat reasonable natural frequency for the system problem mode near 1X; however, the mode shape is not valid. Further, adjusting the system model for all 7 natural frequencies required a highly asymmetric motor and significantly lower natural frequencies. This demonstrates an incorrect motor model assumption may provide a "correct" natural frequency for a single mode, but it may not match for many modes and likely will introduce errors in the system model predictions. This is especially important when simulating modifications because the system model is not accurate.

The real physical motor asymmetric stiffness causes all modes near 1X to be modified with a plane of motion shifted to the discharge plane. The system column/pivot mode shapes are compared for both motor models in Figure 30. The motor reed/bending modes for each motor model are compared in Figure 31.

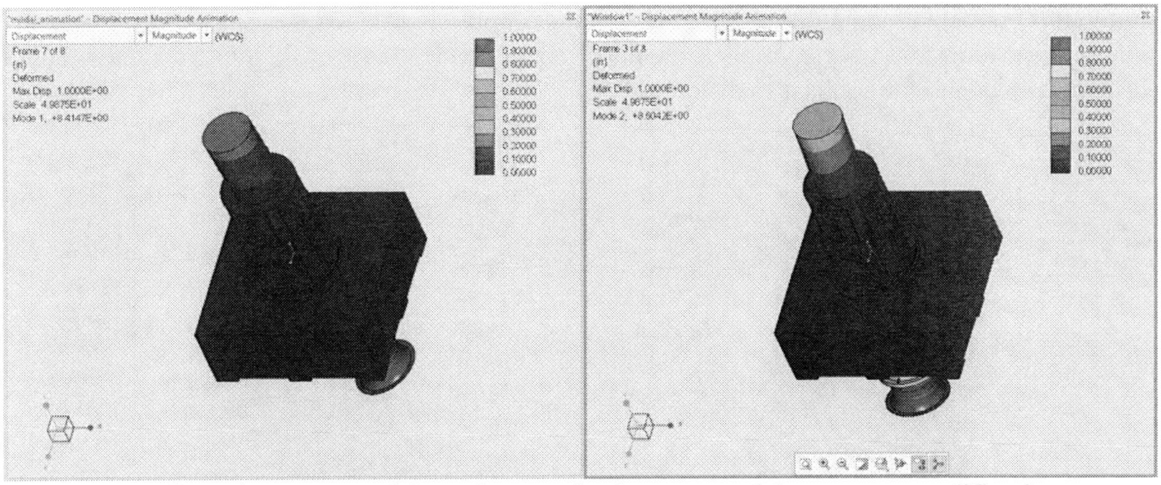

Figure 28. Adjusted system model results; standard motor model; no pump modifications

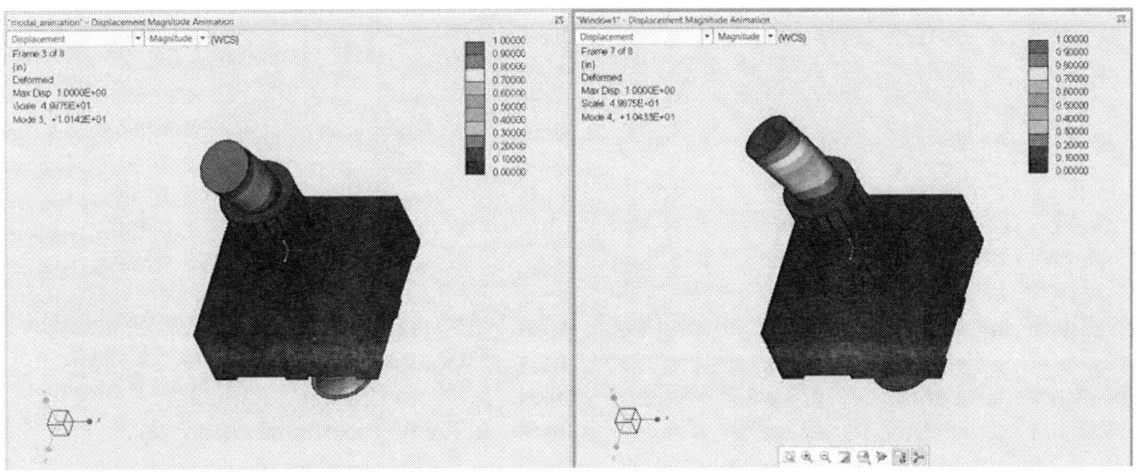

Figure 29. Adjusted system model results; standard motor model; no pump modifications

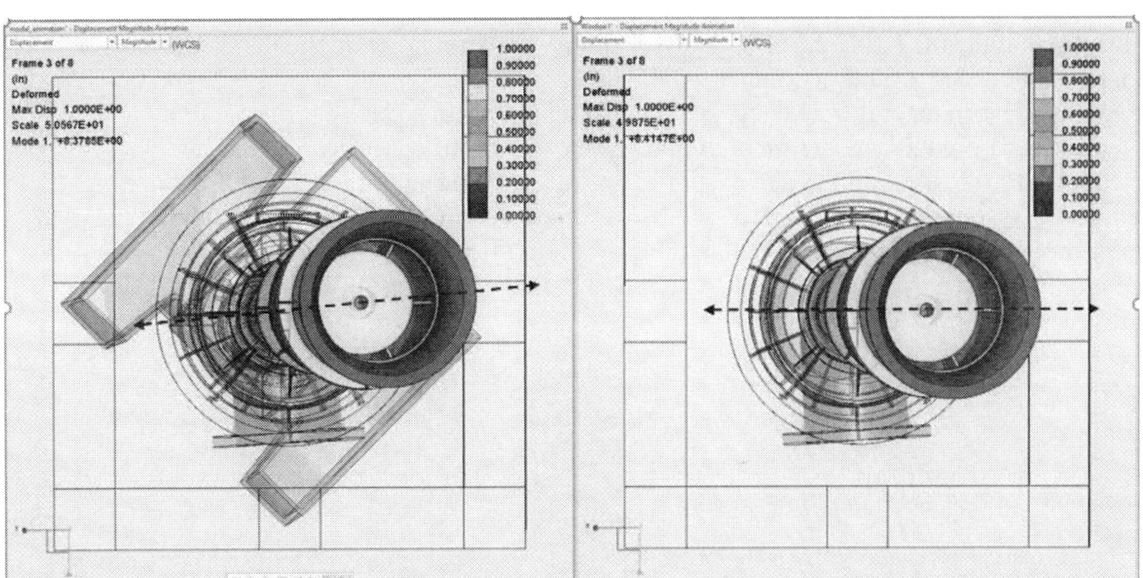

Figure 30. Compare column/pivot modes for detailed motor and standard motor models; no pump modifications; transparent model. Standard motor model has a 1% higher frequency, but the shape or plane of motion is also different.

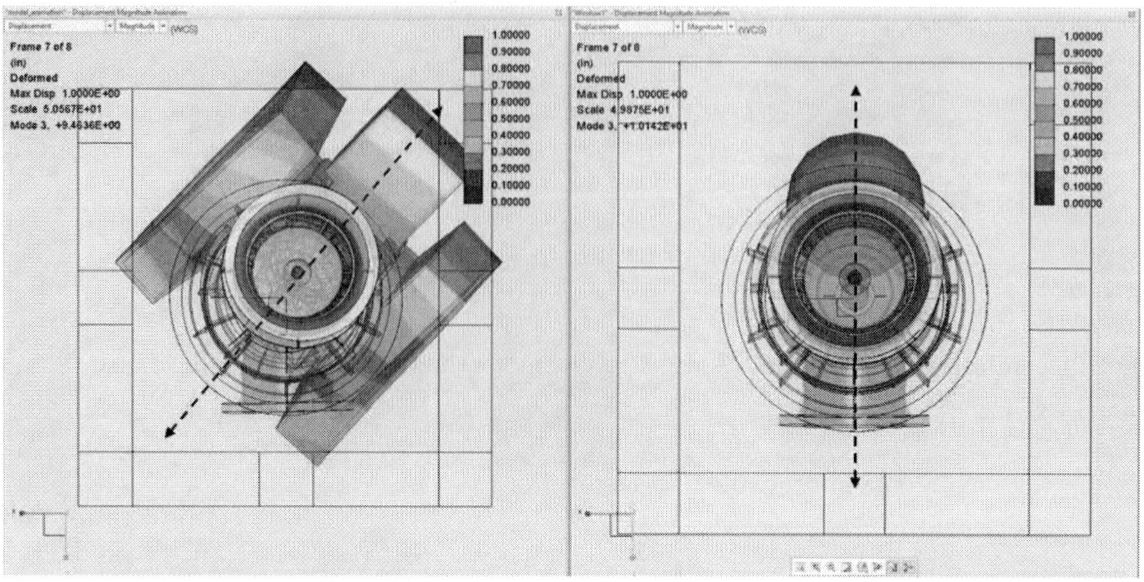

Figure 31. Compare motor/bending modes for detailed motor and standard motor models; no pump modifications; transparent model. Standard motor model has a 7% higher frequency, and the shape or plane of motion is shifted 45 degrees.

The foundation is typical and is somewhat rigid; it is far less than infinitely rigid or fixed. The somewhat rigid concrete foundation allows some transmission of motion/forces between the above and below grade components. This is why the column modes have some angular offset to the discharge plane; the motor stiffness is asymmetric and the major/minor stiffness axes are aligned 45 degrees to the discharge plane.

An extremely important boundary condition to evaluate is the foundation stiffness. The original un-modified and adjusted (tuned) system with the detailed motor model was evaluated with an infinitely rigid or fixed foundation (Figure 32). This is an assumption often used to save time and money in the modeling process. Only changing the foundation in the detailed system model to a rigid boundary assumption increased the problem mode from 8.37Hz to 9.28Hz; this is a 10.8% increase in natural frequency. This modeling error caused by the rigid foundation assumption would indicate marginal frequency separation margin when in fact there is essentially zero margin on the real machine. Installing the standard motor on this same rigid foundation system had little effect on the below grade column modes compared to the detailed motor model because the rigid foundation assumption effectively isolates the below grade column motion from the above grade structure where the motor is mounted (Figure 33).

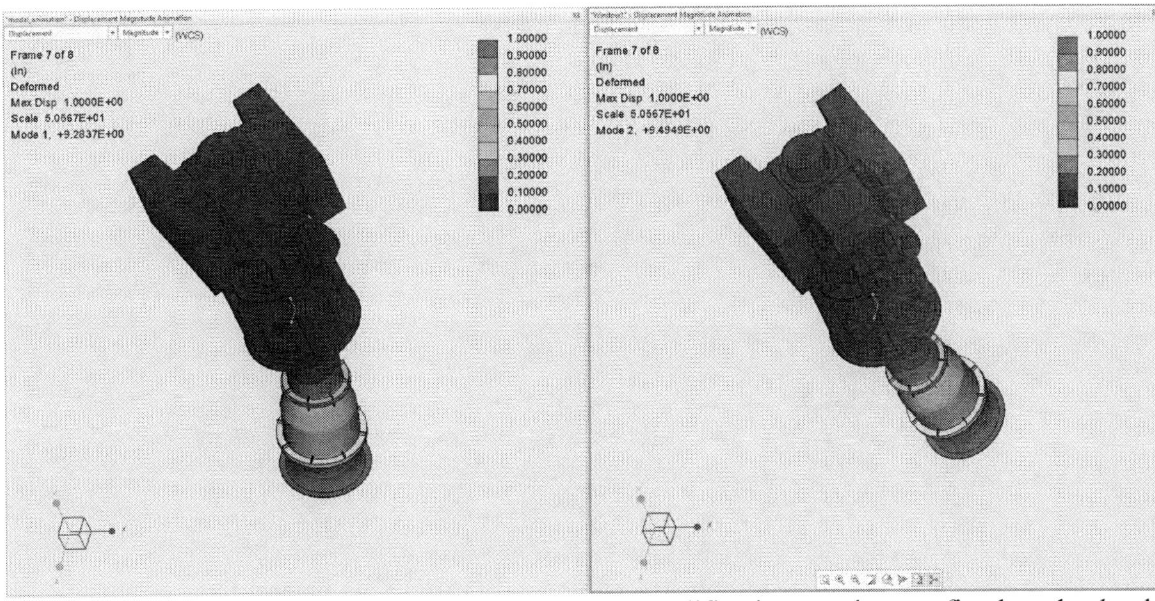

Figure 32. Adjusted system model results; no pump modifications, no braces; fixed or absolutely rigid foundation boundary constraint.

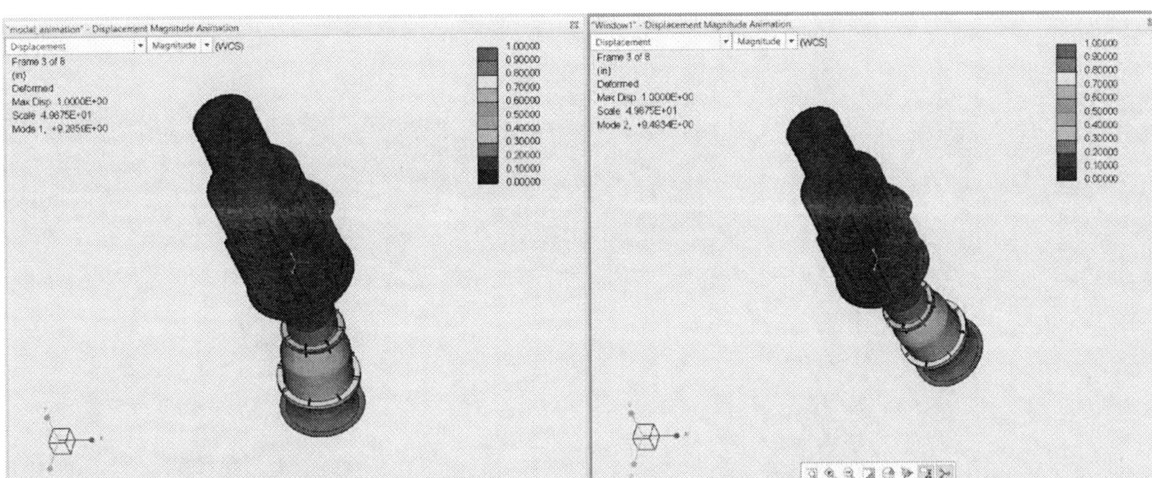

Figure 33. Adjusted system model results; no pump modifications, no braces; standard motor model; fixed or absolutely rigid foundation boundary constraint. This simulates the combined effect of a poor motor model and a rigid foundation boundary constraint.

The above grade motor reed/bending modes are significantly affected by different motor models with the rigid foundation assumption because the motor is a significant component for the above grade modes and the motor stiffness is relevant to these modes. The lower frequency motor reed/bending mode with the detailed motor model (Figure 34) is 8.5% lower than the lower frequency motor reed/bending mode with the standard motor model (Figure 35). Additionally the mode shapes are significantly different because the standard motor model produces motion in-line and perpendicular to discharge and the detailed motor produces motion 45 degrees left and 45 degrees right of the discharge. The standard motor model system is more affected by the small pump stiffness asymmetry (both in-line and perpendicular modes have nearly the same frequency). The detailed motor model system is more affected by the large motor stiffness asymmetry; this is far more consistent with the actual machine vibration shapes and frequencies as compared to the standard motor model system.

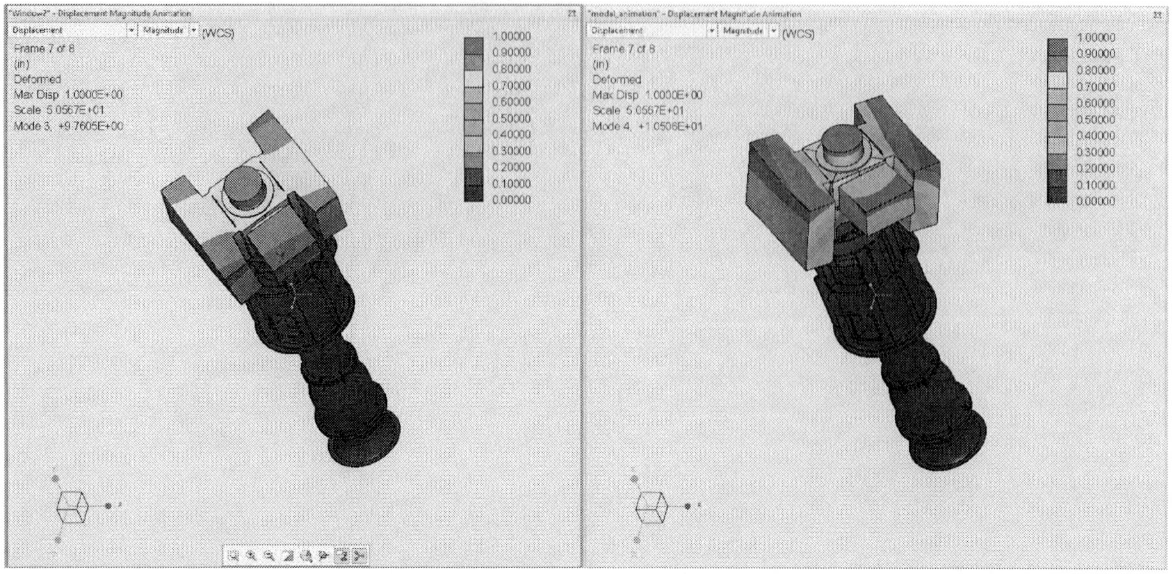

Figure 34. Adjusted system model results; no pump modifications, no braces; fixed or absolutely rigid foundation boundary constraint.

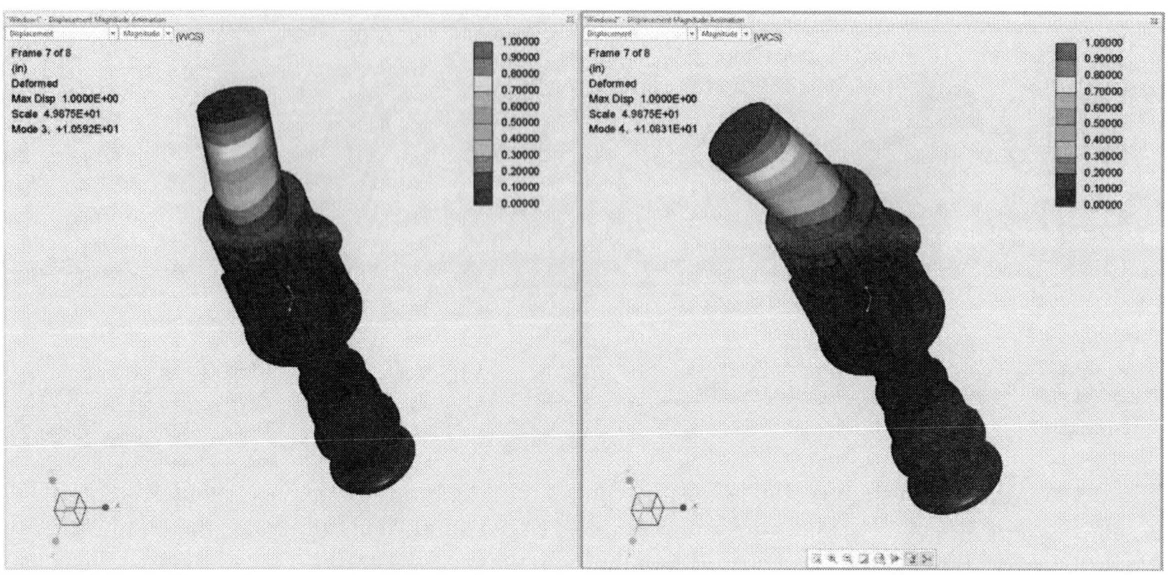

Figure 35. Adjusted system model results; no pump modifications, no braces; standard motor model; fixed or absolutely rigid foundation boundary constraint. This simulates the combined effect of a poor motor model and a rigid foundation boundary constraint.

MODELING CONCLUSIONS:

It is the author's opinion and experience that the assumed boundary conditions and model simplifications are far more significant than is typically expected with regard to natural frequencies, mode shapes, and frequency separation margins. The type of cells in the finite element model and the specific algorithms used are probably much less relevant to model accuracy and the vibration amplitudes of the installed machine than the model assumptions. The case study has shown the effects of two significant modeling approximations on vertical pumps; the simplified motor model and the foundation model or boundary condition. The motor model is often a crude approximation of a real motor structure. For the foreseeable future motor models will have to be approximations of the real motor; however, the key to success is to make the motor structure as realistic and accurate as possible before adjusting the simplified motor model to the motor OEM provided characteristics of mass, center of gravity, and reed critical frequency. The effect of the foundation on modal stiffness and natural frequency varies significantly with the specific machine design and the natural frequency/mode shape of concern. Some foundation model is probably better than no foundation model in most cases. Again, make the foundation model as geometrically and materially accurate as possible. There will always be a compromise between model size and accuracy. Larger models require more time to create and to run simulations but they can provide better results if done well. The size and cost of a vertical pump system does not affect the physics of resonance, but larger or more critical machines may warrant a more expensive analysis. If a high level analysis is desired then motor structure details, foundation details, operating fluid level ranges, and piping details should be provided to the analyst. For a new machine (not built yet) it is also critical that the expectation for these parameters to be included in the FEA is communicated. If an existing machine has a resonance problem then it is imperative that site vibration data is collected; this should include operating data and offline modal data if possible. As in the case study provided it is not always possible to collect acceptable impact test data; usually this is because of excessive background noise. ODSFRF data may be an alternative for modal information in a system with excessive noise or operating vibration amplitudes. Sometimes limited experimental modal or site vibration data must be used with the finite element model to understand the system characteristics. As shown the above grade and below grade components may or may not be isolated through the foundation.

Large and relatively short circulating water pumps typically have relatively large coupling of the above grade and below grade motions. Accelerometers can be installed below grade during repairs and these will allow detailed shape data. It is not always practical to remove the pump to install accelerometers so above ground measurements will have to be used to identify the natural frequencies and the FEA will have to provide some of the shape/mode details. Experience is helpful when adjusting models with limited experimental modal data. Pumps with higher pressure containment often have thicker cross sections, stiffer components, and they may use a containment can. These machines often will have more isolation between the above and below grade components so that motor details and the foundation details may have little effect on the column/bowl modes below grade. Longer columns/bowl assemblies and higher speeds may create resonance at higher modes of the column. A slender column/bowl assembly can have several bending modes in a frequency range that can be excited by machine driven frequencies like mass unbalance, vane-pass, etc. Operating speed excitations typically create large amplitude motions and create most vibration problems experienced by the author on vertical pumps. As noted a vertical pump mounted on a steel structure will likely have significant influence from the structure and it should be modeled in detail. It is important to collect as many natural frequencies and mode shapes as possible to adjust the system model. Just knowing you have resonance at one frequency will likely not provide enough information to validate the FEA modal predictions. Incorrect model inputs may provide one correct natural frequency, but an invalid model.

It is common to assume a stiffer system is best to increase a natural frequency above a problem forcing frequency at resonance, but often a less stiff system is superior. A less stiff system may be easier to achieve, has the same ability to increase frequency separation margin (below rather than above), and it may be less likely to degrade as the machine ages. Machines usually lose stiffness with age and damage rather than increase stiffness. Damage to a "soft" machine structure may just improve separation margin if the system natural frequency is below the forcing frequency at the beginning. Adding stiffening structures also adds mass so the net stiffening effect may be less than expected; this is why an accurate finite element model is extremely valuable to test modifications in the model so the field modifications are efficient and effective.

Frequency separation margin is an important topic that was also demonstrated in the case study. A target of 20% above 1X was estimated in the modified pump model, but the actual realized margin was 10%. The higher anticipated target provided safety margin for unknowns in the analysis. It is believed the last minute motor brace modifications caused most of the deficit in margin, but real world issues are always possible and high margin targets are recommended. This is especially true when trying to increase stiffness to move a natural frequency higher. Bolted joints and low stiffness connections are areas that can be unpredictable. Even with all the care and details used to create a very accurate system model a last minute assumption could have created an unsuccessful modification. A second identical pump was modified with improved motor braces and may prove the conclusion that the motor brace connections caused the margin deficit.

Hopefully this paper has shown many of the areas of potential errors in modeling vertical pumps. New machine evaluations (before purchase and manufacture) are the most difficult because field vibration and modal data is not available. In addition, there are usually different groups and companies involved; there are many places for communication to break down. It is recommended that the purchaser engage the analyst or designer as directly as possible to ensure a FEA will be conducted and that it will include the necessary details. The relevant motor, foundation, piping, and fluid data must be provided to the designer. If data is not provided assumptions and qualifications will have to be made and the analysis may not be accurate enough to prevent vibration problems.

VIBRATION ANALYSIS CONCLUSIONS:

Below grade accelerometers are not common, but they are relatively inexpensive for most water services. They should be considered on all vertical pumps, but especially if a given machine has vibration issues and frequent repairs. Enough sensors need to be installed at anti-nodes to fully define possible modes; typically 1-3 axial stations with two orthogonal accelerometers is sufficient. Proximity probes can be installed near the mechanical seal or packing area without much difficulty; this provides rotor motion information and a good indication of bearing wear (especially wear near the top of the pump). An important aspect of condition monitoring is to have permanently mounted accelerometers (stud or cement mounted) so that trending is effective. Even if data is only collected in a periodic route the permanently mounted sensors will provide repeatable measurements so that small changes in pump condition can be trended. Permanently mounted accelerometers on the above grade motor and pump can provide good trend data regarding the entire pump system. As bearings wear the 1X amplitudes will likely increase and harmonics may increase in number and amplitude. The amplitudes may be small because the sensors are not near the bearings and locations of forces, but consistent data from permanently mounted sensors will still provide trends that can predict condition changes. 1X and harmonic amplitudes can be trended to evaluate rotor/bearing condition, but overall RMS trends can be misleading because many vertical pumps have large random excitation from flow related forces like cavitation. It is important to look at the FFT and/or trend frequency components. Overall RMS trends can miss important mechanical changes that create small amplitude vibration related to 1X and harmonics. Large noise floors and broadband energy may indicate improper flow rate. High flow and low flow relative to the pump's design or best efficiency flow rate can create turbulence and cavitation. These flow related issues create large amplitude random vibration in the time waveform and broad band energy in the FFT. Flow related random forces can impart impulses to the structure and excite system natural frequencies.

A large circulating water pump at low flow has been observed to have much higher vibration amplitude at the natural frequencies as compared to the 1X component.

Condition monitoring of vertical pumps is difficult because most of the pump is usually not accessible for vibration measurement. Permanently mounted accelerometers can provide accurate and consistent trends that improve condition monitoring. It is important that components or frequencies related to mechanical condition be trended rather than overall RMS trends. Overall RMS trends can be significantly affected by flow rate or pump operation; turbulence/cavitation can excite natural frequencies and create large overall RMS amplitudes. A change in broad band energy and random signals in time waveforms may indicate an operational change or flow problem.

REFERENCES:

[1] Vibration Institute Machinery Vibration Analysis training material.

VIBRATION ISSUES WHEN OPERATING CENTRIFUGAL PUMPS AT OFF-DESIGN CONDITIONS

Author: Andrew H. Lerche
and William D. Marscher

VIATC VIBRATION INSTITUTE ANNUAL TRAINING CONFERENCE & EXPO 2022 | MFPT ANNUAL CONFERENCE

2022 CONFERENCE PROCEEDINGS

TRAINING CONFERENCE & EXPO
SAVANNAH, GEORGIA · 2022

VIBRATION ISSUES WHEN OPERATING CENTRIFUGAL PUMPS AT OFF-DESIGN CONDITIONS

Andrew H. Lerche
Senior Staff Engineer
Mechanical Solutions, Inc.
andrew.lerche@mechsol.com

and

William D. Marscher
CEO & Senior Consultant
Mechanical Solutions, Inc.
11 Apollo Dr.
Whippany, NJ, 07981 USA
www.mechsol.com

Mr. Lerche has more than 20 years of professional experience in the areas of design, testing, analysis (fluid and structural), and troubleshooting of mechanical equipment and machinery such as pumps and turbomachinery. He is experienced in the areas of turbomachinery, high pressure fluid control equipment, and shipboard equipment. He has extensive experience with flow-induced vibration of turbomachinery blades, as well as shock and vibration testing and analysis of fixed mechanical and electrical equipment. Mr. Lerche has authored and presented several technical papers in the area of turbomachinery and holds three patents related to industrial compressors and valves. He holds a B.S. and M.S. in Mechanical Engineering from the University of Texas at Austin and the University of Texas at San Antonio, respectively.

Bill Marscher performs vibration analysis and testing on pumps, fans, compressors, and turbines throughout the world. Previously, he worked for Honeywell Bendix, Pratt&Whitney, Worthington Dresser, and Concepts NREC. He obtained his BSME and MSME degrees from Cornell University and a Masters in Engineering Mechanics from RPI. He is a member of the TAMU Pump Advisory Committee, and is past President of the Society of Tribologists & Lube Engineers and of the Machinery Failure Prevention Tech Society of the Vibration Institute. He is voting member of the ISO TC108/S2 standards committee for machinery vibration, and is Vice Chair of the Vibration Committee of the Hydraulic Institute, where he is also a Standards Partner. He has authored twelve handbook chapters, and is co-author of the Oxford Press book Centrifugal Pump Design and Performance.

ABSTRACT:
This presentation will discuss the serious vibration problems that can result from operating a centrifugal pump at flows that deviate from the BEP Design Point. For example, intuition suggests that reduced flow should result in less energy being expended, and therefore inexperienced operators expect that vibration will decrease as flow is throttled. However, at off-design, the poor angle-of-attack of the flow entering the pump impeller typically causes vibration to increase instead, and even may lead to unexpected additional shaking frequencies due to stalling of the pump impeller blades. Such flow-induced vibration from off-design operation can be mitigated by using a VFD rather than simple throttling of a discharge valve. However, as both speed and load are changed versus the full-speed Design Point, resultant changes in the bearing load capacity can limit bearing life. More importantly, speed change leads to changes in the rotordynamic coefficients of the bearings and of the wear ring annular seals (due to Lomakin Effect) which in turn can result in rotordynamic critical speeds becoming unexpectedly resonant. In some types of pumps, flow phenomena can couple with system dynamics to result in more complicated problems, e.g. in vertical turbine pumps, vertical end suction pumps, and pumps with specialty wear rings. In all cases, both common or complex, careful evaluation of the rotor system and its support components can anticipate harmful issues so that design or operational modifications can be made to avoid reliability problems. This presentation will discuss, for example, how to evaluate particular pumps and systems for such issues by using methods outlined in either the API Rotordynamic Tutorial RP-684 or the ANSI/HI 9.6.8 Guideline for Dynamics of Pumping Machinery.

KEYWORDS: Best efficiency point; Lomakin effect; off-design operation; pumps; rotordynamics; suction recirculation; variable speed; vibration

INTRODUCTION:
Both fatigue and rubbing wear in pump components are most commonly caused by excess rotor vibration. One unfortunate, and often avoidable, root cause of excess vibration is operation of the pump far away from its design point, either in terms of flow rate being too far away from the Best Efficiency Point (BEP), or the speed range being too aggressive. Running away far from BEP results in excessive hydraulic force such as from suction recirculation stall or vane pass pressure pulsations. Running over too wide a speed range can lead to bearings running too hot or with too thin a lubricant film, or can cause otherwise acceptable vibration levels to be amplified by a natural frequency resonance.

DISCUSSION:
Pump users typically require a centrifugal pump to operate over a range of discharge pressures and flow rates ("capacities"), not only at a single operating point. Pressure is typically converted into "head" when characterizing a pump, since for a given speed head is independent of fluid density, while pressure is not (in fact pressure is density times head, times a units constant). At any given speed, a pump's head versus flow rate is determined by its characteristic curve, which is a function of the impeller and flow path design, and is available from the manufacturer. This pressure versus capacity roughly follows half of an inverted parabola, with peak head at zero flow, and zero head at maximum flow as shown in Figure 1. The relationship of head to speed is also a parabolic curve, but in this case an upward rising one, such that head is zero at zero speed, and head at any

given speed is approximately proportional to speed squared.

Therefore, as illustrated in Figure 1, speed, head, and flow are all inter-related in a centrifugal pump. There is one more important issue however, and that is the head versus flow characteristic of the system that the pump is sending its flow through. This is typically also a roughly parabolic curve, in this case of head vs. flow, where head is approximately the square of the flow, sometimes with an offset in head at zero flow depending upon whether or not the discharge is different than suction pressure at zero flow. This is shown as the dotted line in Figure 1. The most important concern relative to this is that since the pump must run on its curve, and the system must run on its curve, the only way for the pump and system to operate together is for the pump to operate at the intersection of the pump and system curves, again as illustrated in Figure 1, by the intersection that the red arrow points to. This is the operating point for the illustrated pump flowing through the illustrated system (e.g. a pipe and valve).

If the operating point pressure and/ or flow needs to be changed, there are two options. One option is to change the speed at which the pump is operating, in the case of achieving more flow thereby shifting its characteristic curve up and to the right. It will then intersect the system curve differently, also up and to the right. The other means of changing the operating point is to "throttle" the discharge line, such as with a valve. This shifts the system curve to the left, without changing its head at zero flow. By changing a combination of speed and valve throttle position, any point of the head-capacity map can in principle be operated at, if it is below the full speed pump curve.

However, some points are much safer to operate than others. Typically, the zones of reasonably high efficiency have the lowest internal pressure pulsations, and associated vibrations. The superior point in this regard at any speed is known as the Best Efficiency Point, BEP. There is a zone of roughly +/- 15% BEP to the left and right of the BEP point, where pulsations remain relatively quiet, and this is known as the Preferred Operating Range, POR. Outside of the POR, the pulsations and vibrations pick up considerably, but are still considered acceptable by the OEM, in that operating life, while reduced, is still considered acceptable. This zone is called the AOR, and depending upon pump design typically extends from about 70% to 120% of BEP.

The changing of flow rate by throttling tends to increase radial loads on the bearings, both steady as well as pulsation loads. This will increase bearing temperature, as well as bearing wear rate, and may lead to more rapid wear of wear rings, balancing devices, and mechanical seals. The degree to which the steady and oscillating loads vary with flow are dependent upon pump type and manufacture, but are shown for a typical case in Figure 2.

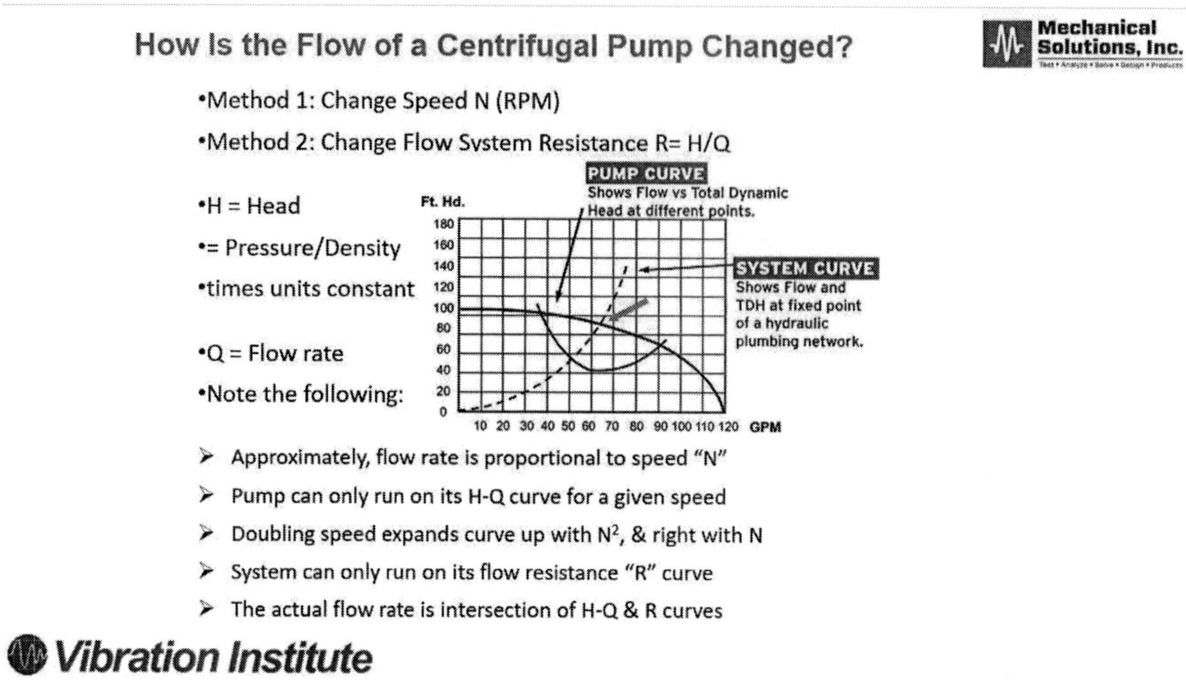

Figure 1: Alternative means of changing the flow rate of a centrifugal pump

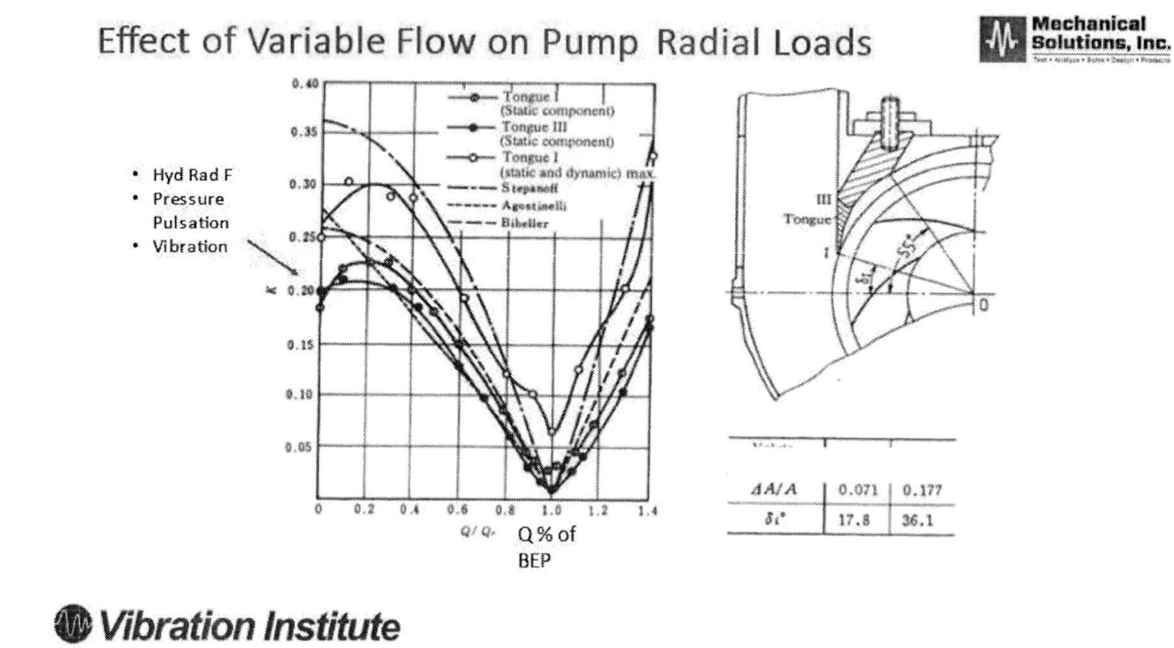

Figure 2: Effect of flow rate on hydraulic radial loads

When speed is changed, any flow throttling occurs as the pump curve shifts downward, so that it intersects the roughly parabolic system curve at a lower head and flow. Since the plot of BEP flow on the pump H-Q curve is also roughly a parabola, this tends to not change the percentage of BEP that the pump is running at, which means the flow throttling in this manner leads to lower vibration and general punishment of the pump.

However, changing speed has other effects as well. The first is that, if the bearing is of the journal type, as is the case for most high value pumps, the bearing film is affected in a manner illustrated in Figure 3. The best way to characterize this is with something called the Sommerfeld Number, or Bearing Characteristic Number, which is basically speed times bearing oil viscosity, divided by load on the bearing. As speed increases, the bearing stiffness actually decreases if the radial load and viscosity stay the same, although they probably won't. Generally, pressure changes with the square of the speed, ad with the increased pressure comes increased radial load. Also, as speed increases, more energy is dissipated in the bearing (typically proportional to about the square of the speed), leading to higher oil temperature, and resulting rapid decrease in oil viscosity. So with respect to speed increase, bearing behavior is a bit complicated. The main takeaway in this regard is that the ability of the pumps bearings to handle speed changes should not be taken for granted, and should be vetted with the OEM.

Another type of (in terms of practical behavior) bearing in most pumps is the wear ring and/ or balance device (often a close radial clearance bushing or drum), through something called the Lomakin Effect, illustrated in Figure 4. The Lomakin bearing stiffness is typically only about 1/10 the stiffness of the journal bearing, but it is located centrally in the pump, where it has more leverage versus the most dangerous (i.e. simplest shape) rotordynamic modes. The Lomakin stiffness is not directly sensitive to speed, but it is directly proportional to discharge pressure, and since discharge pressure changes with roughly the square of the speed, there is a speed effect through that avenue.

The primary potential problem with the changing of bearing stiffness is that, as the rotor support stiffness (especially Lomakin in a multistage pump) changes, the lower rotordynamic modes change in frequency. If one of these natural frequencies drifts into running speed, or some other strong excitation, a strong resonance can occur, with associated excessive vibration levels.

Returning to flow change by throttling the discharge, this can happen (perhaps accidentally) by means other than partial closure of a valve, or adding extra resistive components to the discharge line. Figure 5 illustrates how, when two pumps are operated in parallel with each other, i.e. both discharge into the same system, the effective resistance (head vs. flow ratio) increases because only half as much cross-sectional area of the pipe is available for the pump to discharge into. This can lead to a pump that would have operated near BEP to now be throttled back, perhaps out of the AOR range, as shown in Figure 5. When this happens, the angle of incoming flow (now coming into the impeller much slower for the given spin speed of the impeller) results in a poor angle of attack on the pump impeller vanes, stalling them. Thus, operating outside of the AOR (even for single pump operation) leads to flow stall and separation from the vanes, causing vortices (like miniature tornadoes) to form. These cause severe pressure pulsations and shaking of the pump, as discussed earlier. In the case of two pump operation such they are both pegged at the same discharge head, the weaker pump may be forced to run even further below BEP, or possibly even at dangerous "dead head" (no flow).

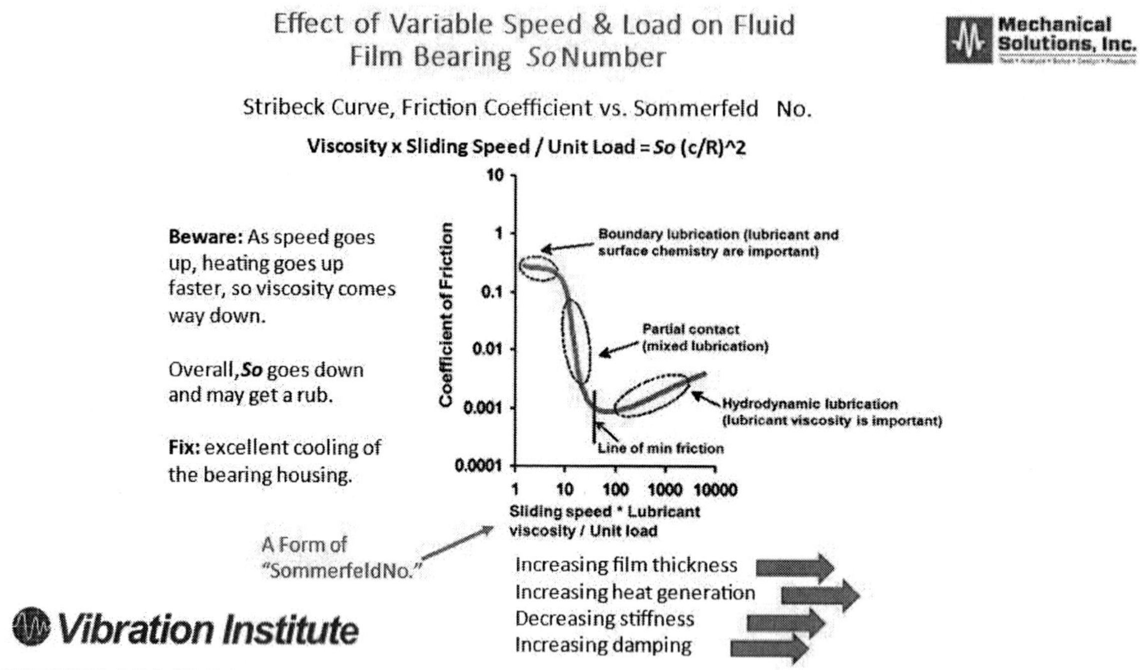

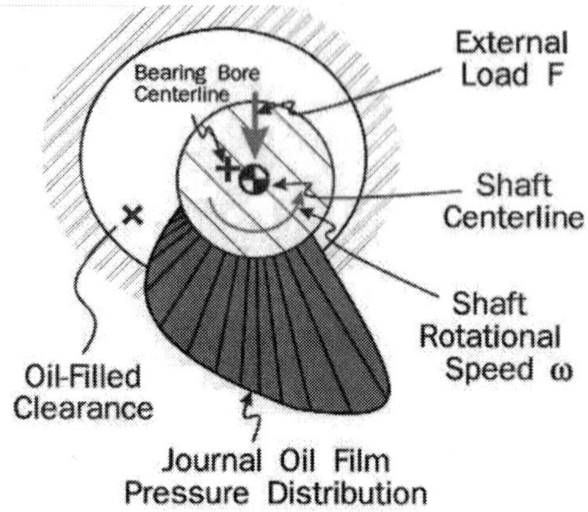

Exaggeration of hydrodynamic journal bearing eccentricity vs. load, and Pressure (Red Zone) within Bearing Film.

Figure 3: The effect of speed and bearing load on bearing behavior

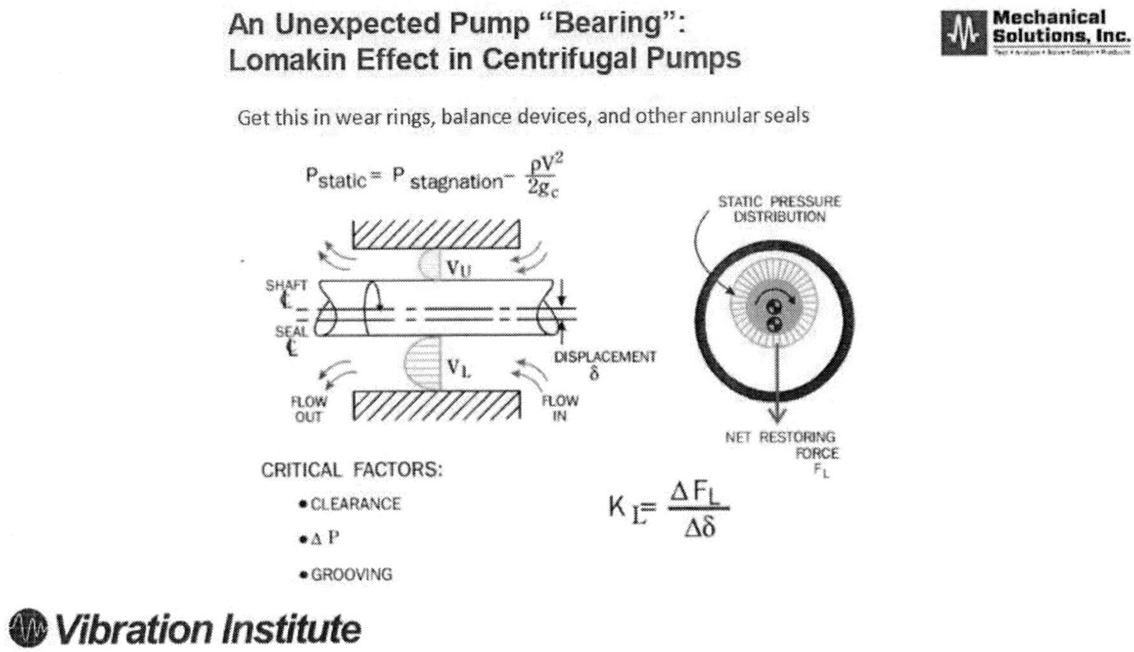

Figure 4: Lomakin effect: an unexpected "virtual bearing", highly dependent on discharge pressure, which depends on flow rate and speed. Does not depend on speed if speed does not change pressure (it normally does, by speed^2).

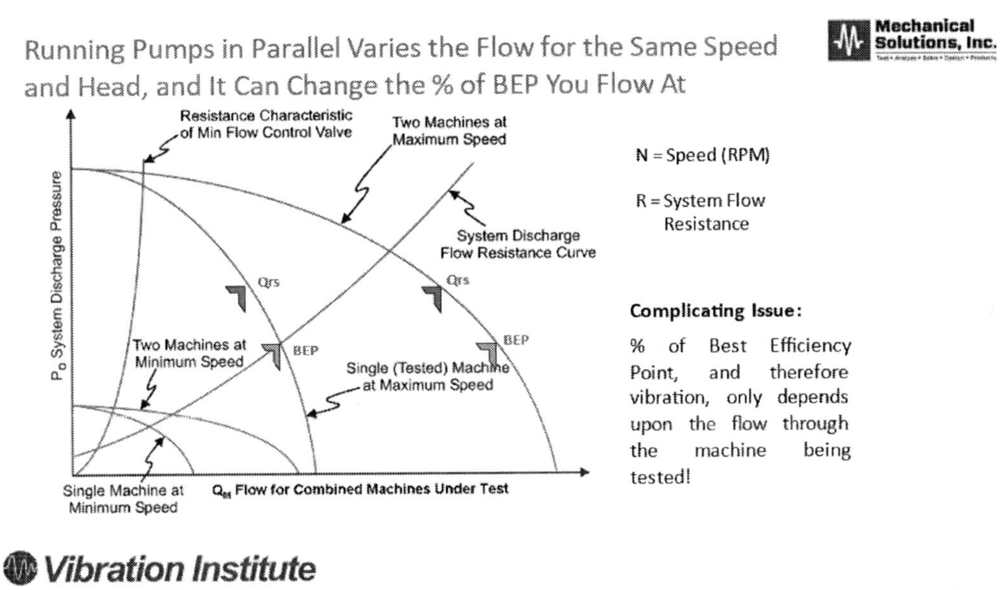

Figure 5: Example of operation of two pumps in parallel. At the same speed that a single pump was at BEP, now both pumps run at unacceptably low flow!

One style of centrifugal pump is particularly unique in its configuration, and in how potential problems with flow and/ or speed changes can affect those problems. This is the vertical turbine pump, shown in Figure 6. Figure 6 is a typical (good) design for such a pump, often used in deep well service. The flexibility of both the pump column, as well as the spindly lineshafting that attaches the impeller rotor to the motor rotor far above, sets the stage for some unique issues, as listed in Figure 6.

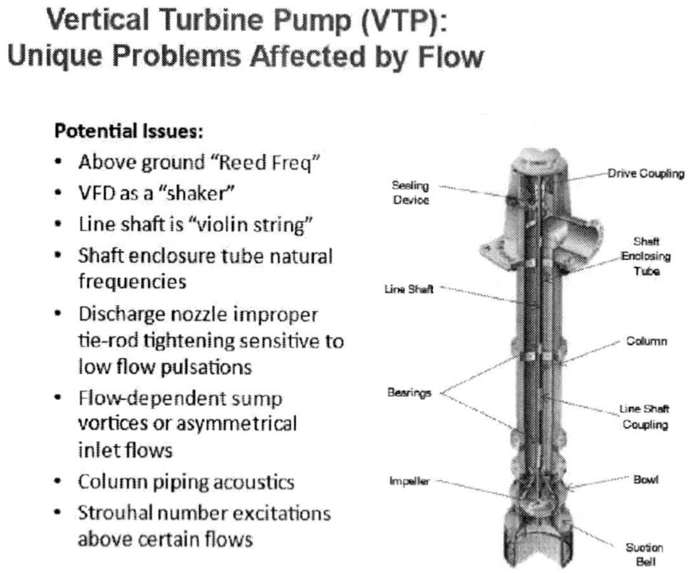

Figure 6: Issues particular to vertical turbine pumps

So, operating centrifugal pumps over a range of head and flow conditions is more complex than simply installing a quality discharge valve and VFD. Figure 7 lists what the authors intend that the main take-aways from this paper should be. Issues listed that were beyond the scope of our discussion, but are covered in the references, are cavitation (typically occurs at flows above or below the AOR), and fluid whirl (again, typically occurs in industrial pumps outside of the AOR).

Wrap-Up Take-Aways

Mechanical Solutions, Inc.

1. VFD Variable speed affects more than just the pump running speed!
   - Bearing film temperature, less film, possible rubs
   - Natural frequency resonances
   - Whirl instabilities
2. Variable flow can lead to problems, maybe severely reducing life.
   - Suction or discharge recirculation
   - Cavitation
   - Higher bearing loads
   - VTP lineshaft low tension "jumprope" whirl
3. Seals and bearings are at risk, unless speed and flow variation are "engineered".
4. **Analysis and testing tools exist to evaluate variable flow operation so any problems can be identified and mitigated (for example, see HI 9.6.8 procedures).**

Vibration Institute

Figure 7: Why operators should exercise care during any type of pump throttling, by either closing discharge valve or changing speed with a VFD.

CONCLUSIONS:

- Whenever possible, run pumps at or near their Best Efficiency Point. If this is not possible, try to keep the flow range within the POR. If this is not possible, do not allow any part of the pump flow range to be outside of the AOR, except for very brief (less than two minutes) start-up or operational load transients.
- Be very careful about the size of the pump purchased versus what is truly needed for the plant process pumping system. Buying an oversized pump will run well below BEP, maybe outside of its AOR. If future demand grows, then re-fit (or replace) the pump at that time.
- Be careful about running multiple pumps in parallel, especially if they are not nominally identical, or if one is significantly more worn than the other. One pump feeding a given discharge system may run at BEP, but because of the extra flow losses in the same size pipe with two pumps running, the pumps are both forced to run well below BEP, maybe even out of the POR or AOR.
- Operating pumps outside of their AOR will likely lead to suction recirculation, and possibly discharge recirculation as well. This recirculation can be envisioned as a tornado at the inlet of the pump, and typically causes severe shaking forces and vibrations, usually at frequencies below running speed. Decreased reliability, and perhaps greatly shortened life, can be expected in such circumstances.
- Concerning speed as opposed to flow, a very wide continuous speed range can lead to bearings running too hot, or with too thin a lubricant film.
- Another effect of a wide speed range is to make it more likely to encounter excessive vibration levels because of otherwise acceptable vibration being amplified by a natural frequency resonance, e.g. when running speed passes through a natural frequency.

NOMENCLATURE
BEP= best efficiency operating point of the pump
POR= preferred operating range
AOR= allowable operating range

REFERENCES & BIBLIOGRAPHY
Agostinelli, A., Nobles, D., Mockridge, C.R., 1960, An Experimental Investigation of Radial Thrust in Centrifugal Pumps (at Off-Design), J. Engrg. for Power, Trans ASME, v.82.

API 610, 12th Ed., 2021, Amer. Pet. Instit., Washington D.C.

API-RP-684, 1st Ed., 1994, Amer. Pet. Instit., Washington D.C.

Arndt, N., Acosta, A. J., Brennen, C. E., Caughey, T. K., 1979, "Rotor-Stator Interaction in a Diffuser Pump," NASA Tech Briefs, MFS-26092 .

Black, H.F., 1979, Effects of Fluid-Filled Clearance Spaces on Centrifugal Pump Vibrations. 8th Turbomachinery Symposium, Texas A&M Univ.

Bolleter, U., 1988, "Blade Passage Tones of Centrifugal Pumps, Vibrations, V. 4, n. 3, pp. 8-13.

Bowman, D., Marscher, W., and Reid, S., 1990, Pump Rotor Critical Speeds: Diagnosis and Solutions (for Variable Speed), Proc. International Pump Symposium, Texas A&M University.

Fraser, W.H., 1985, Centrifugal Pump Hydraulic Performance and Diagnostics, *Pump Handbook,* McGraw-Hill.

Hergt, P., Krieger, P., 1970, Radial Forces in Centrifugal Pumps with Guide Vanes at Off-Design), Proc IMechE, v.184 Pt 3N.

HI/ANSI Standards 9.6.4, 2020, The Hydraulic Institute, Parsippany NJ.

HI/ANSI Standards 9.6.8, 2019, The Hydraulic Institute, Parsippany NJ.

Kovats, A., 1962, Vibration of Vertical Pumps, ASME Journal of Engineering for Power.

Marscher, W.D., 1986, The Effect of Fluid Forces at Various Operation Conditions on the Vibrations of Vertical Turbine Pumps, Proc. IMechE, Radial Loads and Axial Thrusts on Pumps.

Marscher, W.D., 1989, Analysis and Test of Multistage Pump Wet Critical Speeds, ASLE/ASME Joint Tribology Conf., Ft. Lauderdale, FL.

Marscher, W.D., 1990, The Effect of Variable Frequency Drives on Vibration Problems in Vertical Pumps, Proc. Water & Wastewater Conference, Barcelona, Spain.

Marscher, W.D., 1999a, Bearing Application and Lubrication Chapter, Modern Marine Engineer's Manual 3rd Edition, Cornell Maritime Press: Centreville, MD.

Marscher, W.D., 1999b, The Determination of Rotor Critical Speeds While Machinery Remains Operating through Use of Impact Testing, IMAC Conf. Orlando FL, SEM.

Marscher, W.D.; Onari, M.M.; Boyadjis, P.A., et al, Pump Mechanical Behavior & Vibration, Chapter in Centrifugal Pump Handbook, 4th Edition, McGraw-Hill, c. 2008. Edited by Paul Cooper et al.

Massey, I., 1985, Subsynchronous Vibration Problems in High Speed Multistage Centrifugal Pumps, Proc 14th Turbomachinery Symposium, Turbomachinery Laboratories, Texas A&M Univ.

Ohashi, H., Sakurai, A., Nishihama, J., 1988, "Influence of Impeller and Diffuser Geometries on the Lateral Fluid Forces of Whirling Impeller, N89-22907, TAMU Instab. Conf., pp. 285-305.

ACKNOWLEDGEMENTS

The authors would like to thank Mechanical Solutions, Inc. for its support of the author in the construction of this paper. In particular, Maki Onari, Paul Boyadjis, Dan Owens, and Erik Fillebrown contributed portions of the analysis results. Special thanks to Mike Marscher and Brian Fletcher who authored much the graphics. Many thanks also to the manufacturers and users of pumps worldwide who have employed MSI in the pursuit of the many interesting design challenges, analyses, and troubleshooting testing that provided valuable background and experiences.

DYNAMICS OF ROTARY LOBE BLOWERS

Author: Walter Kelm, P.E.
and Ray Kelm, P.E.

Dynamics of Rotary Lobe Blowers

Walter Kelm, P.E.

Ray Kelm, P.E.

Kelm Engineering, LLC

Introduction

Rotary lobe blowers are commonly used in industrial applications to either provide a vacuum source or to move air or other gases at relatively low pressure. The description of rotary lobe blower implies that it is a rotating machine (opposed to reciprocating), it has "lobes" (meaning the rotor has a non-circular shape), and it is a blower (relatively low-pressure application).

The typical blower will have two rotors with the drive rotor having a shaft extension for connecting to a motor either directly coupled or with V-belts. The second rotor is driven by the main drive rotor using timing gears so that both rotors operate at the same speed.

The rotors are either 2-lobe or 3-lobe with examples of each shown in figure 1. The 2-lobe blower is the oldest design and likely the simplest function. The 3-lobe design produces less pulsation and noise as well as vibration for the same flow blower.

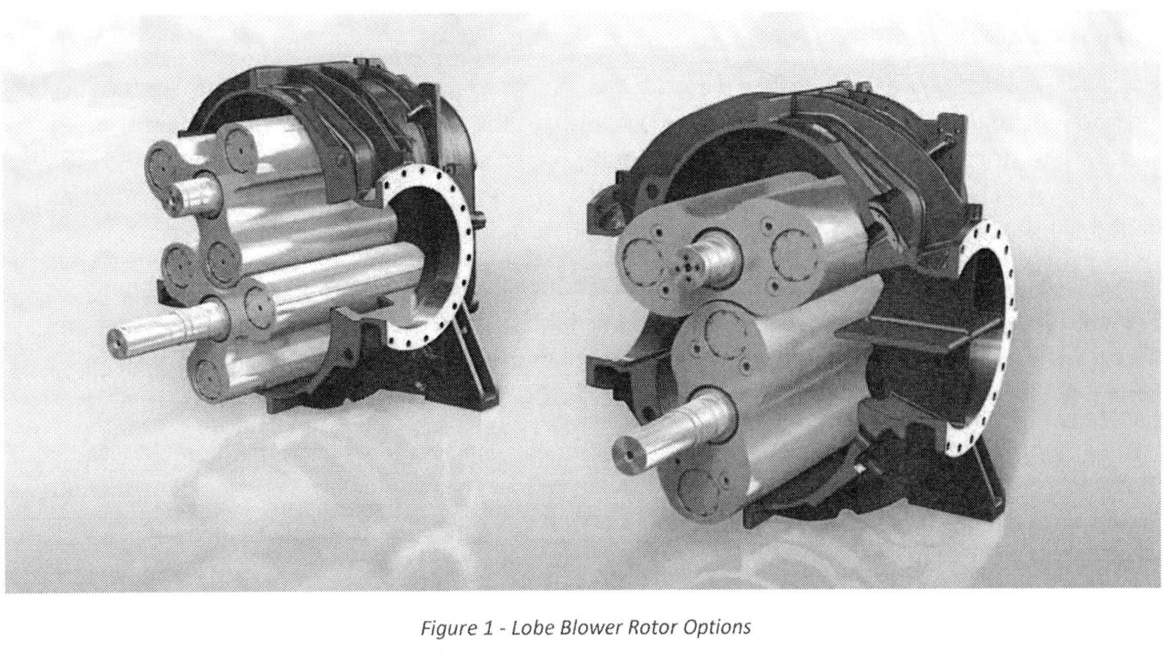

Figure 1 - Lobe Blower Rotor Options

For the drive configuration, it is possible to have the top or bottom rotor be the drive rotor and to drive off either the timing gear end or the opposite side. Various design considerations may dictate the specific drive shaft end orientation.

Both rotor designs will have timing gears on one end of the rotors to keep the rotors phased together properly. The machine efficiency is dependent on the clearance between the rotors and casing. The rotors will not contact each other during normal operation and maintain close clearance without touching due to the timing gears. The rotor profiles will vary between manufacturers, with the cross section and sealing strips impacting performance.

Figure 2 - Rotors for 2 lobe and 3 Lobe Designs

Typical Uses for Lobe Blowers

Lobe blowers are commonly used in the following services:

- Sewage aeration
- Material conveying
- Vacuum trucks
- Low pressure gas pumping

Since the rotor design is a positive flow type, the inlet flow is fixed based on the geometry of the rotors and operating speed. In contrast with a centrifugal blower, the flow will remain constant at fixed speed where the centrifugal blower flow will vary depending on the pressure ratio across the unit. The lobe blower flow will decrease slightly with higher pressure ratio across the blower, but the flow reduction is limited to leakage ("slip") from one rotor pocket to the next depending on pressure, clearances, etc.

The sewage aeration application normally requires producing air flow for the digesters with enough pressure to overcome a liquid head of maybe 6' or so to provide oxygen to the "bugs" processing the sewage. The typical pressure range will be up to about 15 psig discharge pressure with atmospheric air feeding the blower.

Material conveying with lobe blowers is common for moving plastic pellets and other granular products. The blowers will provide pressurized air that is used to "blow" the product from one location to another. For this application, the inlet is atmospheric air and discharge pressure less than 10 psig or so.

Lobe blowers are the normal source of vacuum on vacuum trucks that are used to suck liquids from various locations. In this service, the blower will still pull a fixed volume from the inlet due to the positive displacement design. As the higher vacuum is reached, the mass flow of air reduces due to the lower density due to vacuum.

For vacuum service, there can be very low flow though the blower which can result in elevated operating temperature. One option used for vacuum service is to have an atmospheric air inlet to allow some "cool" air through the blower particularly at high vacuum for rotor cooling. This feature will limit the vacuum level that can be reached while providing cooling so that these designs can normally be operated with the inlet line completely blocked off (as for maximum vacuum).

Blower Operation

The general function of the blower is detailed in figure 3 showing the various steps in the compression process for a 2-lobe blower.

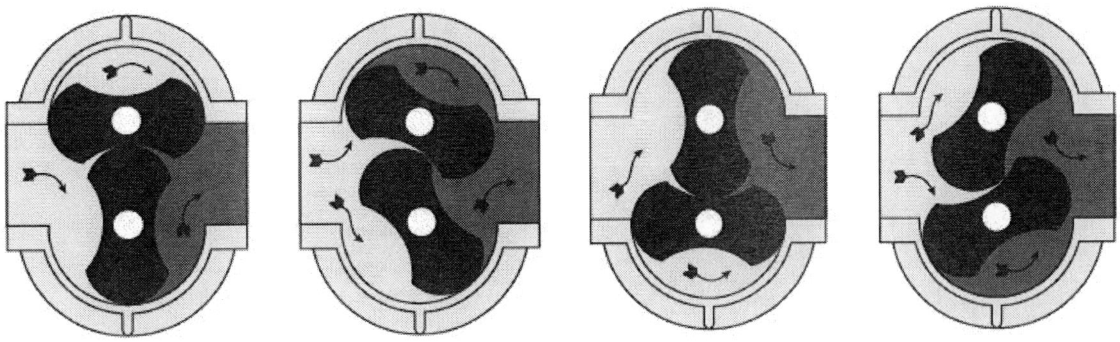

Figure 3 - Blower Function

In figure 3, the four steps are as follows:

1. Top rotor traps a volume of inlet air as bottom rotor pushed flow out
2. Top rotor pocket opens to discharge port and is pressurized by backflow from outlet while bottom rotor completes outlet flow cycle
3. Top rotor pushed flow out while bottom rotor traps inlet air volume
4. Bottom rotor pocket opens to discharge port and is pressurized by backflow from outlet while top rotor completes outlet flow cycle

From the description of the flow cycle, there are some obvious points in the rotation where high flow pulses and vibration are likely. When the captured inlet volume opens up to the outlet port pressure, there will be a sudden backflow to pressurize the captured volume. Once the pocket is pressurized, the flow toward the outlet then reverses to push the now pressurized pocket out.

The frequency that the pockets push flow out will be at 2 x # lobes x RPM. The primary pulsing at twice the rotor lobe pass frequency is due to each rotor producing a pulse for each lobe and with two rotors, the final pulsation frequency starts at twice the lobe pass frequency.

As with any short duration jolt (in force, flow, etc.) that occurs periodically, the resulting frequency content for pressure pulsations will be harmonics of the primary jolt frequency. For the lobe blower, that means that there will be pressure pulsations at the outlet at twice the lobe pass frequency with harmonics sometimes as high as 6-8x. These harmonics can produce high vibration and noise during blower operation and are commonly very high at some blower speeds particularly when operated on variable frequency drives.

Typical Application

Figure 4 shows a typical arrangement for a lobe blower in air service with pressurized outlet. For the atmospheric air inlet, there will be an inlet filter as well as an inlet silencer to reduce inlet noise. In some cases, the inlet silencer may be omitted, and the inlet filter connected directly to the blower inlet.

The outlet will generally have a discharge silencer to attenuate pressure pulsations and to help reduce the noise.

The blowers are commonly driven by belts or directly coupled to a motor. Operating speeds will depend on the size of the blower with maximum speeds normally around 2,000 RPM for truck mounted units. Some of the very large rotary lobe blowers (up to 10' long rotors) may only operate at 200 RPM or so.

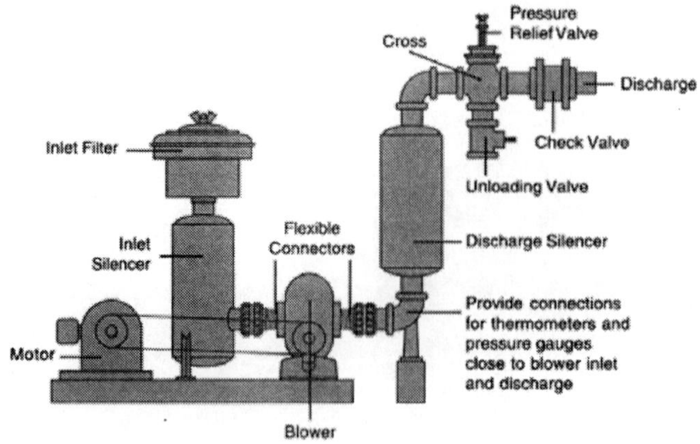

Figure 4 - Typical Arrangement

Application Concerns

Lobe blowers are inherently noisy and produce elevated vibration. Some applications can produce higher vibration and noise in some instances such as:

- Variable speed applications
- Changing gas conditions

Since the machines can produce high amplitude vibration at many harmonics, operating them on variable frequency drives virtually guarantees that you will operate exactly on resonance at some speed(s). Resonant vibration on the blowers can produce very high casing vibration. Resonances on the inlet/discharge silencers and piping are common and can easily lead to catastrophic failures.

For applications on gases other than air, the pulsation characteristics of the blowers will change depending on the molecular weight of the gas being pumped. The mole weight impacts the density of the gas but also changes the speed of sound within the gas stream. Changes in speed of sound in the gas can have a huge impact on the pulsation amplitudes observed. Operating with an acoustical resonance on the silencers or on interconnecting piping can be very destructive.

Other concerns include possible liquid injection and debris. Since these are positive displacement blowers with rather tight clearances, they don't take to liquid very well. Small amounts of water injected into the inlet can improve the operating efficiency of the blower by helping minimize the "slip" or leakage between rotor passages and can also help reduce noise. However, a large slug of liquid can easily destroy the blower. Debris can also be very destructive due to the close clearances between the rotors and casing.

Sources of Vibration

Rotary lobe blowers will have the following contributors to vibration:

- Unbalance of the rotors – 1xRPM (normally not an issue)
- Possible rotor rubs (not present for "healthy" blower)

- 2xRPM and harmonics for 2-lobe blower
    - 3xRPM and harmonics for 3-lobe blower
- Timing gears – in the range of 60-80 teeth (gear mesh and harmonics)
- Antifriction bearings (4 total) – typical bearing vibration
- Pulsation induced vibration – twice lobe pass frequency and harmonics
- Shaft drive oil pump vibration
    - Many larger blowers have oil pump mounted on the non-drive side of blower
- Coupling issues and misalignment – 1x and 2xRPM
- Belts (if present) – normal belt vibration

Lobe Blower Testing

Testing of lobe blowers will normally be like other machinery in a plant. The typical approach is to use portable vibration monitoring equipment with magnetic mounted accelerometers on each end of the blower with measurements taken near all 4 bearings.

Acceptable vibration is difficult to define, with the best approach being to look for vibration change more than absolute values. Manufacturers should supply acceptable vibration limits for specific blower designs, but limits as high as 0.7 in/sec peak are not uncommon.

Example vibration from a 2-lobe blower is shown in figures 5 and 6 with figure 5 on the drive end and 6 on the non-drive end. Each plot shows vertical, horizontal, and axial. This blower was operating at 825 RPM. The maximum vibration observed on the drive end is at 2xRPM (27.5 Hz). No rotor rubs identified. Elevated vibration at 2xRPM on drive end expected to be due to significant coupling misalignment.

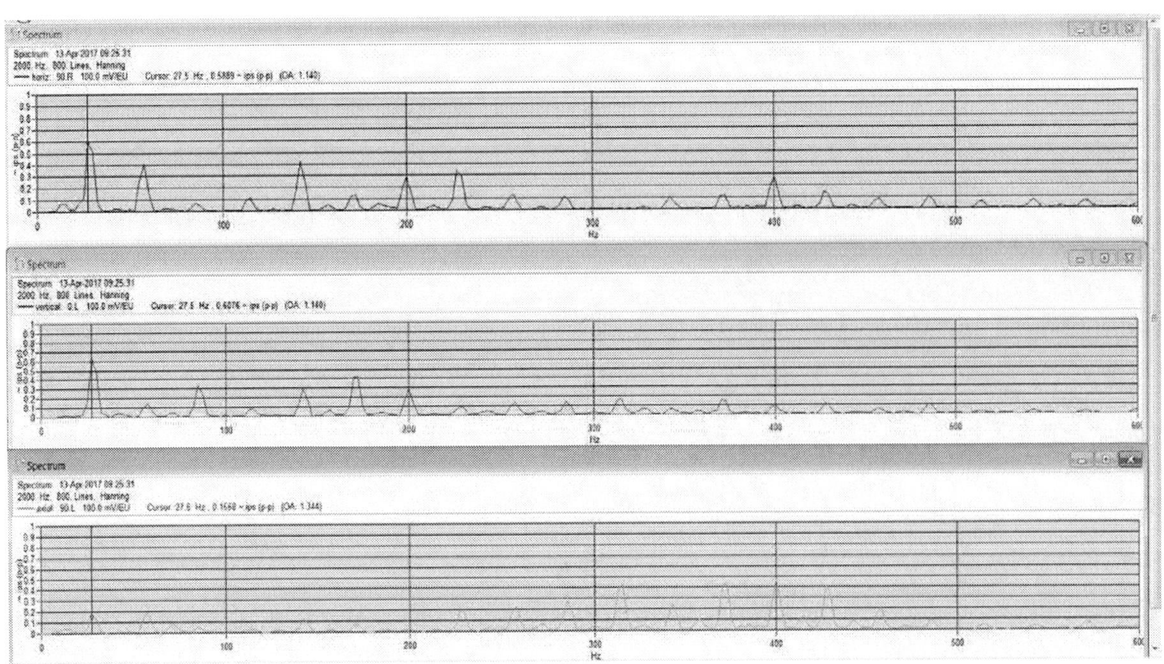

Figure 5 - Drive End Vibration

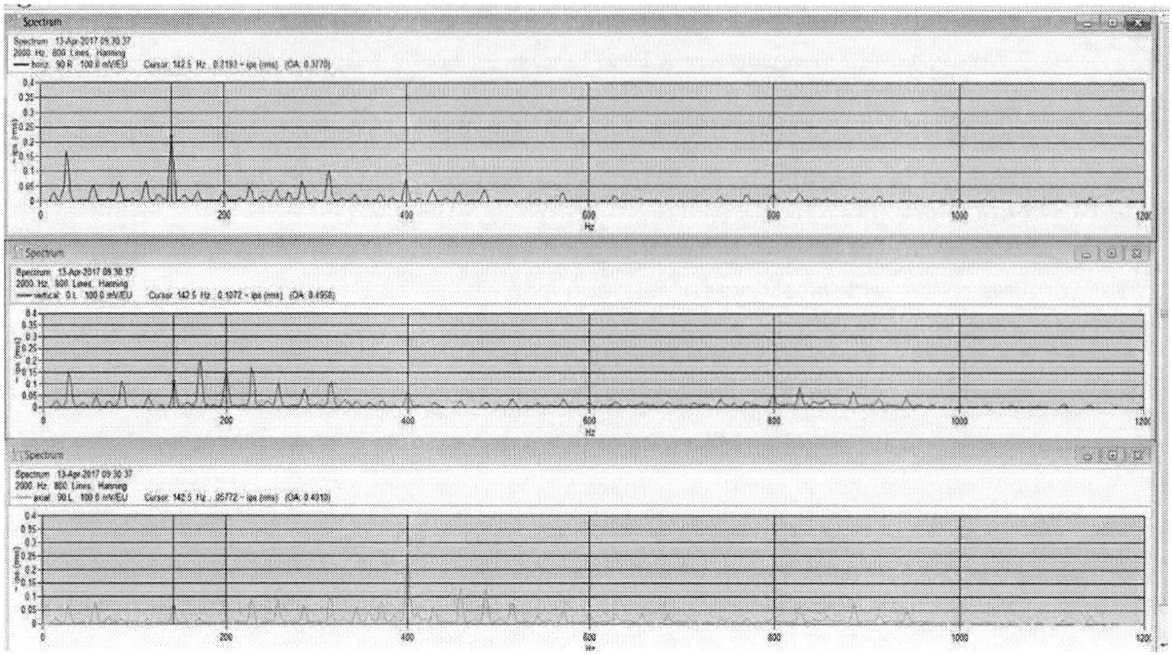

Figure 6 - Non-drive End Vibration

Additional harmonic content in the vibration signatures is observed at 10xRPM and a variety of other harmonics. This sort of vibration is typical, with clear justification for each peak not possible in general. The non-drive end vibration of almost 0.5 in/sec rms is high but often not beyond OEM limits.

Vibration on the silencers can be extremely high and will sometimes lead to silencer failures. Vibration in excess of about 1 in/sec rms can easily produce cracking and internal damage to silencers. An example of elevated silencer vibration that resulted in a silencer failure is shown in figure 7. Overall vibration was 1.9 in/sec rms with dominant vibration at 12xRPM or three times the lobe pass frequency. A silencer resonance was present that resulted in a crack developing through the silencer shell due to excessive vibration on the shell.

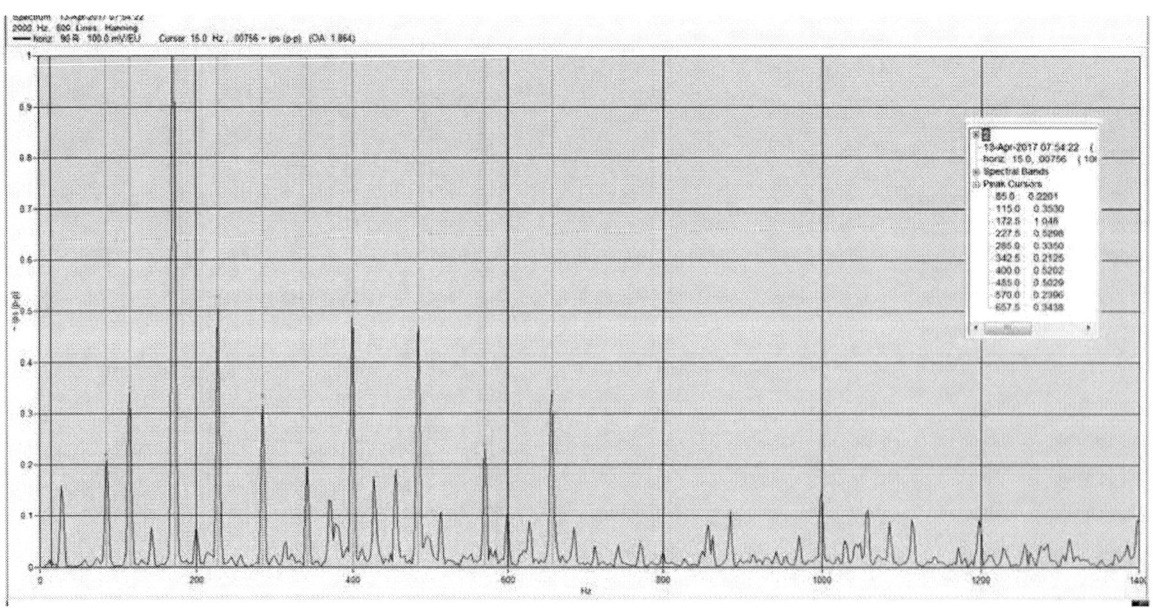

Figure 7 - Silencer Vibration

In addition to typical vibration measurements, it can also be helpful to measure pressure pulsations at the blower inlet and discharge flanges. Many machines have threaded pipe taps located there that are used for control or pressure/temperature sensing that can be tapped into for field testing. Pulsation amplitudes may be reviewed by the manufacturer to determine if it is acceptable for the blower design. Elevated pulsations can lead to rotor shaft end failures as well as fatigue damage to piping or silencers.

Noise measurement is becoming more common and can certainly be used to assist with evaluating the blower. Sound spectrums can be used to help identify surfaces that are producing the elevated sound by matching the sound peaks to vibration peaks on silencers, etc. The typical causes of very high noise will be larger surfaces (silencer shell) vibrating at the same frequency as the observed noise. Noise can be reduced by lagging (covering with some sound damping materials) or stiffening the vibrating surfaces.

Torsional vibration can also lead to shaft failures and will normally have the corkscrew shape shaft end fracture shown in figure 8. The blowers will produce high dynamic torque at lobe pass frequency and harmonics. Torsional vibration can be measured using strain gauges mounted on the drive shaft or coupling spool.

Figure 8 - Torsional Failure of Shaft

Summary

Rotary lobe blowers are common machines in a variety of services. These machines tend to vibrate higher that most common machinery due to the positive displacement rotor design that will generate a lot of harmonic vibration. The blowers are generally designed to tolerate the elevated pulsation and vibration.

Elevated vibration is common at blower harmonics with various resonances being the most common primary cause. Frequencies of concern start at 1xRPM and increase up to about 30xRPM. Very high vibration at 16x, 20x, 24xRPM can be common particularly on the piping or silencers attached to the blowers.

Operating at variable speed will always be difficult since it will be guaranteed that operation will be exactly at resonance at some speed. Operating exactly at resonance can produce elevated vibration with amplification factors of 100 or higher for silencers and piping with low damping. Variable speed operation may require selecting a limited number of fixed speeds that are jumped between opposed to continuous speed variation.

Vibration monitoring and analysis is normally reduced to looking for common bearing fault evidence while looking for vibration changes opposed to absolute vibration limits. Continuous operation at vibration levels significantly higher than vibration on pumps and motors is common.

ELECTRICAL SIGNATURE ANALYSIS APPLICATIONS IN WINDPOWER AND SOLAR UPDATE

Author: Howard W Penrose, Ph.D., CMRP

VIATC | TRAINING CONFERENCE & EXPO
SAVANNAH, GEORGIA · 2022

ELECTRICAL SIGNATURE ANALYSIS APPLICATIONS IN WINDPOWER AND SOLAR UPDATE

Howard W Penrose, Ph.D., CMRP
President, MotorDoc® LLC
Lombard, Illinois

Howard W Penrose, Ph.D., CMRP is the president of MotorDoc LLC, a veteran owned small business, vice chair technical standards American Clean Power (ACP) and powertrain committee ACP, standards development IEEE PES Materials, and past-chair SMRP and SMRP government relations cybersecurity, infrastructure and smart-grid. Dr. Penrose also serves on the Business Leaders United and ACP government relations workforce and CTE programs including presentations to Congress. His areas of research include data science and machine learning for prognostics and Time to Failure Estimation, electric machinery applications and materials, energy/environment, and a specific focus on Electrical Signature Analysis. He is a past adjunct professor of industrial engineering for UIC and a past senior research engineer for the UIC energy resources center. Howard also received five UAW-GM People Make Quality Happen awards for corporate reliability and energy programs from 2004-2009 as part of the UAW-GM QNPM best practices programs. He has collaborated on flywheel machinery design, wind turbine fleet failure analysis and mitigation methods, LVAD heart-pump, the hybrid Tahoe and Volt motors, and the John Deere 644 and 944 hybrid tractor motors and generators.

Abstract: Electrical Signature Analysis has been adopted within the windpower and solar industries through industry guidebooks for prognostics and continuous monitoring of complete powertrain through supply transformers. In this presentation we will discuss the advances in research and practical applications over the past decade for the application of ESA for electrical and mechanical monitoring and prognostics. Several case studies with findings will be included in the presentation. The work has practical applications within other industries.

Key Words: Current signature analysis; electrical signature analysis; windpower; solar

INTRODUCTION

Onshore and offshore windpower and solar applications have specific electrical and mechanical reliability concerns related to the cyclical nature of their applications. For windpower, these conditions start with the step-up transformer, the generator mechanical components, the electrical components and controls, coupling, three stages of gearbox (for asynchronous designs), main bearings, and blades. This does not include the blade pitch gearbox motors, yaw motors, oil pumps, fans, and other electric machine-driven equipment. In solar, the conditions range from the panels, two or three axis systems, converters, and step-up transformers. In this paper we will focus on these specific areas for both windpower and solar, starting with challenges associated with each and then how Electrical Signature Analysis (ESA) is used to evaluate the condition of these machines either through data collection or continuous monitoring.

ESA is a combination of power quality, overall raw data, and spectral analysis of voltage, current, power (watts), and airgap torque associated with an electric machine. The transducer used is the magnetic field between the rotor and stator, although many mistakenly consider the rotor or stator as the transducer. The distinction is critical for proper understanding of the data being analyzed. For instance, in DFIG (Doubly-Fed Induction Generators) machines, a 6-pole generator may be turning at 1440 RPM mechanically which would generate 72 Hz power (120*72Hz)/6poles = 1440 RPM. However, we need to produce 60Hz +/- ~2Hz, which is accomplished by using a special variable frequency drive to inject voltage into the rotor at an angle that will generate a counter rotating field that will negate the difference to bring us close to 60Hz output. A similar process is used in low wind speed conditions to speed-up the rotating field, resulting in a relatively constant >1200 RPM rotating magnetic field. While vibration will utilize the 1440 RPM mechanical speed, ESA uses the ~1200 RPM magnetic field to evaluate the condition of the wind turbine system.

In low wind conditions the controls are also used to correct for power factor in the system. This is done in one of several ways. For both DFIG and SFIG (Singly-Fed Induction Generators), the controls are used to inject capacitance into the system 'stiffening' the electrical system towards unity power factor, which impacts the local grid. This has the additional effect of generating an electrical sub-synchronous resonance into the system, which is a fault driver, but also generates some unusual voltage and current patterns that must be discerned. Once these signatures are identified and a relatively clean signature can be gathered or filtered, then an analysis of the complete powertrain can be performed.

In solar farms, the DC voltage developed by the panels is fed to an inverter to convert the DC voltage to AC voltage. The converters are normally in parallel with two feeding a single step-up transformer (two lower voltage primary and a high voltage secondary). Unbalanced feeds into the transformer can generate circulating currents that cause excessive heating, as well as Ferroresonance signatures in the transformer cores due to the firing frequencies (also found in wind applications). The frequencies associated with the drives, in all cases, tend to be in the 2000 Hz range.

For analysis of a generator, or voltage supply (upstream), the primary spectral signatures are in voltage. In the case of windpower, the combined use of voltage and current is necessary as in some part of the operation the rotor becomes a load, requiring current signatures. The result is that proper analysis requires the use of both voltage and current spectra. To obtain a full analysis of a system several other key capabilities are necessary including an FMAX of at least 5kHz (300,000 CPM), at least a 10kHz sampling rate, 3 averages, and an FFT block of at least 65,000. As we will discuss, these are necessary in order to separate peaks associated with blades, pole pass frequency, and gearbox bearings, as well as keeping the data collection time low to avoid noise due to wind gusts.

One of the interesting features of the airgap magnetic field transition from mechanical to electrical speed, is the feature where electrical and mechanical signatures throughout the system will also translate to the magnetic field speed. Therefore, if the field speed is 1208 RPM, then all of the multipliers used through the system will be based on 1208 RPM (20.133 Hz) which has a slip of 0.133 Hz and a resulting Pole Pass Frequency (PPF) of 0.798 Hz (6 poles * 0.133 Hz). The blades will have a blade pass related to the tower pulse (as a misaligned blade passes the tower) of 14-20 RPM, or 0.233-0.333 Hz. Some planetary and main bearing frequencies will also be found relatively close to these frequencies. Along with the frequency translations will be a dampening of the signatures as measured in dB, requiring tighter alert and alarm tolerances. These conditions are generally unique to wind turbines.

WIND OPERATING CONDITIONS

The power output from a wind turbine is dependent upon the wind speed. As shown in Figure 1, the curve starts at ~3.2 m/s, which is referred to as 'cut-in' and peaks at ~11.5 m/s through to a forced 'cut-out' at 25 m/s. This design turbine will attempt to continue generating to the low-speed cut-out of ~2.8 m/s through the controls. The mechanical speed of this turbine, which will be used as our case study, will start as low as 1175 RPM and a peak of 1440 RPM, depending on how the controls are set. The field rotating speed will range from 1205 RPM to 1220 RPM, also depending on how the output frequency is programmed.

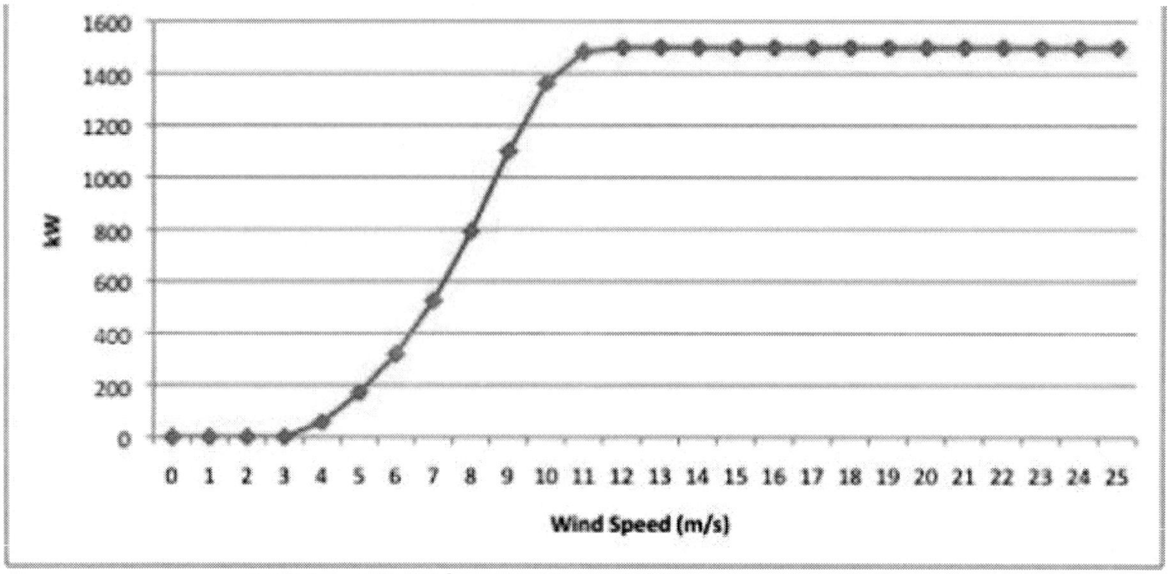

Figure 1: power curve for a 1.5 MW, 1440 RPM asynchronous wind turbine.

The machine being used for this evaluation is configured as shown in Figure 2, where the main shaft rotates between 14 and 20 RPM, feeds a planetary gear into three gear-sets that bring the speed up to the operating speed of the generator. The generator stator is connected directly to the pad-mount oil-filled step-up transformer at 575 to 690 Volts up to either the site voltage of 35kV or 69kV. The control uses some of the power to feed an inverter which is connected via brushes to the rotor with a firing frequency ranging between 1800 Hz and 2000 Hz, a random output frequency based on control settings, and a maximum voltage of 1000 Vac. The rotor control circuit varies significantly depending on the wind speed, power factor, and output voltage and frequency of the generator.

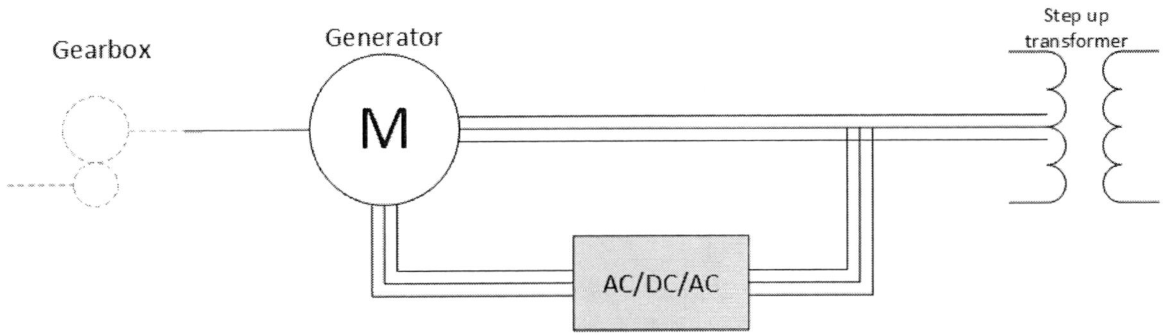

Figure 2: simplified DFIG circuit.

One of the challenges with the design that is considered for each tower is the tank circuit that is formed between the impedance of the generator, the impedance of the transformer, and the induced capacitance of the rotor drive. Much like mechanical resonance, this type of circuit will hit natural electrical frequencies and generate an electrical resonance in the stator and/or magnetic field. These conditions can be detected when they are 'synchronous' through relays that will see variations in output voltage or frequency. However, Sub-Synchronous Resonance (SSR) conditions will also occur and are directly depended upon either the control, known as Sub-Synchronous Control Interaction (SSCI) or from other capacitive sources ranging from local distribution cables to voltage correction in long distance transmission lines, called Sub-Synchronous Torsional Interaction (SSTI). These conditions can affect different turbines, strings of turbines, or a complete site, depending on the location, wind/weather patterns, settings, and type of local or bulk grid.

The larger impact of the operation of the turbine has to do with its operating profile. When the turbine detects wind speed and direction it will turn towards the wind and start turning. Once it sees the minimum speed (i.e.: 3.2 m/s) for more than a minute, depending on the control settings, it will cut-in. At that moment, the complete powertrain and generator, which has been turning at speed and no-load, is shock-loaded. This may occur from 1-2 seconds before the soft start kicks in and slowly loads the generator. In specific low-wind conditions, this can occur from six to twelve times per hour. In all present SFIG and DFIG designs, the main shaft to gearbox and gearbox to generator couplings are rigid and no clutches exist in the system.

SOLAR OPERATING CONDITIONS

Utility-scale solar operations will operate optimally depending on the angle of the sun, time of day, and shading by cloud cover or even overhead passing aircraft. When observing output at one site shade from a cloud drifted across the multi-acre site and the passage was directly observable on the wattage output of the various drives in the system. While the energy density of various panel types continues to increase, the need to maintain the cleanliness of the surfaces and selection of the site locations maintains a concern. There are various single, two and three axis positioning systems for solar panels and the selections depend on the payback for the system. In most cases, the optimal cost-benefit appears to be fixed systems set at an optimal angle for specific seasons.

The conversion system for a solar site is different than most in that the DC power from the panels feeds an inverter which may have an output ranging from 300 volts to over 1000 volts into one of two primaries of a local distribution transformer. Both primaries are stepped from a range from 12.4kv to 69kV for site distribution as shown in Figure 3.

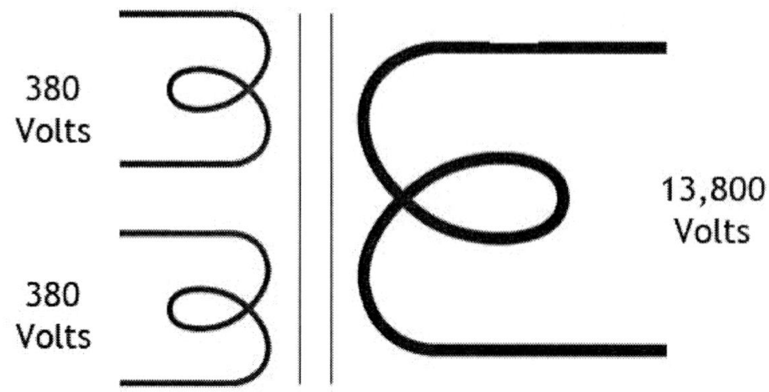

Figure 3: demonstration of primary to secondary step-up transformer used in solar.

The challenge with the solar design is that both primaries must be balanced to avoid circulating currents in the transformer and electrical resonance conditions (Ferroresonance) from a combination of the drives and distribution circuit capacitances. These conditions would cause heating and gassing and are found to be similar in wind applications, which have a single primary. In most cases, the dissolved gasses from both wind and solar are high enough that they are specifically excluded from dissolved gas standards, such as IEEE Std C57.104, and specific standards for the higher gas levels are required. The loading of these transformers is also part of the equation as they will vary in load from ~10% to over 100% depending on the site designers. The higher loads are based upon the idea that lower loads are more harmful and that most wind/solar sites average 35% capacity factor (average output/load of a site).

WIND TURBINE CASE STUDY

The primary defects found in wind turbines relate, in order, as blade degradation, generator defects, the gearbox, and the drive. On average there will be one major gearbox defect/replacement, two major generator defects/replacements, at least one major replacement of blades or main bearings, and countless blade repairs during a planned 25-year project life. This does not include other conditions such as a high rate of yaw or pitch motors and controls, sensor failures, and other electrical or mechanical systems, of which most have an impact measured in hundreds or thousands of dollars and hours of operation. Major failures or defects are measured in $100's of thousands and weeks or months of downtime.

Electrical Signature Analysis (ESA) has been explored for use in wind turbines since the early 2000s and solutions found by MotorDoc in 2003. Most ceased work in the area as data did not translate from laboratory electric motors to in-field turbine generators. The primary cause is that wind turbine generators do not operate the same as standard generators, and the method used for generator testing is voltage signature analysis versus current signature analysis. Most also viewed the speed being analyzed as the mechanical speed versus magnetic speed. In the case of a wind turbine, due to the complexity of the system, resolution is critical as is the use of both current and voltage signature depending on the loading of the system. At various points in the operating setpoints the analyst will be required to look at voltage signature, current signature in other conditions, and both in yet other conditions, and these can change by site and/or manufacturer.

In this case study we will take a worst-case scenario windpower site that has been operating for about ten years. The wind conditions for that entire period show as in Figure 4 and an average year as in Figure 5. The site is hundreds of miles from the nearest heavy loads and it feeds a 500kV bulk power system with both series and parallel VAR (capacitive and inductive voltage control) systems in-place.

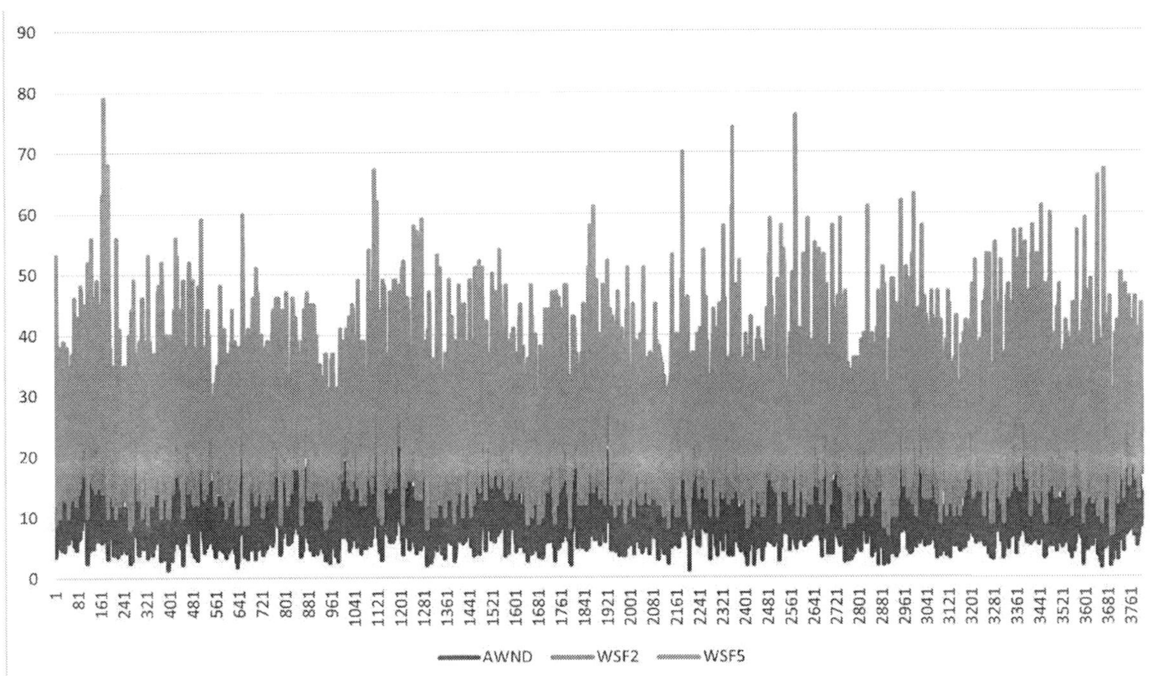

Figure 4: ten years NOAA average and peak 80 meter windspeed and gusts from 2011 to 2021. Blue is wind speed (mph) and grey/orange are gusts (average = orange; peak = grey).

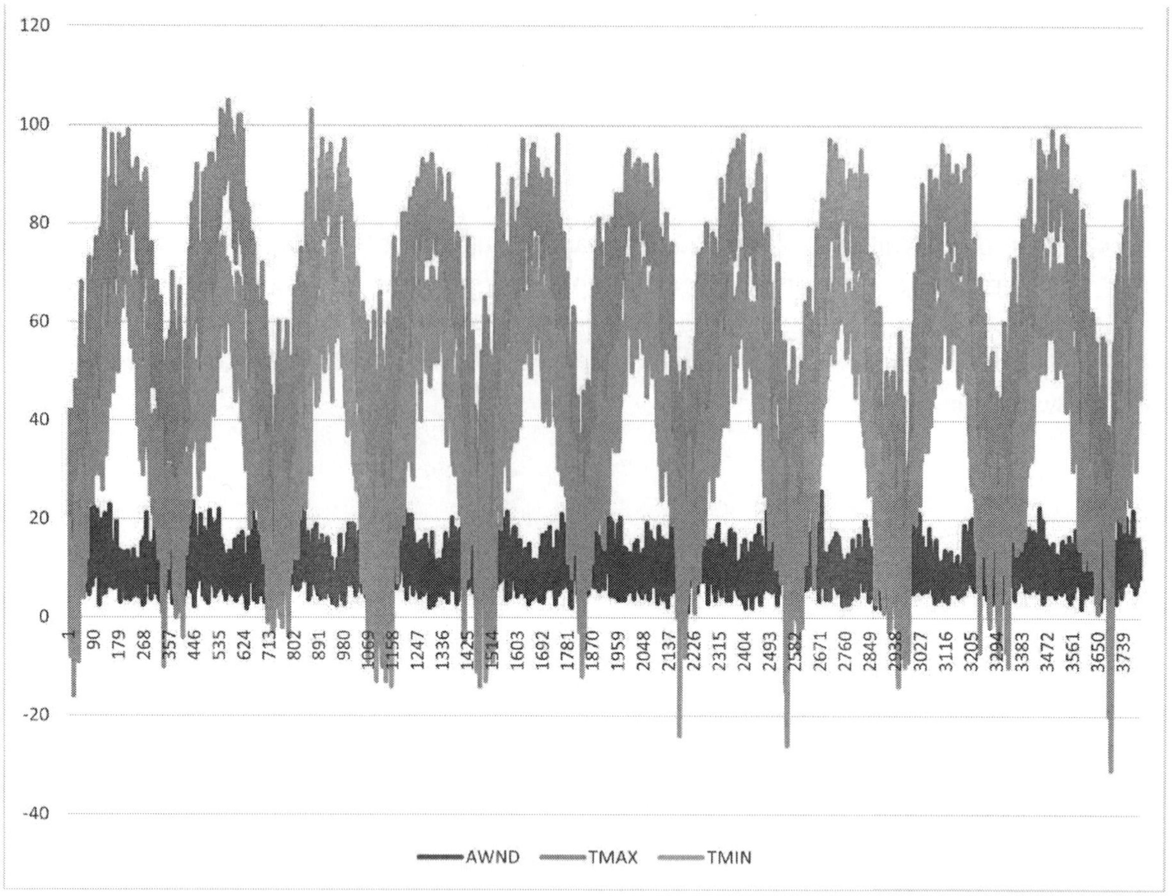

Figure 5: average wind speed (blue – mph) and min/max temperatures (F) by day across 2020 for the site.

Normally connections would be made in the down-tower controls for this type of turbine. However, it was determined that the data would be taken directly at the generator in order to evaluate conditions directly at the terminations.

Figure 6: voltage and current transformers connected at the terminal box of the transformer.

As noted in Figure 6, the current clamps are around one of three cables for each phase. In the system that would be entered as a 3:1 CT ratio. The ESA system was then set up to capture the start for 240 seconds in both voltage and current. The wind speed was measured at 5.8 m/s with gusts to 8 m/s and an ambient temperature of 13C.

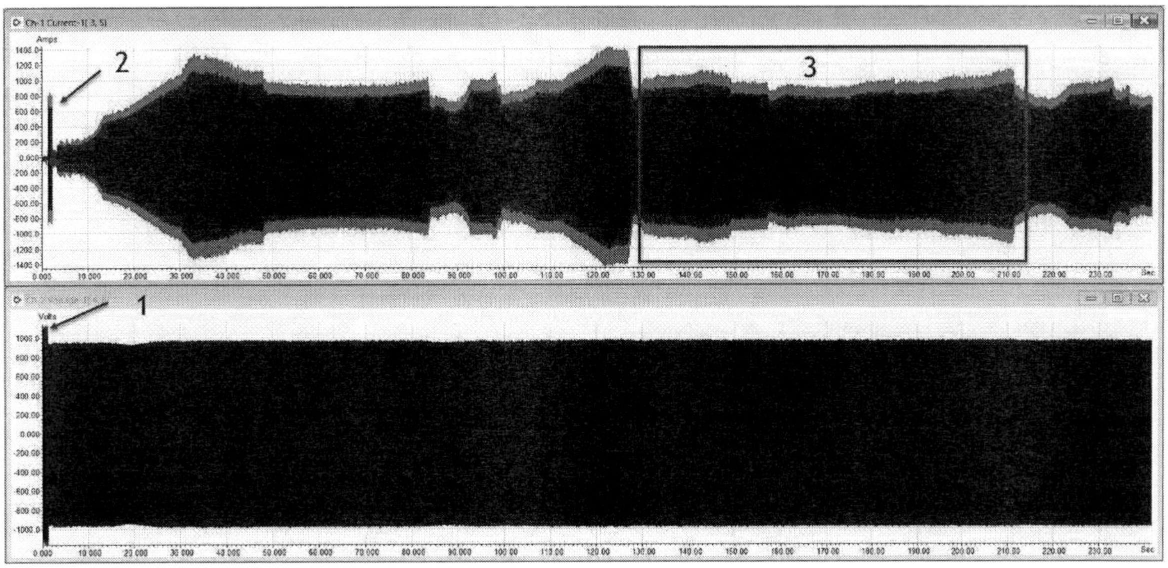

Figure 7: ESA inrush data across 240 seconds. Top is current and bottom is voltage.

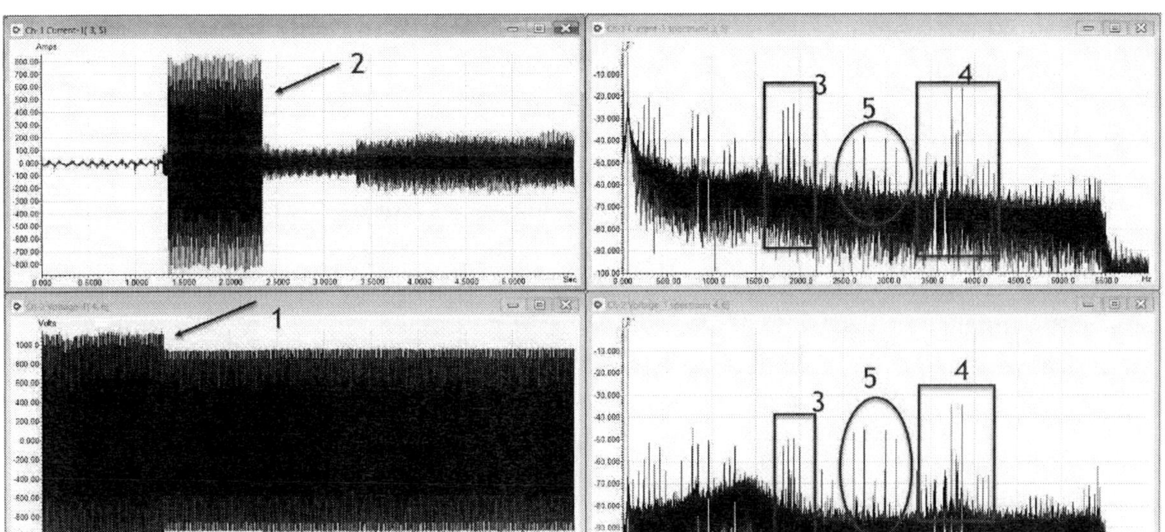

Figure 8: FFT analysis of current (top) and voltage (bottom) during initial closing of breaker (cut-in).

In Figure 8, point 1 is the voltage drop when the generator is brought online, point 2 is the 2-second impulse as the breaker closes and before the soft-start systems kicks in. Point 3 is the inverter signature with multiple sidebands (firing frequency of rotor drive) and point 4 is the second harmonic of point 3. Point 5 was interesting and required some study to discover it was a low-level frequency identifying the Ferroresonance of the transformer. In comparing this signature to a dataset of over 3400 turbines of the ~69,000 turbines in operation at the time, a correlation of number of generator, gearbox and transformer defects was found. As these peaks are higher at a site, the larger number of dynamic defects are found in the generator, transformer and gearbox. This particular site had the highest peaks of those found to date and also had the highest number of defects for this type of turbine at only 10 years operation.

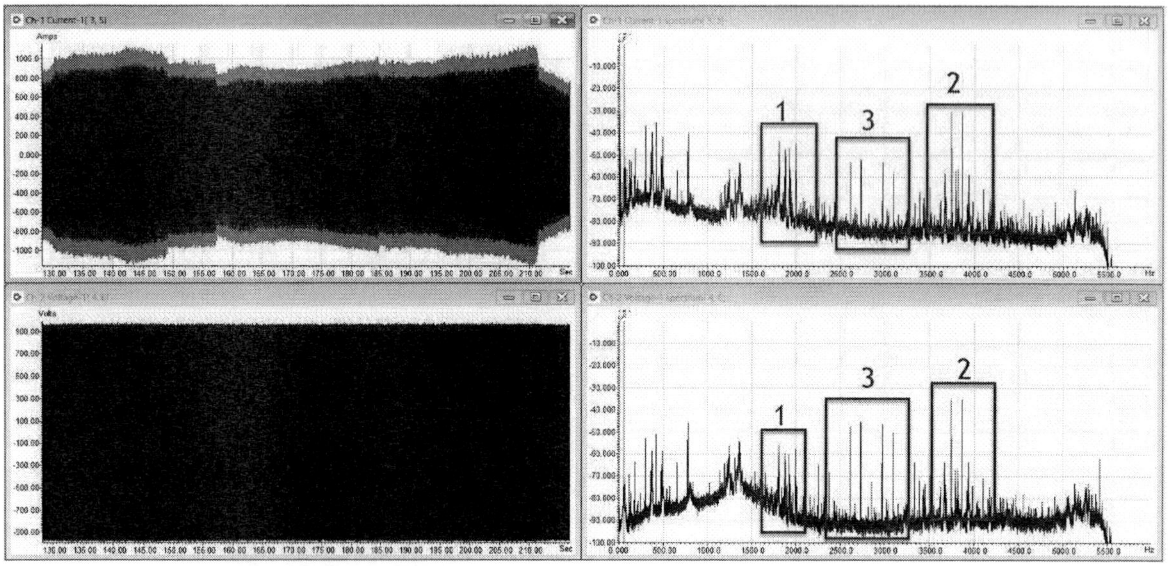

Figure 9: peaks associated with the 'steady state' portion of data (point 3 of Figure 7).

In the steady-state operating portion of the signature, as shown in Figure 9, points 1 and 2 are the drive frequencies and harmonic and point 3 is the transformer. As ESA is an amplitude modulated technology, there are multiple sidebands around the various frequencies that are either speed or line frequency related.

The expected frequencies related to SSR are determined to be ~15 Hz, ~21 Hz, and ~51 Hz in a 60 Hz operating system. In the case of Figure 8, point 5, the center frequency of 2854.0 Hz is 190 times the SSR frequency of 15.02 Hz. In Figure 9, point 3 has shifted to 2852.9 Hz, which is a shift in the SSR frequency to 15.016 Hz. The output from the turbine is 161.75 kW and the losses at the SSR frequency are 1.6 kW, or 1% of the total load. An existing wye ring fracture is showing a loss of 5 kW, or 3% of the total load.

The airgap torque signature is shown in Figure 10. In this case, note the starting torque and spikes from when the breaker closed, then the peak-to-peak torque (ftlb) as the generator comes up to its present load. The oscillation associated with the wide band is at 15.02 Hz. Note that this is magnetic airgap torque, physical shaft torsional analysis would be of a different value.

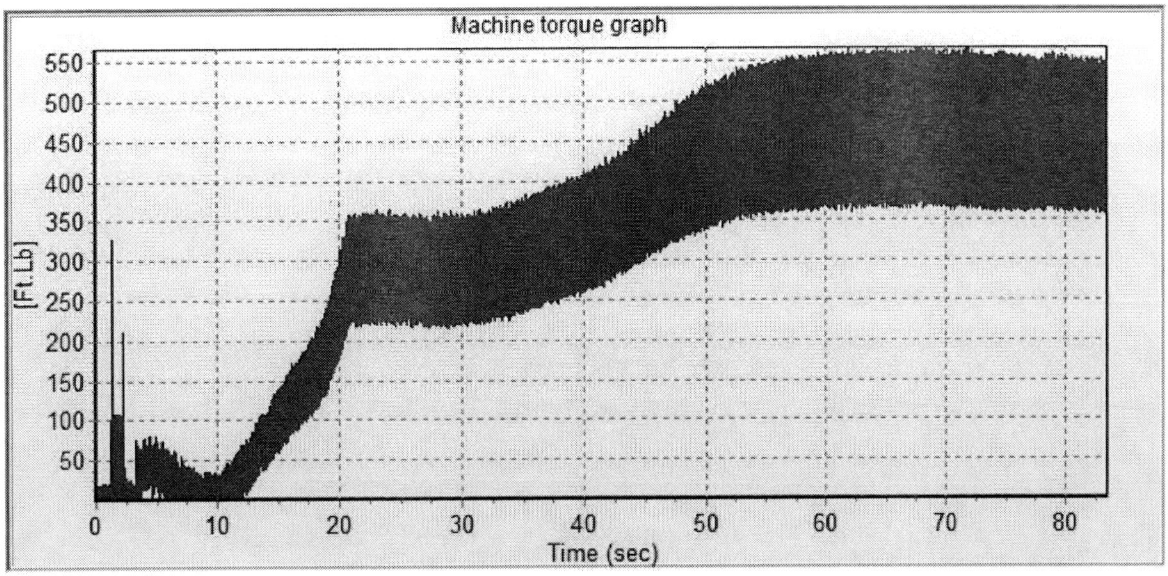

Figure 10: air gap torque spectra across 85 seconds showing an increasing 15.02 Hz air gap torque as the generator loads.

SSR AND ITS IMPACT ON TURBINES

The first exposure to Sub-Synchronous Resonance was in large steam-powered generation at Southern California Edison's Mojave power plant in 1970. The application of HVDC transmission systems, voltage correction systems, and power electronics on the customer-side create potential issues where faults in the bulk or local distribution can initiate low frequency oscillations that break shafts in larger machines. Following several events prior to and one significant event in 2009 in the Electric Reliability Council of Texas (ERCOT) with an SSR event that damaged a windpower site, the study of SSR within wind and solar has been primarily academic. The first measured versus modeled impact of SSR was performed by MotorDoc in 2019 and similar work performed in China starting in 2020.

The above case study is a sample from over 3,400 datasets from multiple wind turbine OEMs in different operating environments through North America with a smaller dataset of 50 Hz applications from Asia. In all cases, similar conditions and frequencies exist and are consistent with related SSR literature. For wind generation, SSR includes Sub-Synchronous Torsional Interaction (SSTI), which is generated from distribution system voltage correction, and Sub-Synchronous Control Interaction (SSCI), which is generated from the turbine controller. The challenge is that while large events have occurred since 2009 on wind farms, the general consensus was that the values were generally under 10% of the force required to break a shaft, which is the common result of SSR in large turbines. Instead, normal operating oscillations generate a fatiguing condition with larger, sudden, events being the result of short circuit conditions in the distribution system.

The primary areas for impact by the conditions outlined in this paper (both SSR and starting) include the planetary gear set, the high-speed gear set, DFIG/SFIG rotor connections, stator magnetic wedges, and transformer Ferroresonance and heating. It is no surprise to discover that these types of defects are the highest powertrain defects found in wind turbine DFIG and SFIG types.

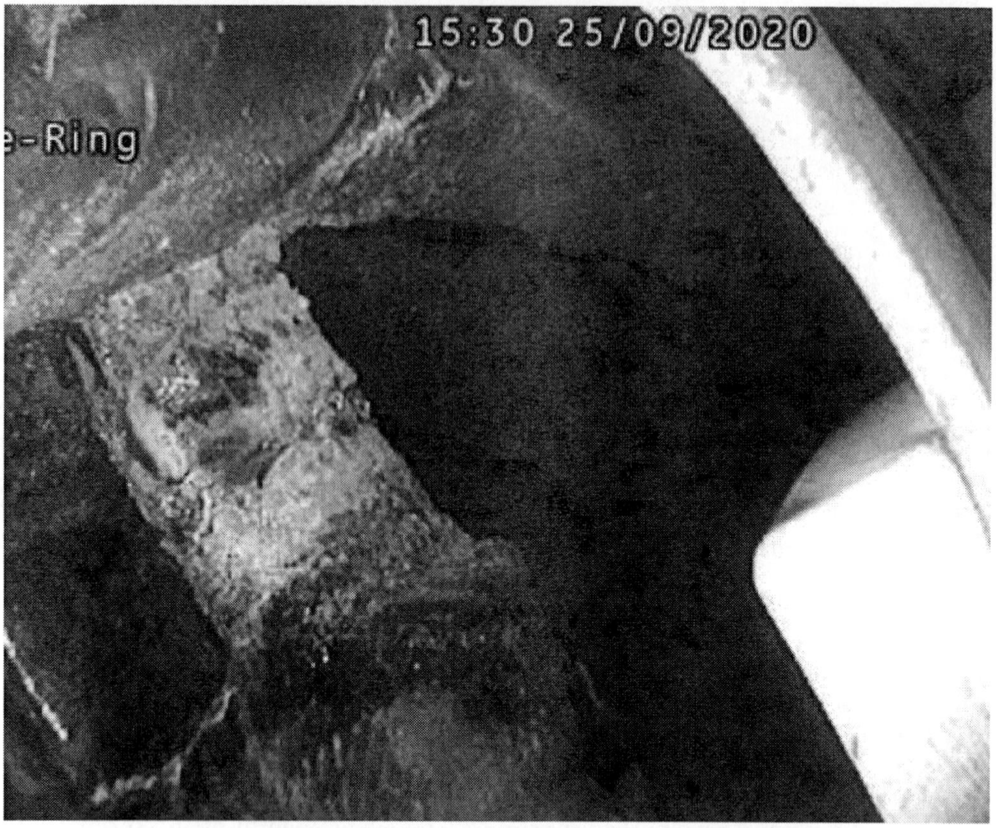

Figure 11: wye ring fracture: most common generator defect in aging turbines over 10 years old outside of bearing lubrication issues...

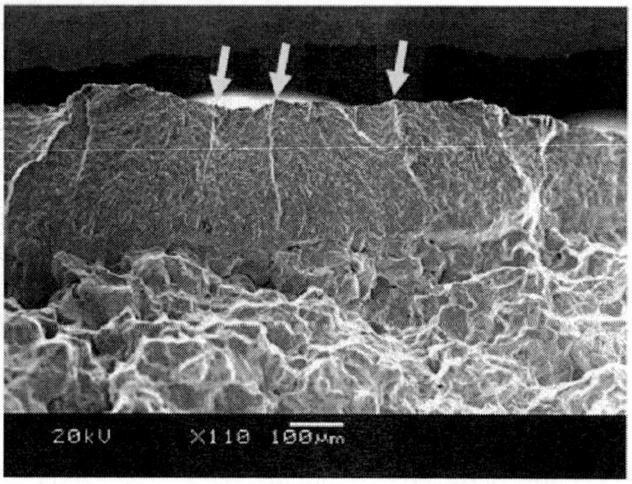

Figure 12: pre-wye ring fracture fatigue findings.

Figure 13: results of catastrophic wye-ring failure.

The impact of the type of failure shown in Figures 11-13 is a minimum of $160,000 USD and over a month of downtime versus an average cost of $10,000 and a day of downtime for an early detection and up-tower (in-place) repair. With the type of fatiguing failure, let alone from the flexing by number of starts, this type of repair would happen on almost all DFIG/SFIG at least once in a 25-year period. The average USA wye ring failure occurs within 15 years of operation depending on the operating conditions.

Figure 14: stationary ring gear of planetary gear set.

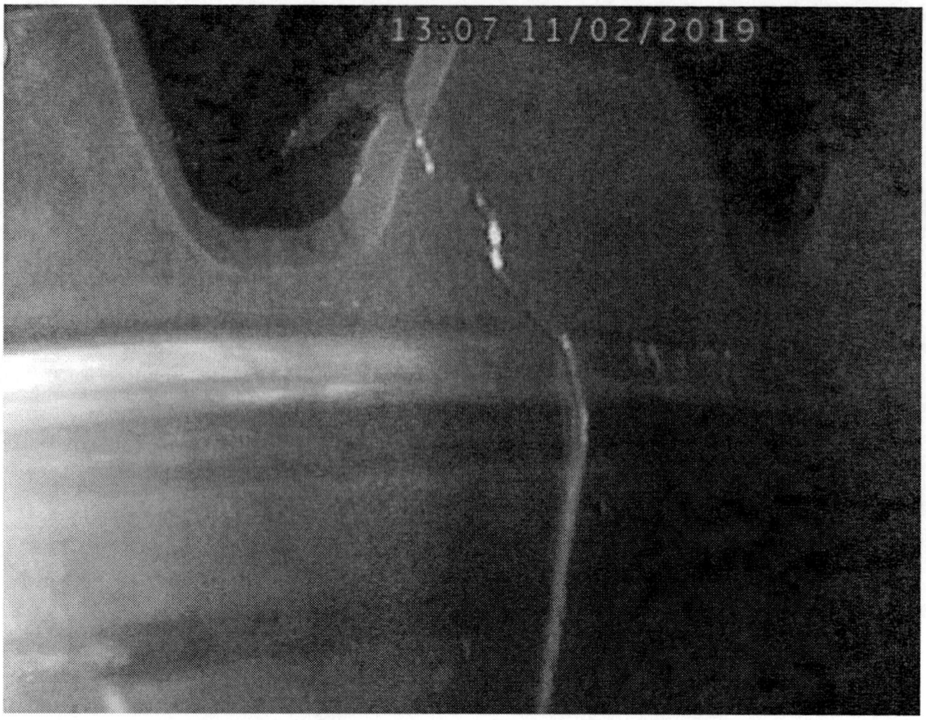

Figure 15: planet gear fracture.

Typical failures of a planetary ring gear and planet gear can be found in Figures 14 and 15. The most common type of failure in these applications are fractures of the high-speed bearings.

SSR AND IMPACT ON SOLAR

While we don't have rotating generators associated with solar, the transformer is the point of stress as discussed. In this case, the primaries of the site have unbalanced supply voltage and power.

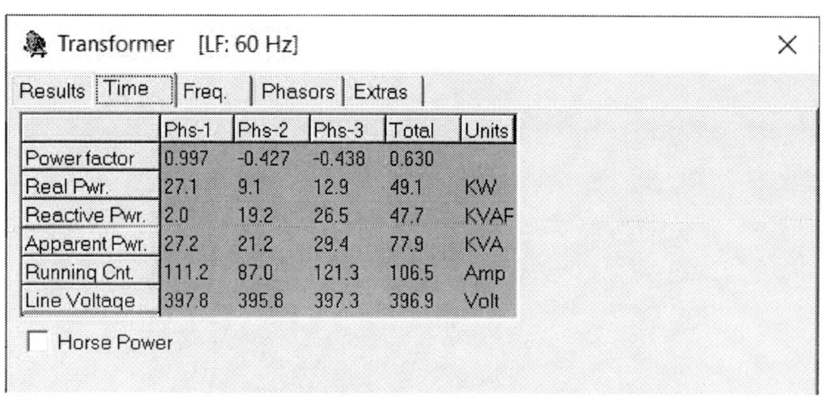

	Phs-1	Phs-2	Phs-3	Total	Units
Power factor	0.997	-0.427	-0.438	0.630	
Real Pwr.	27.1	9.1	12.9	49.1	KW
Reactive Pwr.	2.0	19.2	26.5	47.7	KVAF
Apparent Pwr.	27.2	21.2	29.4	77.9	KVA
Running Cnt.	111.2	87.0	121.3	106.5	Amp
Line Voltage	397.8	395.8	397.3	396.9	Volt

Figure 16: primary 1 inverter to transformer.

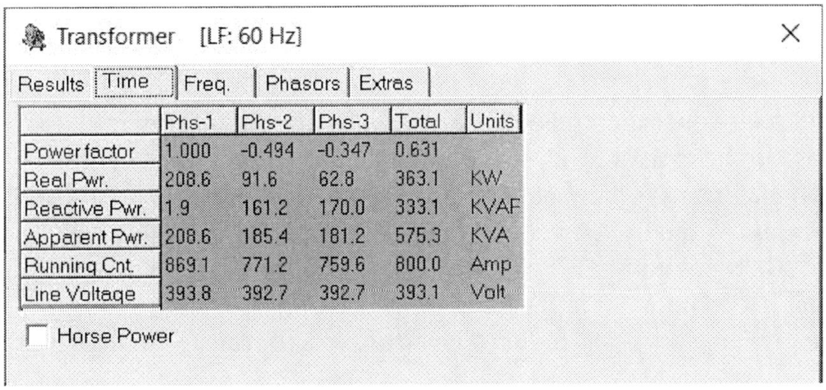
Transformer [LF: 60 Hz]

Results | Time | Freq. | Phasors | Extras

	Phs-1	Phs-2	Phs-3	Total	Units
Power factor	1.000	-0.494	-0.347	0.631	
Real Pwr.	208.6	91.6	62.8	363.1	KW
Reactive Pwr.	1.9	161.2	170.0	333.1	KVAF
Apparent Pwr.	208.6	185.4	181.2	575.3	KVA
Running Cnt.	869.1	771.2	759.6	800.0	Amp
Line Voltage	393.8	392.7	392.7	393.1	Volt

Horse Power

Figure 17: primary 2 from inverter to transformer.

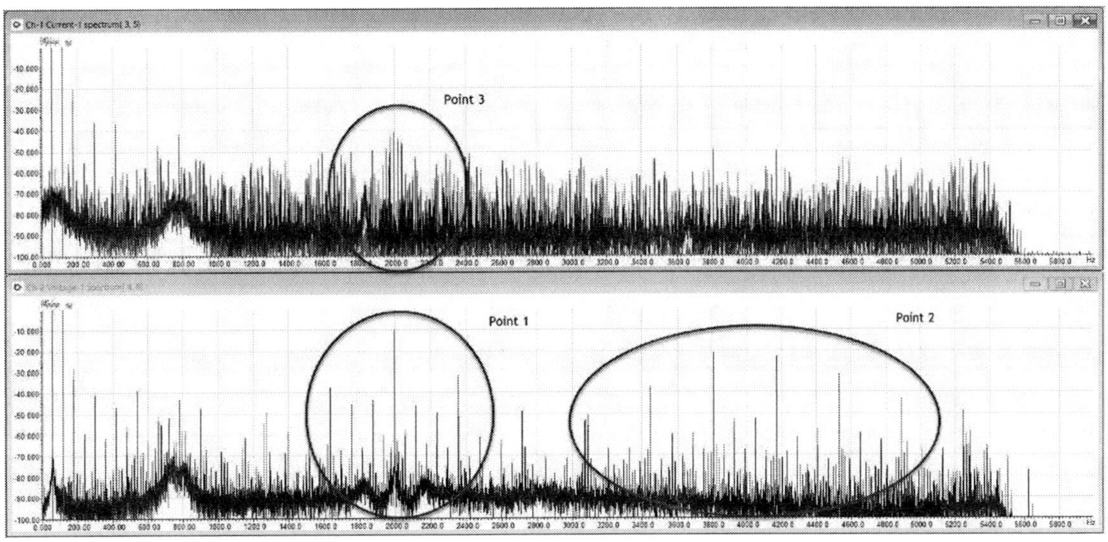

Figure 18: inverter signature (firing frequency) and harmonic in voltage between primary 1 and inverter.

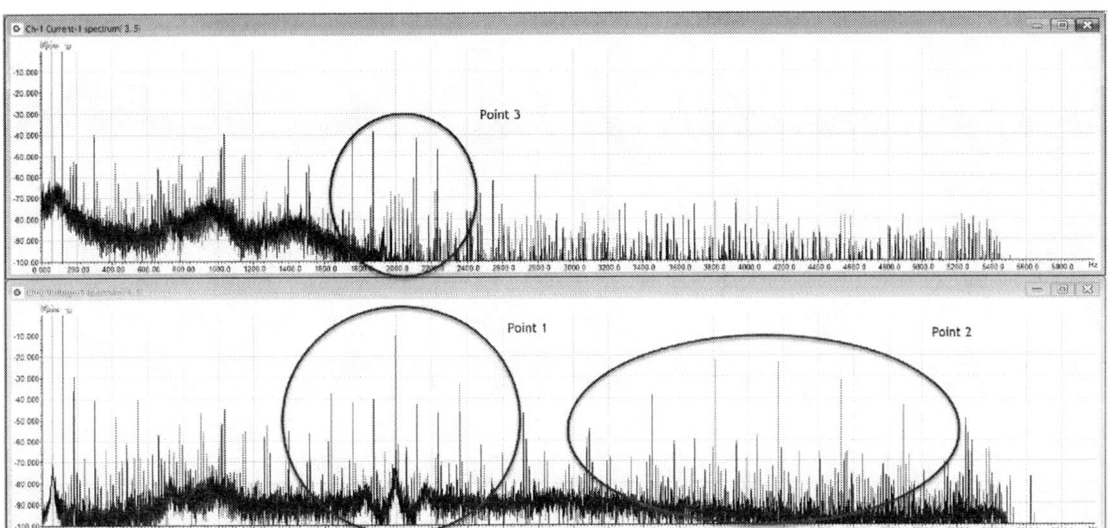

Figure 19: inverter signature and harmonic in voltage between primary 2 and inverter with signficant signature in current (transformer).

The result of these unbalances in the transformer is heat.

CONCLUSION

Wind turbine and solar generation reliability are directly affected by operating profiles and the control strategies. In the case of wind, SSTI and SSCI (SSR), number of starts, and environmental conditions have a direct correlation to the reliability of turbine components. The greatest stresses related to SSR occur at lower wind speeds and lighter loads where power factor correction strategies initiate air gap torsional oscillations. These generate long term rotor fatigue, gearbox component fatigue and transformer gassing and aging. In solar applications, the primary concern is reduced inverter life and transformer gassing and aging.

For wind generation, there are multiple corrective actions to be considered to extend the operating life of wind turbines. These also require a review of the transmission and distribution system as well as turbine control strategies:

- Existing short-term reliability improvements include stiffening systems such as thicker wye rings and modifications to gearbox design. For transformers, certain transformer oils including plant-based esters which have high thermal breakdown.
- Modifications to long-distance transmission and distribution systems for VAR correction. Observations have shown that series and parallel VAR strategies have the same impact on wind and solar reliability.
- De-tune SSR frequencies by turbine. Each turbine, depending on its location in the wind site systems and operating environment, is affected differently. The use of active or passive filtering strategies at different operating points can eliminate the driving forces or improve reliability.
- Changes to control strategies including soft-start and starting/loading characteristics. Modification to the rotor control firing frequencies can also provide improvement.
- A clutch between the generator and gearbox, or main shaft and gearbox, to reduce impact on the gearbox and shocks to the generator.

The immediate corrections for solar include load balancing and the longer term are similar to wind in the case of distribution system improvements.

BIBLIOGRAPHY

- Shuangwen Sheng, “Gearbox Reliability Database: Yesterday, Today, and tomorrow,” NREL Wind Turbine Tribology Seminar, Chicago, Illinois, October 29-31, 2014.
- G. Jose and R. Chacko, "A review on wind turbine transformers," 2014 Annual International Conference on Emerging Research Areas: Magnetics, Machines and Drives (AICERA/iCMMD), 2014, pp. 1-7, doi: 10.1109/AICERA.2014.6908172.
- Hossein Ali Mohammadpour and Enrico Santi, Analysis of Sub-Synchronous Resonance (SSR) in Doubly-Fed Induction Generator (DFIG)-Based Wind Farms, University of Nebraska, Lincoln: Morgan & Claypool, 2015.
- X. Dong, J. Hou, Z. Wang, J. Pi, X. Tian and Y. Jin, "Study on the subsynchronous resonance mechanism of DFIG based on the analysis of stator and rotor torques," 2016 China International Conference on Electricity Distribution (CICED), 2016, pp. 1-6, doi: 10.1109/CICED.2016.7576188.

- X. Dong, W. Zang, C. Dong and T. Xu, "SSR characteristics study considering DFIGs at different locations in large wind farm," 2016 China International Conference on Electricity Distribution (CICED), 2016, pp. 1-6, doi: 10.1109/CICED.2016.7576178.
- Hossein Ali Mohammadpour and Enrico Santi, Analysis of Sub-Synchronous Resonance (SSR) in Doubly-Fed Induction Generator (DFIG)-Based Wind Farms, University of Nebraska, Lincoln: Morgan & Claypool, 2015.
- J. Samanes, E. Gubia, J. Lopez and R. Burgos, "Sub-Synchronous Resonance Damping Control Strategy for DFIG Wind Turbines," in IEEE Access, vol. 8, pp. 223359-223372, 2020, doi: 10.1109/ACCESS.2020.3043818.
- Shuangwen Sheng, "Wind Turbine Gearbox Reliability Dataase, Operation and Maintenance Research Update," NREL Drivetrain Reliability Collaborative Meeting, Feb 21-22, 2017.
- K. Alewine and H. Penrose, "Field Experiences Utilizing Electrical Signature Analysis to Detect Winding and Mechanical Problems in Wind Turbine Generators," 2020 IEEE Electrical Insulation Conference (EIC), 2020, pp. 246-248, doi: 10.1109/EIC47619.2020.9158573.
- E. M. Hope, T. A. Bellei and M. Reyes, "Wind turbine generator step-up transformer failure investigation," PES T&D 2012, 2012, pp. 1-7, doi: 10.1109/TDC.2012.6281584.
- M. K. Siahpoosh, D. Dorrell and L. Li, "Ferroresonance assessment in a case study wind farm with 8 units of 2 MVA DFIG wind turbines," 2017 20th International Conference on Electrical Machines and Systems (ICEMS), 2017, pp. 1-5, doi: 10.1109/ICEMS.2017.8056530.
- X. Xie, X. Zhang, H. Liu, H. Liu, Y. Li and C. Zhang, "Characteristic Analysis of Subsynchronous Resonance in Practical Wind Farms Connected to Series-Compensated Transmissions," in IEEE Transactions on Energy Conversion, vol. 32, no. 3, pp. 1117-1126, Sept. 2017, doi: 10.1109/TEC.2017.2676024.
- Hossein Ali Mohammadpour and Enrico Santi, Analysis of Sub-Synchronous Resonance (SSR) in Doubly-Fed Induction Generator (DFIG)-Based Wind Farms, University of Nebraska, Lincoln: Morgan & Claypool, 2015.

MOTOR FAULTS THAT VIBRATION DATA DOESN'T ALWAYS DETECT

Author: William Kruger

VIATC VIBRATION INSTITUTE ANNUAL TRAINING CONFERENCE & EXPO 2022 | MFPT ANNUAL CONFERENCE

VIATC TRAINING CONFERENCE & EXPO
SAVANNAH, GEORGIA · 2022

MOTOR FAULTS THAT VIBRATION DATA DOESN'T ALWAYS DETECT

William Kruger
ALL-TEST Pro
PO Box 1139
Old Saybrook, CT 06475
wkruger@alltestpro.com

William Kruger joined ALL-TEST Pro, LLC as the Technical Manager in 2004. Since joining ATP, Bill has traveled the world teaching the Theory and Application of Motor Diagnostics, helping Fortune 500 Companies implement Predictive Maintenance Programs. He has also spent over 20 years training in the field of vibration. In 2010 Bill's proven instructional techniques earned him the new title of Training Technical Manager.

ABSTRACT:
Vibration analysis is a very useful tool in identifying faults in rotating equipment and can easily separate rotor faults from stator faults, however beyond that the faults are left up to the electrical team to "sort out" electrical vibration calls. Traditionally the electrical technician grabs a megohm meter and multimeter to sort out the electrical vibration. How often does this result in an "All's OK" response? This paper presents why this approach doesn't work and what faults the vibration data does identify and what faults it doesn't. More importantly it provides some techniques and guidelines to enable the electrical group to identify and correct most motor faults using MCA™ & ESA.

MOTOR FAULTS:
Motors are subject to all of the mechanical forces and faults associated with all rotating equipment. Unbalanced rotors, bent shaft, misalignment, bearing defects all provide vibration forces in a motor at the same frequencies as any other rotating equipment. However, there is a group of vibration faults that are considered "electrical faults" that are unique to AC squirrel cage induction motors.

One of the earliest methods developed for separating "electrical" faults from the mechanical faults that create vibration in these electric motors was to watch the change in the vibration after power was removed from the motor. If the vibration disappeared immediately, it was correctly diagnosed that the fault was electrically related, but, if the vibration decreased as the shaft speed decreased it was determined that it was not electrically related but a mechanical fault such as unbalance or other issues related to the rotor.

Electrical rotor issues were generally easy to detect as the result of the audible "beating or modulating" sound coming from the motor. Confirming this as a rotor issue could be easily done using a tachometer and stopwatch. For example, if the motor was a four-pole motor rotating at

1785 RPM and there were 15 cycles of the audible beat occurring in a 15 second time frame that would equal 1 cycle per second which equates to the number of poles times the frequency of the rotor current. This would strongly confirm the fault is related to the rotor.

SPECTRUM ANALYSIS:

Spectrum analysis, using modern "handheld" high resolution data collectors made it much easier to separate "Electrical" vibration issues from mechanical faults. Stator electrical faults occur at a frequency that is related to the number of poles and the rotating speed of the magnetic field. By mathematical laws, this frequency works out to be 2 times line frequency.

From a practical approach fault such as static eccentricity (unequal air gaps, soft foot), mechanical looseness all occurs at twice line frequency, whereas mechanical faults are related to shaft speed or multiples of shaft speed.

Since most motors in use are squirrel cage induction motors, there must be relative motion between the rotor and the rotating stator magnetic field to induce power onto the rotor to create an electromagnet field on the rotor. Therefore, the shaft speed will normally be less than synchronous speed (SS).

SS = 120 F/P

Where: SS = synchronous speed

F = Frequency in Hz

P = Number of Poles

If a machine driven by a 2 pole AC electric motor had severe misalignment or any other issue that created forces at 2 times running speed, this frequency would be close to the 2X line frequency created by "Static Eccentricity".

In the days of tunable filter analyzers, the customary practice was to remove power from the motor to separate mechanical and electrical sources. If the vibration dropped out "instantaneously" the analyst correctly assumed that the fault was electrically related, however, if the vibration decreased along with the speed of the shaft it was them again correctly assumed the fault to be mechanically related.

With the development of the portable "high resolution" data collectors high resolution spectra were on the motor to easily separate the spectral peaks at 2 X running speed from twice line frequency. These peaks at twice line frequency have become known as "Stator Electrical" problems.

STATOR ISSUES (2 X line frequency):

Occurs if the geometric centerline of a "perfectly round rotor is not aligned with the geometric centerline of a "perfectly round stator." This creates a condition where the distance between outer periphery of the rotor and the inner bore of the stator core will not be equally spaced. Therefore the "air gap" between rotor and the stator will be unequal and create one point that has the narrowest gap and directly opposite will be the point with the widest gap. For example, if we index the drive end of the motor with clock positions. And the air gap at the 3 o'clock position is the narrowest, the air gap at the 9 o'clock position will have widest gap.

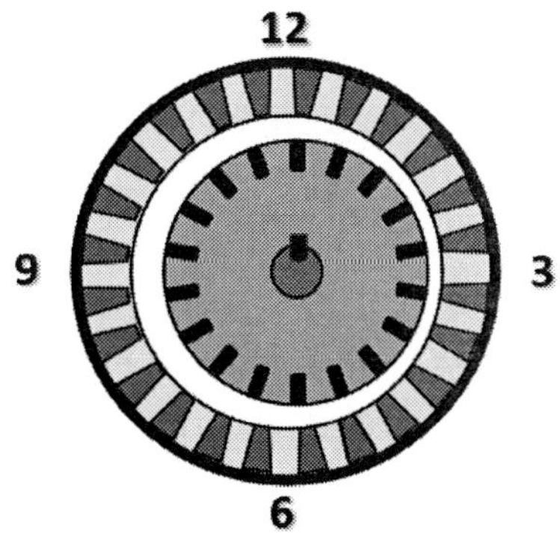

Coulombs inverse square law states that the attractive or repelling forces of electrostatic charges increase as the square of the distance between them decreases. Therefore, since the air gaps between the rotor and stator aren't equal distance, the electrostatic forces will be greater at the 3 o'clock position where two electrostatic field are closest together than the electrostatic fields at the 9 o'clock location, where they are furthest apart.

Applying three phase AC power to the motor's stator windings the magnetic field rotates around the stator at synchronous speed. For a two-pole motor, the rotating magnetic field will have a North pole and 1 south pole which are magnetically 180 degrees apart. The poles rotate around the stator at synchronous speed.

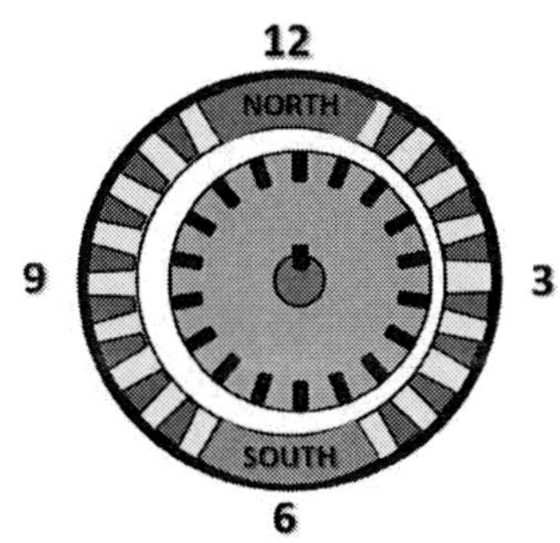

Balanced Forces

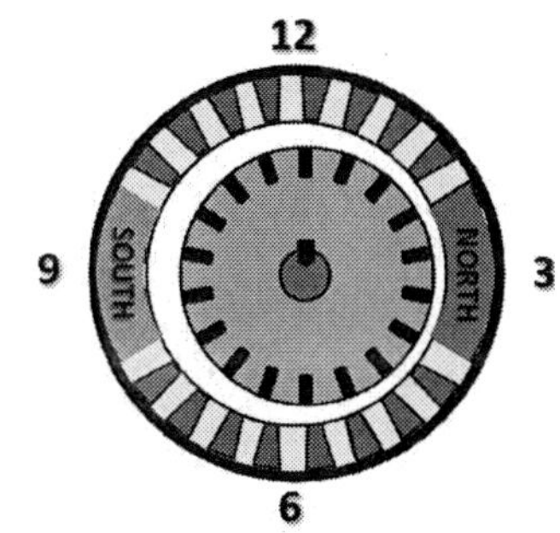

Unbalanced Forces

When the north pole is at 12 o'clock the south pole will be at 6 o'clock the magnetic attraction between the stator and the rotor will be equal and opposite and cancel each other out. However, when the North pole passes the narrow gap at 3 o'clock the south pole will be passing the widest gap at 9 o'clock.

Since the gap between the rotor and stator is narrowest at the 3 o'clock position and widest at the 9 o'clock position the attracting radial forces between the between the rotor and stator will begin to increase as the north pole rotates clockwise toward the 3 o'clock position and weaken on the opposite side of the stator as the south pole rotates toward the 9 o'clock position. The increasing magnetic field at the 3 o'clock accompanied by the weakening magnetic field at the 9 o'clock will pull the rotor from its static position towards the stator at the narrowest point (3 o'clock).

When the north pole passes the 3 o'clock position and the south pole passes the 9 o'clock position the unbalanced magnetic forces will be strongest pulling the rotor towards the 3 o'clock position.

As the poles continue to rotate the south pole will rotate towards the 12 o'clock position and the north pole will rotate towards the 6 o'clock positions. Since at these locations the air gaps are equally spaced the opposing magnetic forces will return to being equal and opposite the rotor will then be pulled back to its static position.

When the South magnetic field passes over the 3 o'clock position again the attracting force will be greatest at 3 o'clock and the north pole will be passing over at the 9 o'clock position and weakest again pulling the rotor toward the 3 o'clock position.

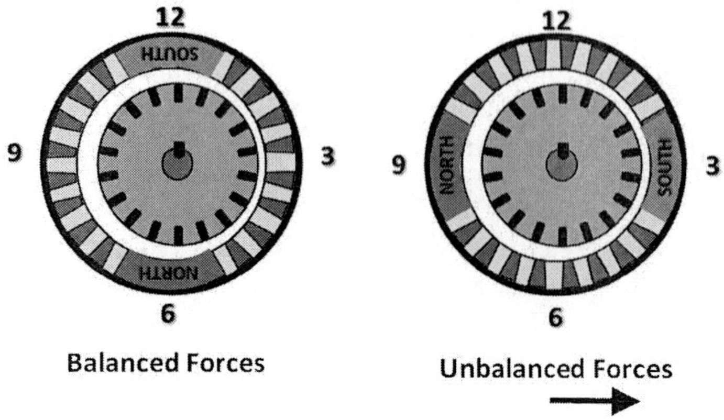

When the south magnetic field pole moves away from the narrow clearance at 3 o'clock and the north magnetic field will move away from the wider clearance at the 9 o'clock position and the opposing magnetic forces again will be equal and opposite position the rotor back to the stator's geometric centerline.

After one complete rotation of the magnetic field the north pole be at the 12 o'clock position and the south pole will again be at the 6 o'clock position.

Following increasing and decreasing radial forces in the horizontal direction, since there are 2 magnetic poles in a 2-pole motor that rotate around the stator at synchronous speed, therefore the rotor will be pulled away from its static position towards the 3 o'clock position every time a magnetic pole passes the narrow clearance. This means that on a 2 pole motor these forces will occur twice per each revolution of the magnetic field. On a 4-pole motor this will occur 4 times per rotation of the magnetic field, on a 6-pole motor it will occur 6 times per rotation of the magnetic field as the number of poles increases the number of tugs on will rotor per each rotation of the magnetic field will increase respectively.

The following table shows the frequency of unbalanced forces created by unequal air gaps for supply frequencies of both 50 and 60 Hz. It's easy to see that the results of multiplying the number of times synchronous speeds will always equal twice line frequency.

Number of poles	Synchronous Speed in RPM 60 Hz	# Poles X Synchronous Speed in CPM	Synchronous Speed 50 Hz	# Poles X Synchronous Speed in CPM
2	3600	7200	3000	6000
4	1800	7200	1500	6000
6	1200	7200	1000	6000
8	900	7200	750	6000
10	720	7200	600	6000

For a 60 Hz input frequency the synchronous speed is 3600 RPM so this will create the radial vibration force at 7200 CPM, which is twice line frequency. For a 50 Hz input frequency the frequency of the vibrating force will be 6000 CPM, which is also twice line frequency.

Most vibration source charts; frequency tables and diagnostic charts classify these faults as electrical problems. So, the "electrical" are passed on to the electrical team. But these frequencies occur as a result of mechanical issues and are undetectable using traditional electrical testing methods. In most cases this leads the technician make a call of an electrical fault and will contact the Electrical department.

More importantly this means that its possible and usually probable that electrical problem are developing that won't create any vibration until in the latest stages of the fault! This is when the smoke comes out.

STATIC ECCENTRICITY:

Some unequal air gap issues are caused by incorrect machining of the rabbet fits on the motor frame or register fits on the end bells. It's also possible the bearing housing bores aren't perfectly centered in the end bell. These machining errors prevent the rotor from being centered in the stator core. Or it's possible the stator core may not be concentrically placed in the motor enclosure.

It's important to remember that the shape of an inductors core shapes the magnetic field. Therefore, if the core of the inductor is perfectly round the magnetic field will be perfectly round. Since the stator is a three-phase inductor, the shape of the stator core shapes the stator magnetic field. If the stator core is distorted (eccentric) the magnetic field will then be distorted.

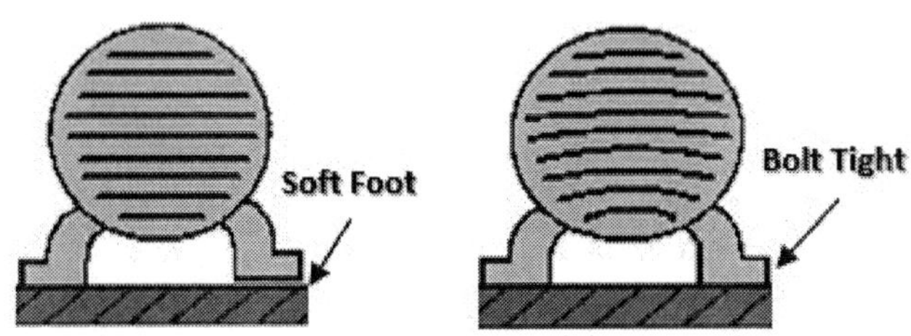

So, if all the machined fits are correct it's possible the stator core is eccentric or becomes eccentric with a load applied to the motor. Mechanically the most common cause of an eccentric or distorted stator core is soft foot on the motor. A soft foot condition occurs if all of the feet of the motor are not in same plane, or the base of the motor isn't level. If a soft foot condition is uncorrected, tightening the hold down bolts will distort the frame of the motor and since the stator core is mounted in the motor frame this will cause the stator core to distort. Since the stator core has been physically distorted, the stator magnetic field will also become distorted and create unequal air gap clearances.

Other cases of vibration occurring at twice line frequency also occur because unequal air gaps caused by distorted magnetic field. Among them are winding shorts, or shorted stator laminations. In both situations the electrical shorts create excess heat that creates localized heating or "hot spots' in the area of the stator where the fault exists. This hot spot causes the stator core to distort and again distort the stator magnetic field. Again, creating the unequal air gap condition.

MECHANICAL LOOSENESS:

If there are loose windings in the stator slots or the stator is insufficiently tight in the motor frame the rotating magnetic field poles will cause the windings or stator iron to move as the magnetic poles pass the loose component or windings it will also cause vibration at twice line frequency.

As previously stated, when the power is removed from the motor the magnetic field will collapse immediately and these unbalanced electrically forces will also disappear immediately.

ROTOR ISSUES (1 X running speed with PPF side bands):
Faults that occur in the rotor usually occur as a mechanical issue associated with the rotor. The most common and usually severest rotor issues are cracked or broken rotor bars. This condition creates unbalanced current flow through the rotor bars. A less common problem but with similar symptoms are casting voids or air pockets in the rotor bars or end rings. Unbalanced current flow through the rotor creates a non-symmetrical magnetic field around the rotor.

One of Maxwell's equations known as Ampere's Law states that the strength of a magnetic field in tesla's (T) equals ($B = \mu_0 I/2\pi r$) where: I = current. Since the rotor bars are electrically in parallel the voltage across each rotor bar will be the same. Ohms law states the I = E/R. Therefore, if one or more of the bars presents a higher resistance resulting from cracks or voids in that bar, the current through bar will be reduced, thereby reducing the magnetic field surrounding that bar. Open rotor bars will create dead spots on the rotor. In the figure below, the rotor bar in line with the shaft key has a high resistance therefore, following Ohms law the current flowing through that rotor bar will be lower creating a smaller magnetic field around that rotor bar. This creates a non-symmetrical magnetic field around the rotor core.

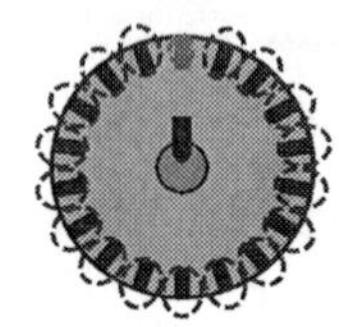

When a rotating magnetic pole, passes over the weaken portion of the rotor magnetic field it creates unequal magnetic attracting forces between the rotor and the stator and will cause the rotor to be pulled towards the strongest portion of the rotor magnetic field, which is directly opposite the shaft key. When magnetic fields are located at any other portion of the rotor these unbalanced forces will disappear.

Since both the rotor and the stator are rotating each magnetic pole will pass over the weaken portion of the rotor magnetic field once for each rotation of slip. This creates an electrical unbalance that adds to the mechanical radial forces. This causes the amplitude of the 1 times shaft speed forces to modulate at a frequency equal to # of poles times slip speed. This is referred to as Pole Passing Frequency (PPF). The modulation of the 1 x running speed This creates PPF sidebands around the 1 time running speed spectral peak.

PPF = # poles X slip frequency

Eccentric Rotor (Dynamic Eccentricity):

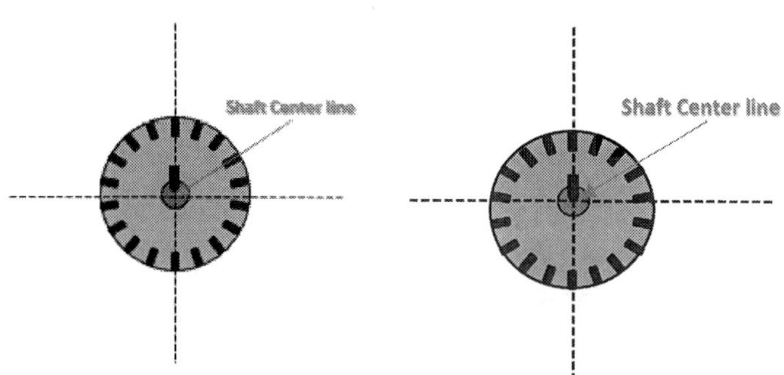

An eccentric rotor occurs when the geometric center of the shaft is "not" in line with the geometric centerline of the rotor core.

This can occur if the shaft isn't located in the center of the rotor core, or if the rotor core isn't perfectly round.

This results in narrower clearances between the rotor and stator cores. When the center line of the shaft is aligned in the centerline of the core and the rotor is perfectly round all clearance will be equal between the rotor and the stator core. However, if the rotor isn't perfectly round then a narrow clearance is created between the "high spot" on the rotor and the stator core. However, since the eccentricity is on the rotor the narrow clearance will rotate with the shaft.

So, when any of the magnetic poles pass by the narrow clearance the rotor will be pulled toward the narrowest gap. In the picture below with the shaft key aligned at the 12 o'clock position the narrow clearance is at 6 o'clock so when either the north or south magnetic field passes by the 6 o'clock position the rotor will be pulled towards the 6 o'clock position, however, when the shaft key is aligned toward the 3 o'clock position the narrowest clearance is now at the 9 o'clock position and as either of the poles passes the 9 o'clock position the rotor will be pulled toward the 9 o'clock position. Since the narrow air gap changes with rotor position this is referred to as dynamic eccentricity. These faults will also cause the vibration amplitude at 1 times shaft speed to modulates at PPF and create sidebands around the 1 times RS spectral peak. As with stator vibration issues these forces will disappear when the power is removed from the motor again prompting the electrical vibration call.

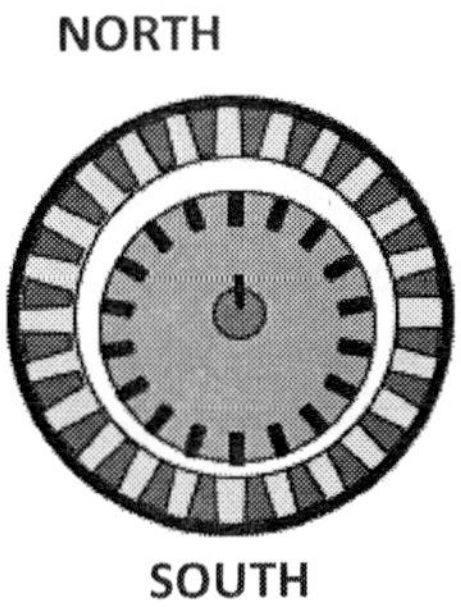

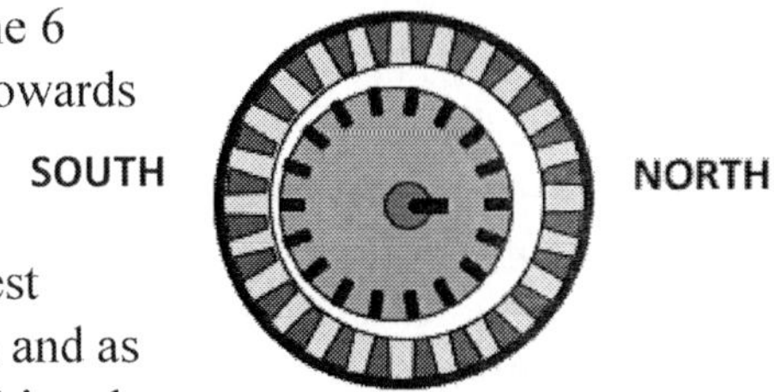

What should the electrical technician look for?
Armed with a "megger" and a multimeter, the electrical technician normally responds with "all's OK" electrically. This doesn't mean the motor is not the problem. It just means that the motor isn't grounded, and the windings aren't open, so why are the vibration tech's calling an electrical problem. To provide the answer as what fault or faults are causing either the rotor or stator problems, additional energized or deenergized motor testing techniques can be implemented.

Perhaps a better question for the electrician is what faults aren't detected by vibration and how do we identify them?
Most electrical failures occur when the electrical insulation fails. Electrical insulation is a material in which electrons don't flow freely. The purpose of electrical insulation is to direct current where it's wanted and prevent it from going where it's not wanted.

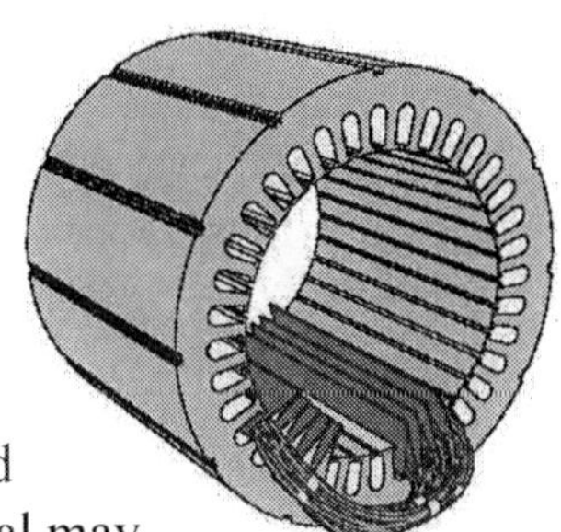

The atoms of an insulating material have very tightly bonded electrons in the outer most or the valance shell. Whereas a conductor has very loosely bonded electrons in the valance has and allows for free flow of electrons. As insulation ages, or becomes contaminated, the atomic structure can change and can become more conductive and less of an insulator. Or the insulating material may wear away, crack or develop other breaks in the material. Generally, motors don't vibrate until complete insulation failure occurs and the motor will usually, undergo an electrical trip. So, in reality to determine the electrical condition of the motors it's necessary determine the condition of the motor's insulation. To begin this, it's important to remember that motor stators have 2 separate and independent insulation systems.

1) The ground wall insulation (GWI) isolates the energized portion of the stator windings, line leads, and connections from the casing or the motor frame.

2) The winding insulation system is the polymer coating surrounding winding conductors that directs the current through the conductors making up the windings, instead of short cutting around the coiled conductor.

The twice line frequency stator vibration faults are the result of unequal air gaps and will not be detected using the "megohm tester" since this tester only provides a measurement of the weakest point in the ground wall insulation. Faults that create the twice line frequency vibration regarding the air gaps, soft foot, looseness, shorted stator iron or windings are undetectable by measuring the value of the insulation resistance to ground. Although this is the most common and a very important measurement since its main purpose is to ensure the resistance of the GWI is sufficient to keep the voltage applied to the motor windings from leaking to or flowing to an exposed part of the motor frame.

Additionally, since there are no electrical connections between the stator windings and the rotor, the rotor faults that create vibration will not affect the IRG reading or the winding resistance.

So how do electrical technicians identify motor faults for predictive maintenance and troubleshooting?

The two most common methods are Motor Circuit Analysis™ (deenergized) and Electrical Signature Analysis.

MOTOR CIRCUIT ANALYSIS™ (MCA™):

Is a field proven method deenergized motor testing method that uses a series of patented low voltage AC and DC measurements to completely evaluate both the ground wall and winding insulation systems. Most MCA™ tests can be performed from a motor controller (MCC) or local disconnect. If all the measured variables are balanced from the MCC that confirms all components in electrical circuit, cabling, and motor are balanced. However, if an unbalance exists from the MCC then a complete set of data can be performed directly at the motor to systematically isolate the faults' location.

GROUND WALL INSULATION:

Capacitance occurs a dielectric material is placed between two plates or conductive materials. Since both the ground wall insulation and the winding insulation are dielectric materials when placed between conductors, they create capacitance.

The ground wall insulation system creates a natural capacitance between the motor frame, stator, and the winding coils.

A capacitor is an electrical device that stores an electrical charge and opposes a change in voltage. When an AC voltage is applied to a capacitive circuit most of the current is stored on the plates and isn't consumed. This stored current is the capacitive current I_c while a small amount of current that flows across the dielectric is the resistive current I_r that creates heat. The current that flows across the dielectric is the leakage current and determines the electrical strength of the dielectric material.

The most common tests for the GWI are the Insulation Resistance to Ground test. Remember the main purpose of this test is to confirm the motor is safe to energize. This test applies a DC voltage to the one or more of the winding conductors and measures how much current leaks through the insulation to ground. If the IRG values are too low, it's possible that the frame of the motor or other exposed parts of the motor frame can be energized up to the full supply voltage which creates a hazard to both plant and personnel.

One of the basic laws of electricity is that current takes the path of least resistance, so the IRG measurement doesn't provide an indication of the overall condition of the ground wall insulation but only determines the resistance of the insulation at its weakest point. There are additional AC measurements of the GWI provide a better indication of the overall condition of the GWI.

DISSIPATION FACTOR (DF):

The dissipation Factor DF measures the ratio of the I_r to the I_c and measures the loss rate of the insulating material. New clean insulation typically has a DF of 3 to 5%, if the insulation becomes contaminated or undergoes thermal degradation the DF will increase above 5%.

CAPACITANCE TO GROUND (CTG):

New, clean motors will have a specific value of capacitance to ground (CTG) that should be remain the same throughout the life of the motor. Thermal degradation will cause the dielectric to become more resistive and less capacitive which will cause the CTG to decrease. However, contamination or water ingression will cause the CTG to increase.

Combining the DF, CTG & IRG provides a more complete picture of the overall condition of the GWI.

WINDING INSULATION:

The conductors used to form the stator windings are coated with a polymer film that directs the current through the conductors used to produce the magnetic field when coiled. This polymer film is typically 0.001 to 0.002" thick which provides an electrical strength 4.5 - 5.5 kV/mil when new. As the insulation ages or wears away or is subjected to contamination or heat this electrical withstand value will decrease. Wear reduces the thickness of the insulating film while heat and contamination cause the chemical makeup of the polymer film to change. In any case this process occurs over time.

These slight changes are undetectable using an ohm meter which applies < 10 V DC in an attempt to locate any developing breakdowns or weaknesses in the winding insulation. An additional hurdle to using winding electrical insulation faults is that electrical insulation has a negative temperature coefficient, meaning that as the temperature of the insulation increases the electrical strength decreases. This will create issues especially when a is motor experiencing intermittent tripping. As current flows across the developing the temperature of the insulation will increase and the resistance will decreases causing the current flow to increase generating more heat. As this cycle continues the resistance of the insulation between the conductors decreases enough to allow sufficient current to flow between conductors and cause the motor to trip on overcurrent. When the motor trips the temperature of the insulation will decrease and the resistance will increase sufficiently that it will be undetectable using an ohm meter.

When the electrical insulation degrades it undergoes changes in the molecular structure of the insulation surrounding the conductors. These changes in the winding insulations are undetectable using the traditional electrical measurements of Voltage, current, and resistance. However, these changes affect the ability of the coil to store an electrical charge or a magnetic field and the relationship between the resistance (R), inductance (L) and capacitance (C), of coil.

MCA™ TESTING:

To locate developing winding degradation that result in winding shorts MCA™ uses a series of low voltage AC & DC tests. Three phase motor windings should be symmetrical and balanced. If faults begin to develop in the windings insulation system, the faulty phase coils will respond differently and break the symmetry required for proper, smooth operation of 3 phase motors. Modern MCA™ instruments such as the ALL-TEST PRO 7™ are handheld battery powered instruments that provide step by step testing instructions and provide on screen display of the motors condition. The test results are stored in the instrument and can be quickly uploaded to the MCA™ software for analysis, trending, and reporting.

WINDING RESISTANCE:

The DC test is simply a measurement of the winding resistance. As previously discussed, winding resistance doesn't provide any indication of the condition of the insulation surrounding the winding conductors, so if a resistance unbalance does exist it indicates connection issues in the circuit or circuits under test. This could be loose connectors between motor and line leads in the connection box,

or loose connectors in the motor controller, cold solder joints in the motor windings. Loose or unbalanced resistance will create different voltage drops across the connectors resulting in a voltage unbalance supplied to the motor. IEEE recommends that the maximum voltage unbalance an Induction motor can be safely operated is 5%. And any unbalance up to 5% the motor should be derated at approximately a squared relationship.

AC TESTS:

To evaluate the condition of the winding insulation low voltage AC purely sinusoidal at 5 different frequencies to thoroughly exercise the winding insulation system. If any portion of the winding's insulation has changed chemically it will cause the response to the AC signal to become unbalanced. MCA measures the Inductance, Impedance, Phase Angle and Current Frequency Response at the each of the frequencies.

IMPEDANCE (Z) & INDUCTANCE (L):

The impedance and inductance measurements are used to determine if any unbalances are the result of rotor position. Since an inductance motor can be described as a transformer with a rotating secondary. The primary windings of the transformer are the stator windings, and the rotor bars are the secondary windings. The turns ratio of each phase will be determined by the number of rotor bars positioned under each of the stator poles. When the rotor is in a stationary it is possible that there will be an unequal number of bars which would create an inductance unbalance as well as an impedance unbalance.

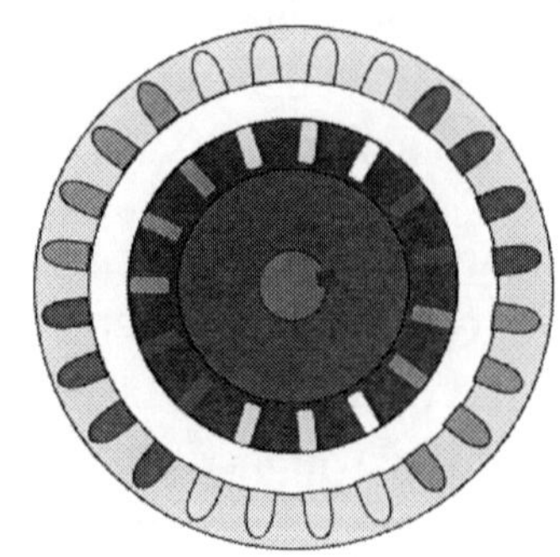

PHASE ANGLE (Fi):
The phase angle will change with the very small changes in C or L that occur in the early stages of a developing winding short. In three phase motors with good winding insulation systems the relationship of resistance, inductance and capacitance should be the same for all three phases. When the insulation begins to degrade this relationship will change in the affected coil or coils creating an unbalance in the phase angle (FI) of the phases. Developing windings faults are present if any FI measures vary by more than 1 degree from the average of the 3 phases.

CURRENT FREQUENCY RESPONSE (I/F):
The current frequency response tests the coils 2 different frequencies, the second frequency is double the initial frequency, for example 100 & 200 Hz; 200 & 400 Hz, by increasing the frequency of the applied frequency the overall impedance increases, which causes the current through the coil to decrease. The I/F response is the ratio of the change in current by doubling the test frequency. On three phase motors all phases should have the same ability to store a magnetic field (inductance) or an electrical charge (capacitance), if any coil or coils in a phase begins to degrade it will change that phase's ability to store an electrical charge or a magnetic field and create an unbalance in the phases.

Therefore, these two variables are the best indication of developing winding faults. Developing winding faults indicated if any Fi or I/F value deviate by more than 2 units from the average of the three phases.

ROTOR ISSUES:
Rotor Issues can be identified by measuring the inductance of all three phases as the rotor is repositioned at equal angular positions through one rotation of the shaft. The results of the inductance are used to create a 3-phase line graph that plots the measured inductance of each phase at each rotor position. This is referred to as a Rotor Inductance Test. Fault free rotors will have symmetrical shapes through one complete rotation of the shaft. Distortions or non-symmetrical graphs indicate rotor eccentricity, unequal air gaps, casting voids, cracked or broken rotor bars.

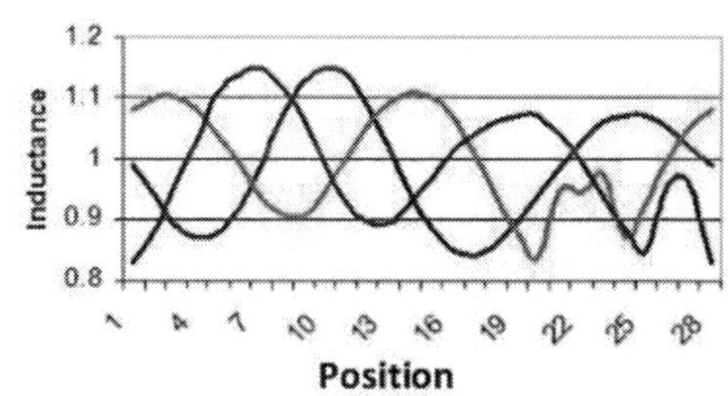

ELECTRICAL SIGNATURE ANALYSIS (ESA):
Is rapidly gaining increased usage as more practitioners become familiar with the capabilities of ESA. Vibration technicians for many years have used current clamps as the transducer to determine the source of the vibration on equipment driven by electric motors. It was learned that the same forces that cause the machine structure to vibrate would cause the motors current to modulate at these same frequencies. Machines with accessibility issues such as the subsurface pump section of vertical pumps or machines in hazardous areas were tested by placing a current clamp around one of the motors feeder leads and a FFT was performed on the motors current. This technique referred to as Motor Current Signature Analysis (MCSA), was very useful in locating and determining the severity of cracked and broken rotor bars on squirrel cage induction motors. Since

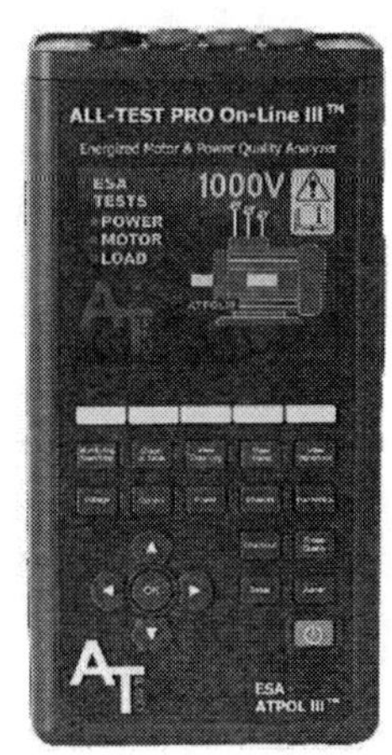

the early days of this practice the technology has advanced to not only perform an FFT on the motor current but the supply voltage as well.

ESA systems range from dedicated computer-based systems which require preconditioning hardware to the handheld battery powered energized ATPOL III™ to quickly and easily perform and store the motor tests in the instruments internal memory or on SD cards for uploaded to a host computers for permanent storage analysis and report generation.

This has greatly enhanced this technology and has been referred to as Electrical Signature Analysis (ESA). ESA now typically encompasses not only capturing and digitizing the voltage and current time waveforms but performs simultaneous capture on all phases of voltage and current to create a power table. This feature provides additional benefits of evaluating the condition of the power being supplied to the motor, including but not limited to voltage and current unbalance, harmonic distortion, motor load, power factor

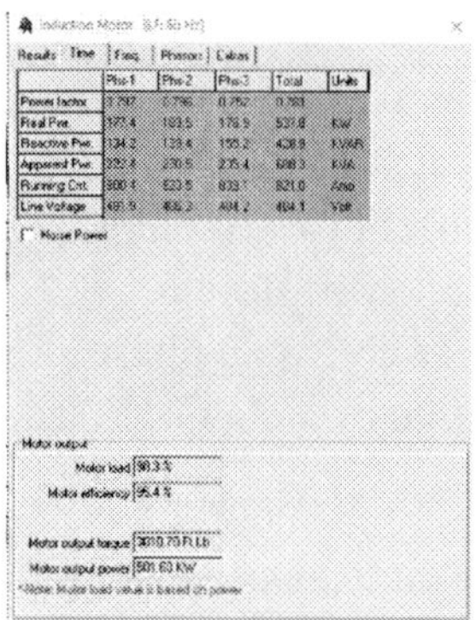

ESA has increased diagnostic capabilities over MCSA. The input voltage acts as a diagnostic filter by eliminating any forces acting on the motor as a result of variances in the incoming power to the motor. If a spectral peak appears in both voltage and current or the voltage alone, it is coming from the power system and doesn't provide any indication of the motor system condition. However, if a peak is only in the current, then the fault comes from either the motor, driven machine, or the process itself.

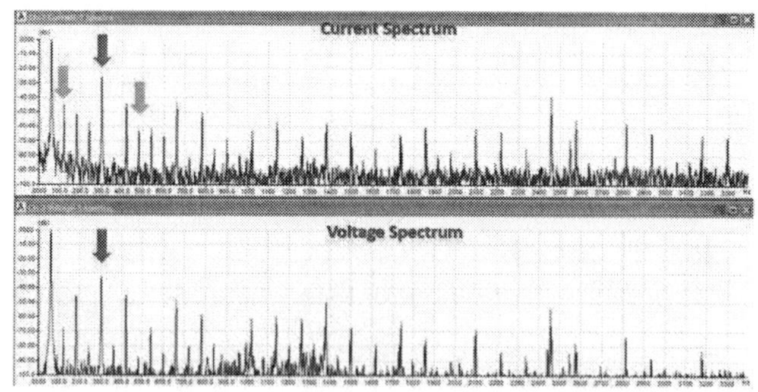

Additional research identified specific spectral patterns that accurately identified faults such as "static eccentricity", "dynamic eccentricity", "stator mechanical" as well as other electrical and mechanical in in their early stages. Additional spectral relationships were developed that can separate casting voids from other rotor issues such cracked or broken rotor bars. ESA uses 5 different measurements to accurately evaluate the rotor bars faults completely missed by vibration.

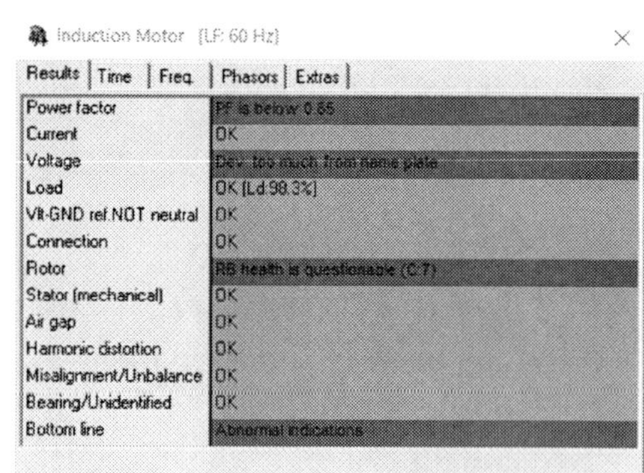

CONCLUSION:

To eliminate an "All's OK", response from the electrical team when informed of an "Electrical Vibration's" diagnosis it's a good idea for the vibration technician provide the electrical group with some additional guidance as to what possible faults they should test for and what available technologies are available to assist them in their endeavor.

VIBRATION ANALYSIS OF BELT CONVEYORS

Author: Tony Wofford

VIATC VIBRATION INSTITUTE ANNUAL TRAINING CONFERENCE & EXPO 2022 | MFPT ANNUAL CONFERENCE

2022 CONFERENCE PROCEEDINGS

VIATC TRAINING CONFERENCE & EXPO
SAVANNAH, GEORGIA · 2022

VIBRATION ANALYSIS OF BELT CONVEYORS

Tony Wofford
Clearwater Paper Corporation
Idaho Pulp & Paperboard Division
PO Box 1126 803 Mill Road
Lewiston, Idaho 83501
tony.wofford@clearwaterpaper.com

Tony Wofford is a Reliability Technician with Clearwater Paper Corporation, a pulp and paperboard manufacturer, in Lewiston, Idaho, and is certified as a Vibration Analyst Category III. He has been involved in vibration analysis for over 23 years and was in the electrical apparatus service and repair industry prior to that.

ABSTRACT:
Belt conveyors are simple machines that have been commonly used for material handling in industry since the early 1900's. However, little seems to be formally documented or communicated about performing thorough vibration analysis on a belt conveyor beyond the electric motor and gearbox traditionally used to drive it. Unfortunately, failure of a belt conveyor can be due to many possible causes and result in significant production loss and costs. This paper explores common terminology used for various components and the information necessary for accurate vibration analysis. It also presents case studies of various belt conveyor issues.

KEYWORDS: Cleat passing frequency; feet per minute (FPM); high frequency envelope (HFE) data; idler; linear speed; pulley; revolutions per minute (RPM); rotational speed.

INTRODUCTION:
Many vibration analysts are already familiar with belt conveyors and have worked with or around them. However, at least in our facility, typically only the main drive components (motor, gearbox, and the drive pulley) have received much attention–until a component not regularly checked—if ever—fails—and causes significant production loss (a.k.a. $$!). To complicate matters, a belt conveyor system can have many "hidden" vibration sources that may require some investigative work to identify.

It is very helpful to start with understanding the names and functions of the basic components of a belt conveyor, as well as be familiar with typical conveyor configurations.

GENERAL COMPONENTS AND CONFIGURATIONS:

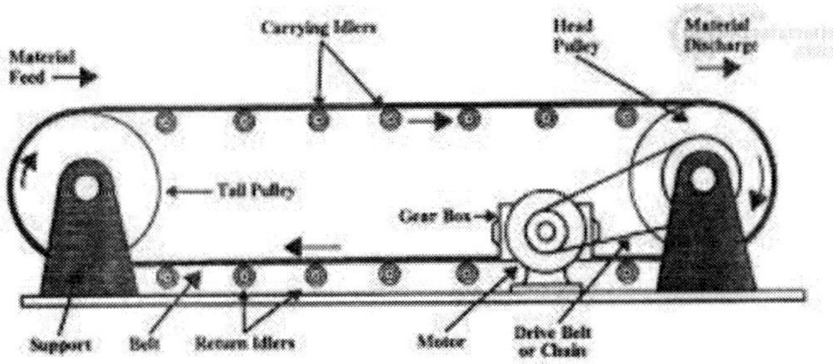

The simplest version of a belt conveyor consists of ***pulleys*** and ***idlers***, which are named according to their position and function, and a flat flexible ***belt*** made of one or more layers of material, with rubber being the most common component (see Figure 1).

Figure 1 – Simple Belt Conveyor

Pulley usually refers to the larger rolls, typically with external bearings, that usually involve some degree of "wrap" of the belt, while ***idler*** applies to the smaller bracket mounted and internal bearing rolls (or "rollers") that carry and guide the belt. Additional names for these pulleys and idlers are assigned according to their specific position and function. In this example, the ***head*** or ***drive*** pulley is typically belt or chain driven by a motor-gearbox combination at the discharge end of the conveyor, while the ***tail*** or ***return*** pulley is at the opposite end from that. ***Carrying*** idlers handle the weight of both the belt and the product, while ***return*** idlers just carry the belt alone after the material is discharged.

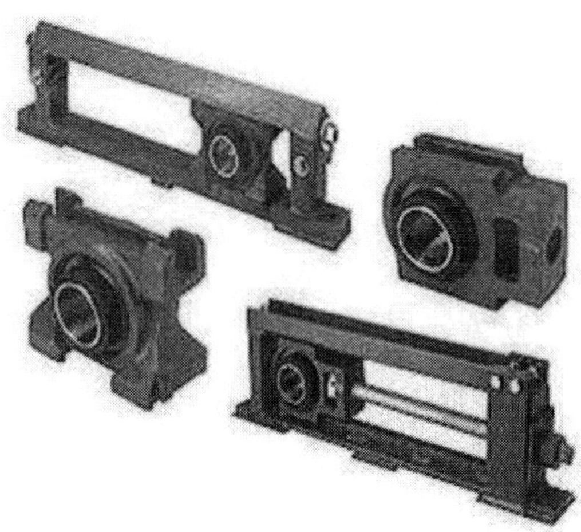

All belt conveyors require the belt to be tensioned to ensure transmission of power, and in this simplest configuration this is usually accomplished at the tail pulley by use of a sliding bearing housing / bracket arrangement called a ***take-up frame*** (see Figure 2).

However, the longer a belt gets, the more take-up is required. This requires additional ***weight*** or ***take-up*** pulleys that are either attached to a relatively large mass or hydraulically pulled. ***Bend*** pulleys are used in conjunction with weight pulleys to change the direction of the belt, and ***snub*** pulleys are used to increase the arc of contact between another pulley and the belt (see Figure 3).

Figure 2 – Take-up Frames & Bearings

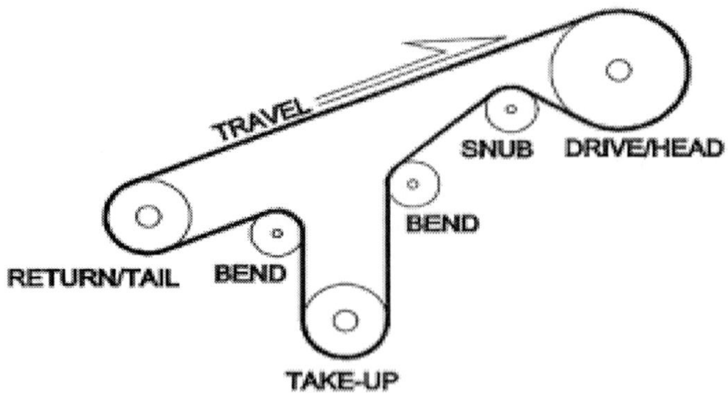

Figure 3 – Bent, Weight & Snub Pulleys

In some belt conveyors, the ***drive*** and ***head*** pulleys may be two different pulleys, with the head pulley always on the discharge end of the belt, but with the location of the drive pulley being fairly flexible (see Figure 4).

Additionally, a belt conveyor may have multiple drive pulleys (not pictured), although this requires some sort of specially integrated control system to avoid belt tension variation between drives.

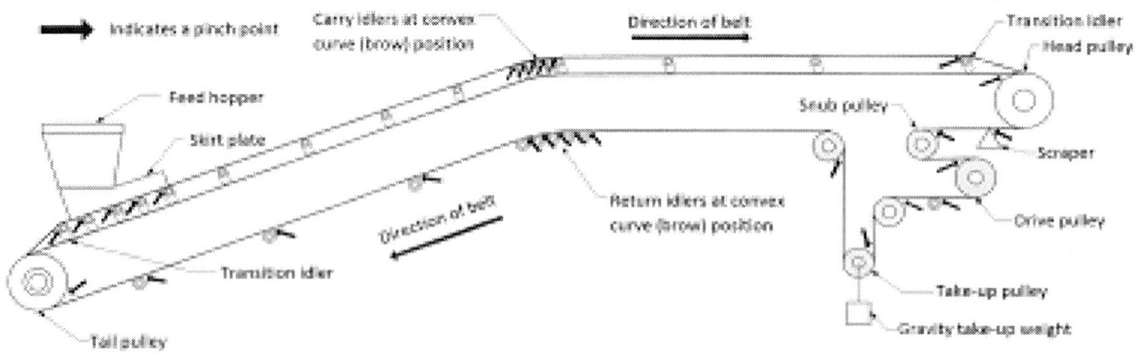

Figure 4 – Separate Head and Drive Pulleys

It is important to understand that the diameter and rotational speed (usually expressed as "Revolutions per Minute" or "RPM") of the drive pulley dictate the linear speed (usually expressed as "Feet per Minute" or "FPM") of the belt over the remainder of the belt system, and that belt speed, in turn, governs the speeds of the remaining rotating components based on their diameters. Therefore, changes in size (diameter) of the drive pulley results in changes to speeds of everything else turned by the conveyor belt; while changes in other pulley diameters results in changes only to the speeds of those pulleys.

Figure 5 – Clutch

A pulley may also be fitted with an anti-reverse mechanism known variously as a ***clutch, clutch brake, back stop clutch,*** or just a ***backstop***. This prevents an inclined belt conveyor from rolling backward when loaded with product but not in operation. These devices may contain a combination of rolling element bearings and cams, and due to their proximity to the pulley bearing on the same shaft, wear or failure of a clutch component can be misconstrued as a pulley bearing problem if not taken into account. It is therefore helpful to include route vibration data collection from these as well as the bearings.

Depending on the nature of the product being moved, ***scrapers*** or ***cleaners*** may be installed that ride against the belt (typically at the head pulley) to prevent excess carryover of product that may adhere to the belt surface. Although not typically considered a "moving" component, scrapers can sometimes significantly contribute to vibration on a belt conveyor under certain conditions if they either rub too hard or "chatter" against the belt (see Figure 6).

Figure 6 – Head Pulley Scraper

Figure 7 – Electric Shaker / Vibrator

A belt conveyor may also have ***pneumatic*** or ***electric shakers*** or ***vibrators*** installed on the outside of a chute into or from the conveyor that may be operating either continuously or cyclically—and can be a significant vibration source (see Figure 7).

Figure 8 – Transfer Junction

Finally, a conveyor belt ***system*** may be composed of many individual belt conveyors (that may be structurally attached) with one or more conveyors discharging product onto another conveyor via an ***infeed chute*** or ***feed hopper***. A structure where multiple belts are "fed" and/or discharge may be called a ***transfer junction*** (see Figure 8). This physical connection of multiple belt conveyors can matter a lot when attempting to find a source of vibration that may or may not be on the same conveyor.

A special instance of integrated belt conveyors is a ***sandwich*** belt conveyor, in which two belt conveyors with separate drives are run in contact with each other with the product "sandwiched" between the two belts. This enables conveying product at very high angles—up to completely vertical--hence another name for the entire system as a ***high angled conveyor (HAC)*** (see Figures 9 & 10).

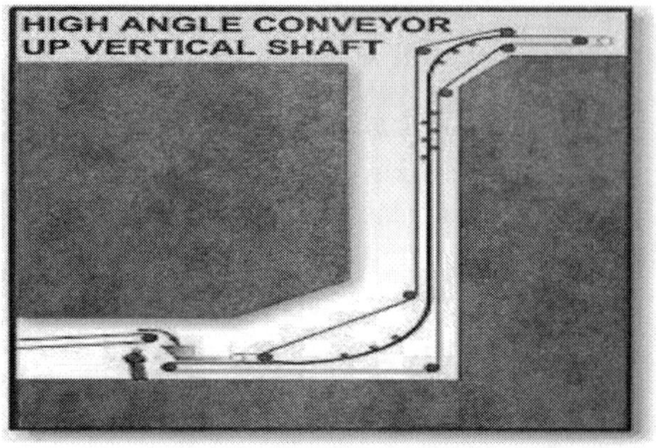

Figure 9 – HAC Layout

Figure 10 – HAC Belt detail

COMPONENT VARIATIONS AND IMPACT ON VIBRATION ANALYSIS:
All of the aforementioned components of a belt conveyor can have varying designs for differing applications—all of which can significantly impact vibration data and troubleshooting.

Pulleys:
Pulleys can have bearings that are mounted either externally, on the structural framework, or internally, with the shaft held on the framework with some type of clamp. Some pulleys are of a one-piece construction with their shaft and some are held on with hubs. Figure 11 shows examples of ***drum***, ***lagged drum***, and ***winged*** or ***clean-out*** pulleys, with and without shafts installed.

A ***drum pulley*** is just like it sounds—essentially a drum with either a full length shaft or two stub shafts. Shaft fits and hubs can come loose, producing classic indications of looseness and/or pulses in high frequency envelope (HFE) data.

Figure 11 -- Pulleys

A layer of rubber or ceramic lagging material attached to the surface of some pulleys for better friction and grip is known as ***lagging***. As this layer wears, the effective pulley diameter changes over time. If this is happening on a drive pulley, belt speed and therefore remaining pulleys' speeds will also change unless compensated for by the drive system.

This becomes very important in the sandwich, or HAC, belt configuration, where unequal wear on the two drive pulleys could result in one belt moving faster than the other, an undesirable situation. This is avoided by adjusting the speed of the drive components (motors, gearboxes, and drive pulleys) on each "component" conveyor to keep the belts themselves at the same speeds.

Finally, ***winged*** or ***clean-out pulleys*** have different kinds of slatted profiles that are meant to keep product from building up on the belt. The passing of the belt over the "wings" of this kind of pulley can result in relatively high vibration at the number of wings times pulley turning speed. Since a winged tail pulley may be in the vicinity (usually below) the discharge, and therefore the head pulley, of another conveyor, vibration may be seen – or even quite high -- on the discharging conveyor that does not relate to any of the components on that particular conveyor belt.

Idlers or rollers can also be of various designs and functions, with the ***trough*** or ***troughing***, ***flat return***, and ***vee return*** being most commonly used (see Figure 12). The typical idler is basically just a tube (or several tubes) with bearings mounted internally and supported by brackets (see Figures 13 & 14). Trough idlers come in various angles based on specific applications and locations on the belt

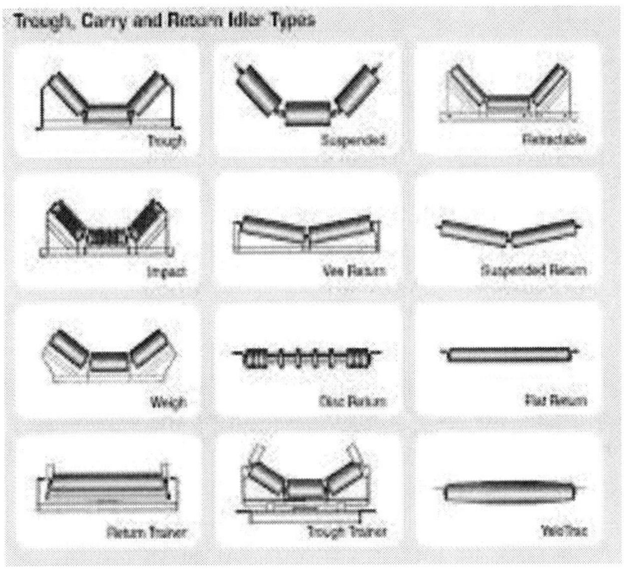

Figure 12 – Idlers

Other types of idlers include ***impact***, designed with multiple discs to minimize belt damage from the impact of product falling feeding/falling onto the belt, ***picking table*** (not shown), which spreads out material on a belt by using a longer center roll, and ***self-training*** or ***training***, which uses a movable bracket frame and two smaller perpendicular rollers to keep the belt centered.

Idlers can often fail without significant impact upon conveyor performance; and are therefore lower priority for data collection, if at all.

A special case of both a pulley and an idler where higher loads are expected is known as a ***live shaft idler*** (not pictured). It is essentially a pulley that is the smaller size of a typical idler but has a pillow block bearing at each end instead of the normal internal bearings and brackets.

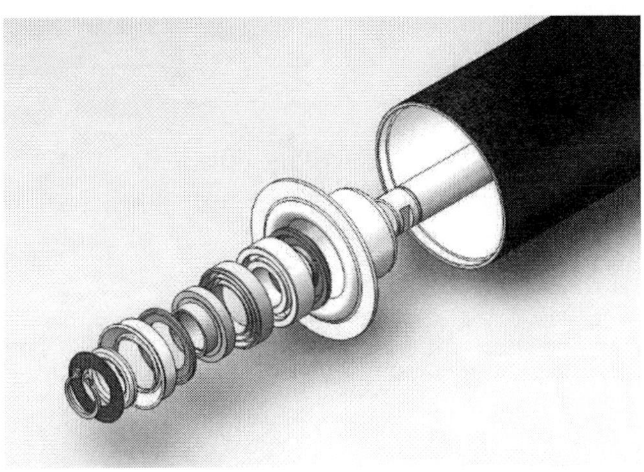

Figure 13 – Idler detail

Figure 14 – Disassembled idler

Belting:
Belting is made up of one or more layers, with the inner layer often a woven or metal fabric to provide strength and outer layers often of rubber or different plastics. The belt can either have a smooth surface or be ***cleated*** – having raised projections that are either molded or fastened onto the belt surface for the purpose of helping product stay on the belt, especially when inclined. The cleats can vary in design, length, width, and height--and can significantly contribute to vibration on conveyor components.

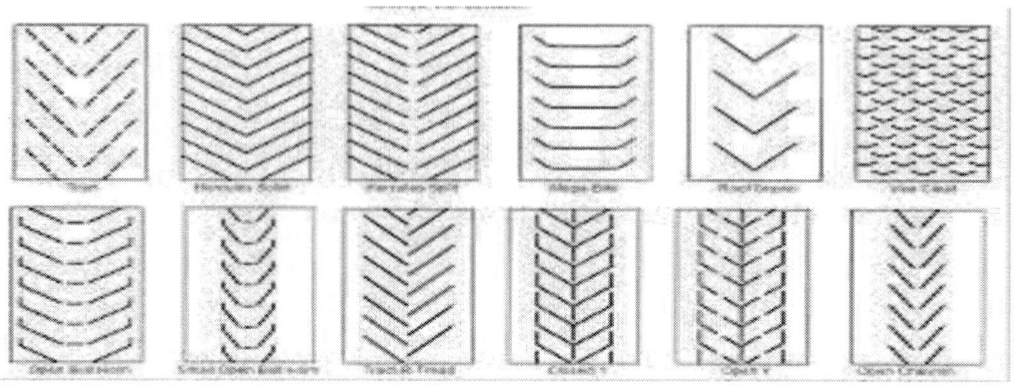

Figure 15 – Belt Cleating

CASE STUDIES AND ANALYSIS:
Case #1: 48" Main Chip Conveyor: High vibration from an unknown source

The 48" Main Chip Conveyor transports wood chips up an incline into our continuous digesting portion of the plant, where they are cooked to produce brownstock pulp, an intermediate product in our paperboard making process.

For years, I had only collected periodic vibration data off the motor, gearbox and drive pulley pillow block bearings. After a recent significant upgrade to this portion of the mill, however, this conveyor became much more critical to our process. As a result, I added the tail, weight, and both bend pulleys into our database and monthly routes based on diameter dimensions found in our CMMS databases. The rotational speed of each pulley is based on calculations from the (known) drive pulley rotational speed using this formula to first obtain the linear speed of the surface of the drive pulley and thus the speed of the belt.

$$FPM = (RPM * Diam'' * \pi) / 12$$

Then, that calculated linear belt speed and the known diameter of each of the other pulleys is used to calculate the rotational speed of each one using the same formula rearranged as:

$$RPM = (12 * FPM) / (Diam'' * \pi)$$

Mathematically, this is just the drive pulley rotational speed multiplied by the ratio of its diameter to each pulley diameter. However, use of the formula using FPM is recommended for reasons explained below.

$RPM_N = RPM_D * (Diam_D / Diam_N)$ *D = Drive pulley; N = Pulley being solved for*

Table I

PULLEY SPEED CALCULATIONS				
Name	**Diam"**	**Calc FPM**	**RPM**	**Hz**
Drive/Head	30	540.51	68.82	**1.15**
Bend	18	540.51	114.70	**1.91**
Tail	24	540.51	86.03	**1.43**
Weight	24	540.51	86.03	**1.43**
Return Idler	5	540.51	412.92	**6.88**

Table II

KNOWN FORCING FUNCTIONS	
Hz	**Description**
44.72	Motor/GB HS
89.44	2X Mtr
1.44	GB LS
160.77	HTD Dr belt tooth freq
1.15	Drive/head pulley

Following my first data collection on the additional pulleys in February 2019, I was puzzled by differences between some of the calculated frequencies and actual ones measured, and even more by the presence and relatively high amplitudes of a peak at nearly 37.9 Hz (and sometimes 2X this frequency) at nearly every pulley--that didn't jive with any of my initially "known" or expected forcing functions (see Table II).

Further investigation resulted in the following observations/findings:

- Underneath a lot of framework and guarding, there was a 14" diameter snub pulley near the drive pulley. This explained some unknown frequencies near the drive components.
- Identified the 5" diameter idlers as clear contributors to the overall vibration picture, so added them to the calculations.
- Although company records showed the drive pulley being a drum style with a 30" diameter, closer inspection revealed that this pulley had 3/8" lagging, making the effective diameter 30.75". Once this corrected diameter was plugged into the "Pulley Speed Calculations," this better explained the initial discrepancy between some of the calculated and actual pulley speeds (by correcting the belt speed)—but still didn't show a clear relationship between any specific pulley and the 37.9 Hz.
-

Table III – Pulley Speed Calcs--revised

Name	Diam"	Calc FPM	RPM	Hz	Act	Ratio to 37.9 Hz
Drive/Head (Lag)	30.75	554.03	68.82	1.15	1.15	33.04
Drive Snub	14	554.03	151.16	2.52	2.46	15.38
Bend	18	554.03	117.57	1.96	1.91	19.84
Tail	24	554.03	88.18	1.47	1.48	25.68
Weight	24	554.03	88.18	1.47	1.47	25.84
Return Idler	5	554.03	423.24	7.05	7.25	5.23

- Also, even with the correct diameter input for the drive pulley, there was still a significant difference between calculated and actual speeds of the return idlers.
- Confirmed that none of the pulleys were winged—which would result in a frequency at the number of wings/vanes X that pulley's speed.
- Finally, closer observation of the belt itself revealed that this belt was cleated, with two rows of chevron shaped cleats 0.25" high on 3.00" centers.

I began to wonder if these cleats could be producing some kind of "cleat passing frequency" and ran the following calculations:

Table IV – Cleat Passing Frequency Calcs--initial

3.000	CC" of cleats	
4.000	Cleats per foot	Cl/F: 12 / CC" of cleats
554.03	FPM at drive pulley	From calcs
2216.12	Cleats/min	Cl/M: FPM * Cl/F
36.94	Cleats/sec (Hz)	Cl/S: (Cl/M) / 60

This 36.94 Hz was close to the dominant frequency of 37.9 Hz, but seemed like it should be closer. This is when I paid more attention to the discrepancy between the calculated and "actual" speed of the return idlers. I realized that, unlike the head and tail pulleys, these were in contact with and being turned by the outside of the belt, which has the cleats.

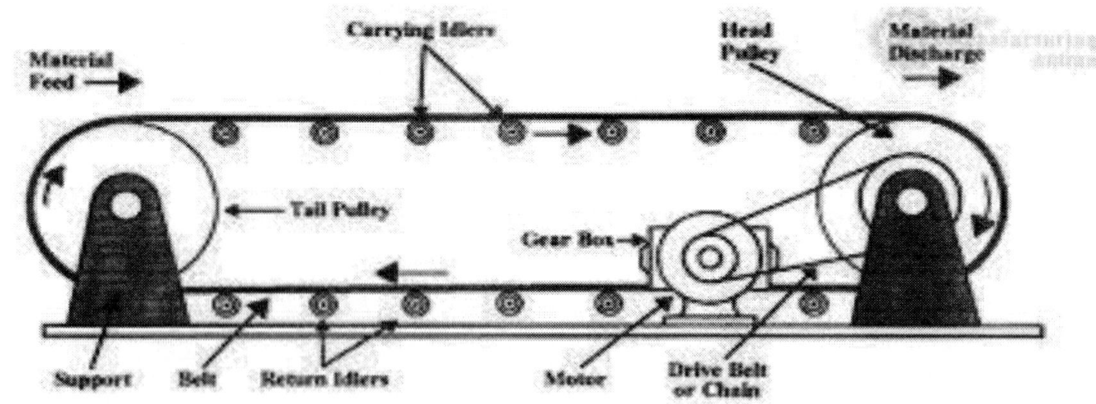

Figure 16 – Return idlers relative to belt

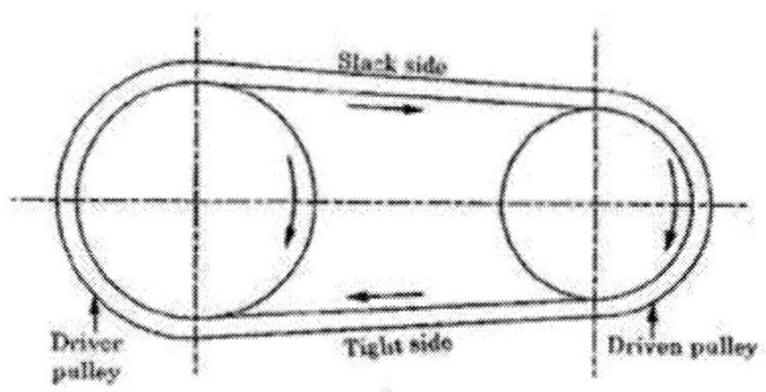

Figure 17 – Belt "diameter" illustration

This is important when you consider that the outside of the belt travels slightly further than the inside of the belt in the same amount of time, with the difference being related to the thickness of the belt. A way to visualize this is to picture the belt as it wraps around the drive pulley. The outside of the belt would be at a slightly greater diameter than the inside, which, as mentioned before, makes the linear speed of the outside of the belt slightly faster than that of the inside (see Figure 17).

To calculate this new/faster outside belt speed, I used the same formula with the diameter of the drive pulley plus the thickness of the belt on the top and bottom of the drive pulley (that is, doubling the belt thickness), with the result of a FPM of 567.54. Using this number to calculate the cleat passing frequency had a result of final confirmation of this being the source of my dominant "mystery" vibration frequency.

Table V

BELT OUTSIDE FPM & DIAMETER CALCULATIONS		
0.375	Belt Thickness	
30.750	Drive Pulley Diam	
31.500	"Effective" DP Diam	
68.820	Drive Pulley Speed (RPM)	
567.539	Belt Outside FPM	

Table VI

CLEAT PASSING FREQ CALCS--revised	
3.000	CC" of cleats
4.000	Cleats per foot
567.54	Belt Outside FPM
2270.16	Cleats/min (CPM)
37.84	Cleats/sec (Hz)

Additionally, using this new/outside belt FPM to calculate the speed of my 5" return idlers resulted in a calculated frequency of 7.23 Hz (433.8 RPM)—far closer to the 7.25 Hz (435 RPM) that showed in my vibration data.

Case #2: 48" Main Chip Conveyor: High vibration on weight pulley

On this same belt conveyor, the highest overall amplitude of vibration on any rotating component was on both weight pulley bearings at over 0.80 ips pk, with the dominant frequency being 3.8 Hz. Based on the calculations mentioned earlier, this equates to the twice the turning speed of the bend pulleys.

Figure 18 – Bend & Weight Pulleys

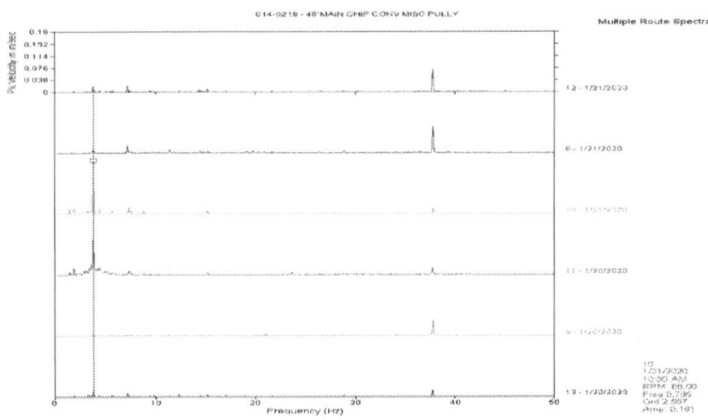

Figure 19 – Spectral Comparison

On closer inspection, it appeared that the south bend pulley (on the left upper portion of the picture in Figure 18) was moving / shaking significantly at twice its running speed—and this was being transferred by the belt to the weight pulley--which is only supported by a hinge and the belt itself. The result is the bouncing action of the weight pulley seen in the vibration data.

Case #3: 48" Main Chip Conveyor: Intermittent vibration source on tail pulley

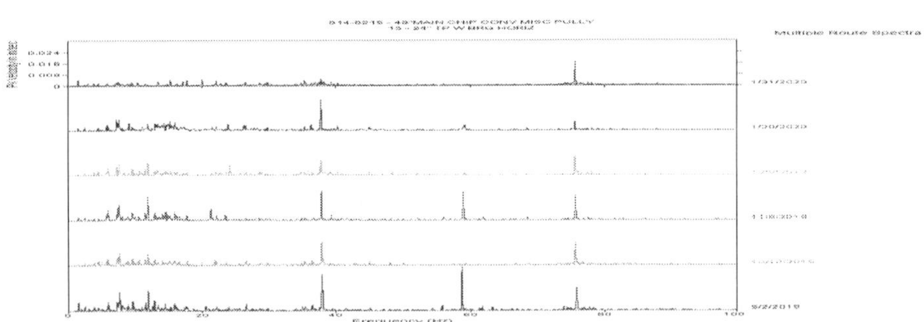

Figure 20 – Monthly spectral data with intermittent peak near 59 Hz

On the tail pulley of this same conveyor belt, I noticed that I was occasionally getting a relatively significant vibration peak at around 59 Hz—that didn't relate to any expected vibration sources in the system--and wasn't consistent every (monthly) survey (see Figure 20).

I knew there were shakers/vibrators in the vicinity—mounted on the bottom of a bin that dropped the chips onto this conveyor belt. But, when I checked, neither were running. After a little more searching, I found this smaller electrical one on the discharge chute of another belt conveyor that also dropped chips on the same belt (see Figure 21)—and noted it that cycled on and off several times over the course of a minute.

Taking a reading from this unit's base yielded 59.1 Hz—with its intermittent nature just due to the timing of data collection relative to its on and off cycle.

.

Figure 21 – Shaker / vibrator

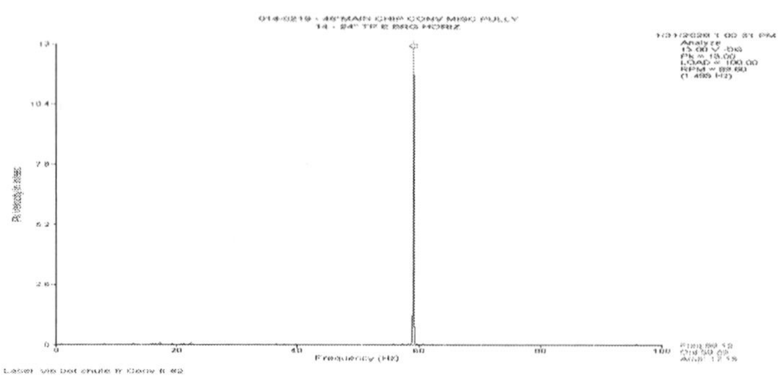

Figure 22 – 59.1 Hz vibration @ 13ips RMS

Case #4: Sawdust Conveyor #2 – Cracked bend pulley

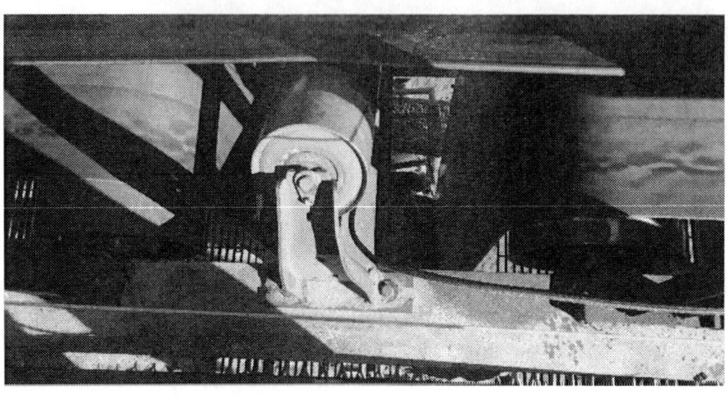

Figure 23 – Bend pulleys & troughing idler

While collecting monthly route data, I heard a "thumping" sound in the vicinity of the bend pulleys and a troughing idler of this conveyor (see Figure 23). Velocity data didn't show any real significant change or difference between the four bend pulley bearings, but HFE data (see Figure 24). indicated the West 12" bend pulley being broken or cracked due to pulses (not sharp impacts) at running speed of the pulley (127.4 RPM). Inspection by maintenance mechanics was requested.

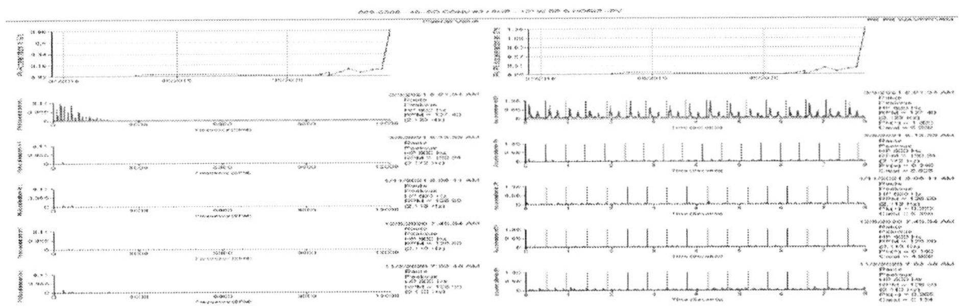

Figure 24 – HFE Vibration data W Bend Pulley S bearing

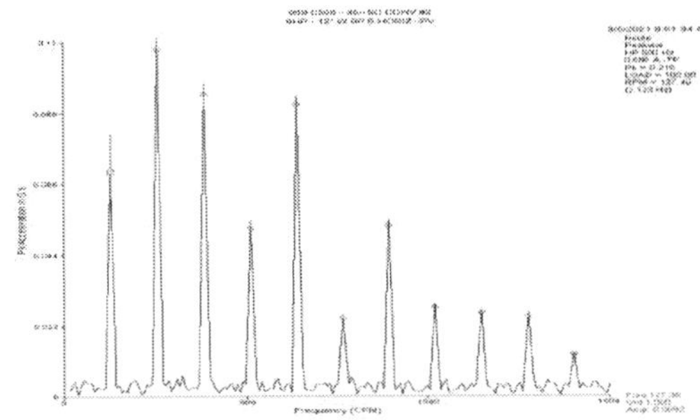

Figure 25 – HFE FFT (detail)

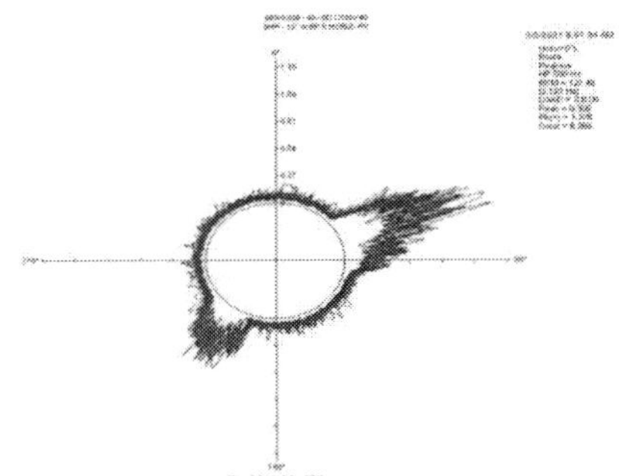

Figure 26 – HFE Circular Waveform

The mechanics looked at while running and felt all the thumping was coming from the 5" idler (turning at around 322 RPM), which is a fairly common occurrence. The bend pulley failed four days later (cracked).

Case #5: Sawdust Conveyor #2 – Carrier (Troughing) Idler

On this same conveyor several months later, again heard a "thumping" sound from the same location. Vibration data this time, however, showed the pulse to be at 5.375 Hz / 322.5 RPM, again mainly in HFE, not regular velocity data. This time, it ***was*** the idler (with failed bearings). I've since added this idler to my monthly data collection route to simplify things here.

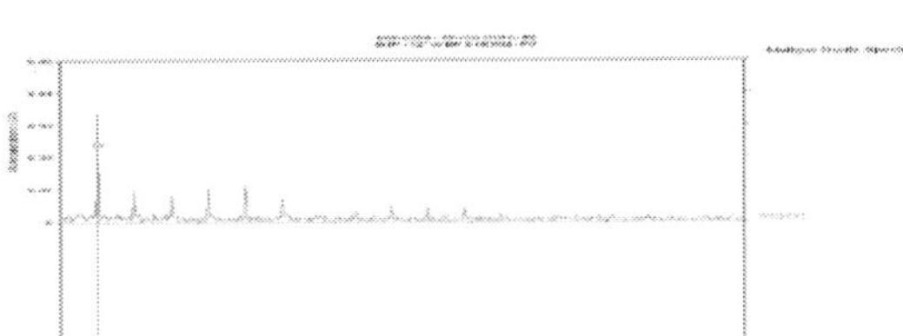

Figure 27 – Vel vs HFE FFT Comparison

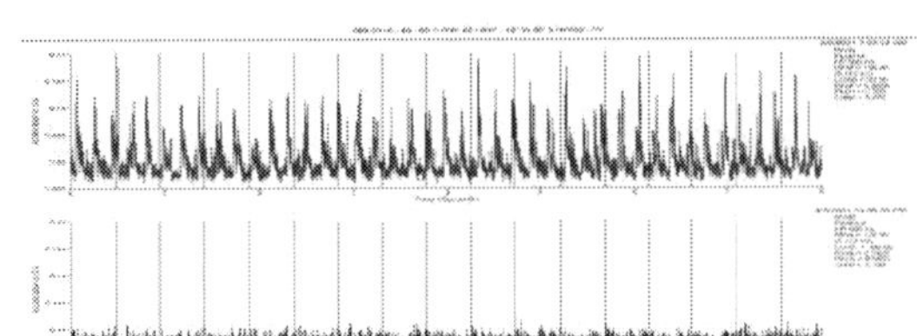

Figure 28 – Vel vs HFE Waveform Comparison

Case #6: South Sawdust Conveyor to Pile: Mystery vibration source

Initially thought source of subsychronous vibration at 16.06 Hz on a 1778 RPM (29.64 Hz) motor coupled to gearbox and chain driving a drag chain (another type of conveyor) was the tooth frequency of its drive sprocket. But, after running calculations, that resulted in a frequency of 1.73 Hz (gearbox output speed X number of teeth)—and it was clear that this was not the source.

Explored a little and discovered the actual source to be the winged tail pulley of the belt conveyor under the drag chain exciting a natural frequency of its own (cantilevered) support structure to amplitudes over 1.27 ips pk. This same frequency can be seen on most components of this conveyor.

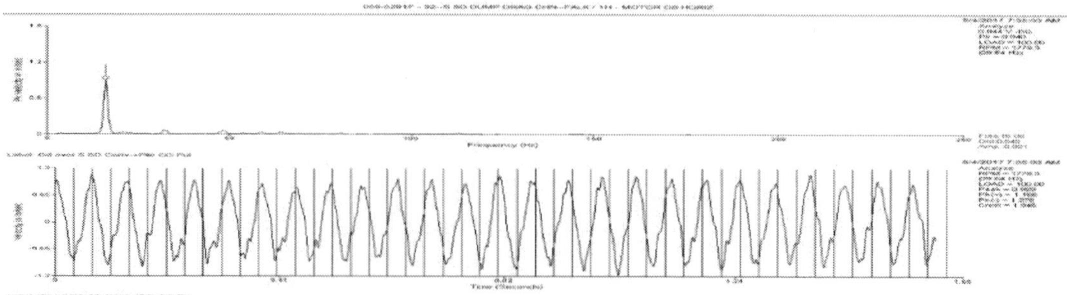

Figure 29 – High vibration from winged tail pulley

Case #7: Chip Conveyor #2 – Fixed shaft tail pulley bearing defect
Both velocity and high frequency vibration climbed on a 20" fixed shaft tail pulley turning at 137.3 RPM. Readings were taken on shaft ends axially and horizontally to compare the two over time. Interestingly enough, overall amplitudes were higher axially in velocity data and horizontally in high frequency energy data.

Figure 30 – Fixed shaft pulley

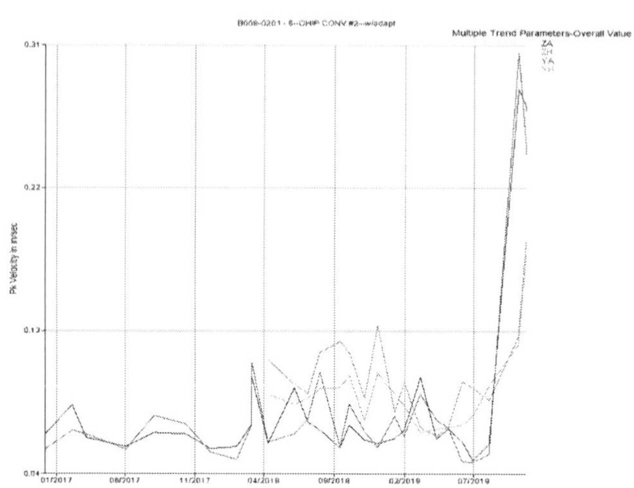

Figure 31 – Overall velocity trends (axial highest)

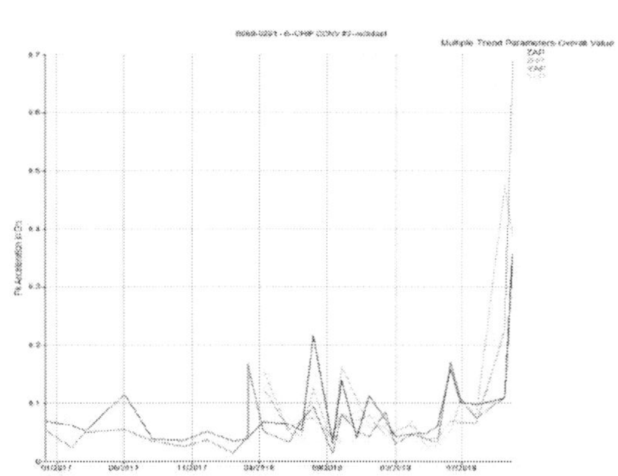

Figure 32 – Overall HFE trends (radial highest)

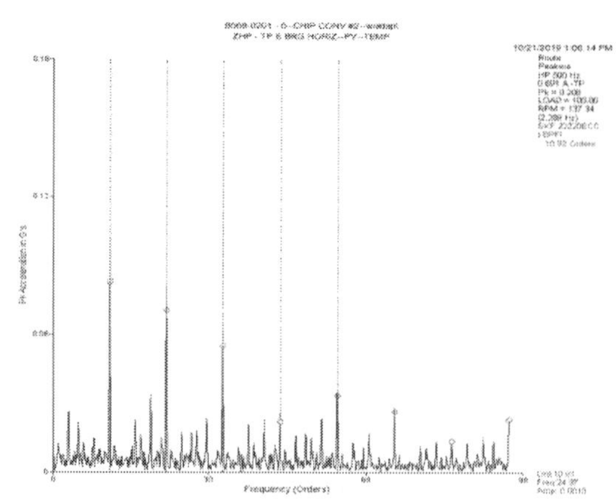

Figure 33 – HFE spectral data

Both types of data showed inner race defect frequencies, which seemed odd because this usually shows up only as impacting at turning speed in high frequency energy. But, this makes sense in this application because the inner race is fixed and the outer race is rotating. That also means that an outer race defect will more likely show up in the high frequency data as impacting at turning speed of the pulley.

Case #8: Chip Conveyor to #2 Silo – Suspected bad (damaged) instrumentation bearing
Vibration data collected off a conveyor drive pulley bearing via a cable and magnet mounted accelerometer was so "noisy" the signal appeared corrupted—and I suspected damaged instrumentation. After deleting the apparently bad data and replacing both cable and accelerometer, I obtained similar data, and wondered if I had lost a bearing!

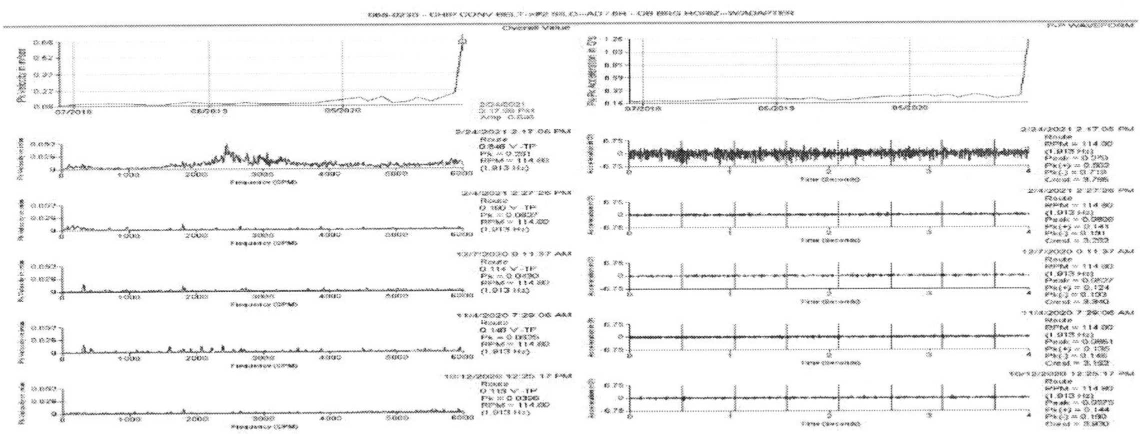

Figure 34 – Vibration data from prior to issue--and after instrumentation replaced

Upon closer inspection, I could see a scraper appearing to be "chattering" against the belt; but couldn't hear it due to noise level in the immediate area. I was able to reach in (safely) and hold my accelerometer onto the framework the scraper was attached to—and got nearly 2.0 ips pk waveform amplitude! This quieted down after mechanics adjusted the spring pressure holding the scraper against the belt.

Case #9: HAC Belts--subsynchronous vibration on motor

The High Angle Conveyor at my facility transfers wood chips from a feed conveyor straight up 135' vertically to the top of a "pre-digesting" vessel. The two drive pulleys on this system are each driven by a shaft-mounted triple reduction gearbox coupled to a motor with a small torque arm under the motor mount.

Figure 35 – Top belt drive assembly

Figure 36 – Both drive assemblies

The two motors are on variable frequency drives and typically run at slightly different, but consistent, speeds that I've become familiar with. However, I still confirm that by using my data collector's "Turning Speed Detection" feature which pinpoints the dominant frequency nearest the one the user inputs. This time, however, the application honed in on a frequency of 27.15 Hz, significantly lower than the typical 31.5 Hz (1890 RPM) that I expected. Not believing that it could be that much slower than normal, I double-checked with a strobe and found actual speed was still indeed around 31.5 Hz.

Back in the office, an expanded look at the lower end of the spectral data showed significant subsynchronous activity, as I expected with everything that is moving up there, but with both this new 27.15 Hz peak and quite a bit of even lower frequency activity.
Expanding that data out even more appeared to show a dominant peak at 6.8 Hz--although pretty poor resolution decreased the degree of confidence here.

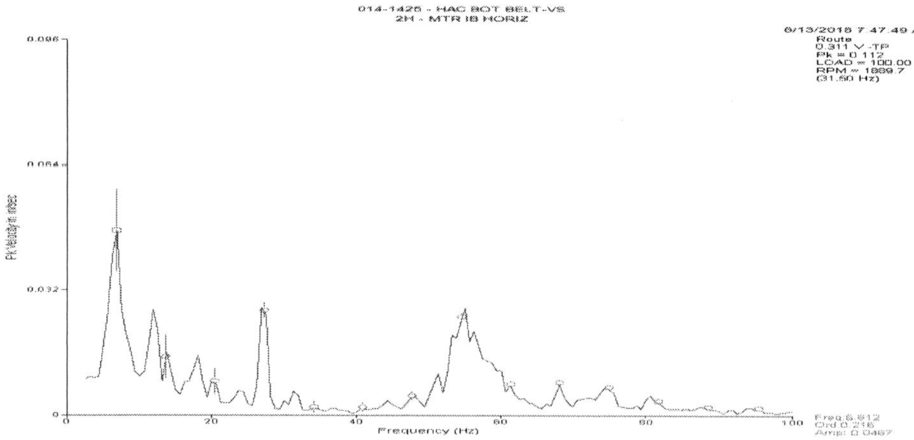

Figure 37 – Subsynchronous spectral data

It's not uncommon for a shaft mounted gearbox to move or "wobble" at the speed of the output shaft. However, when I calculated that speed (by dividing the input speed of 31.5 Hz by the gearbox ratio of 19.82:1), I obtained a speed of 1.59 Hz (95.4 RPM)—clearly not the source of the (apparent) 6.8 Hz peak.

I ran similar calculations as mentioned before, only using a possible speed of 6.8 Hz, a known belt speed of 531.9 FPM, and re-arranging the formula mentioned earlier to solve for roll diameter (in inches): ***D = (FPM * (12 / π)) / RPM***. This yielded a result of nearly 5"—which turned out to be the diameter of the return idlers in the vicinity. This was confirmed several days later when the first of several idlers fell out of the belt conveyor from positions near the drive pulleys. I also increased data resolution on these measurement points to aid in future analysis.

CONCLUSION:

Successful troubleshooting of belt conveyors can involve many challenges for the vibration analyst. Understanding and being familiar with equipment and system configuration; component nomenclature, variation, and function; speed calculation; and attached equipment can all very important for identifying, diagnosing, and communicating a problem.

Additionally, understanding the differences and relationships between linear and rotational speed, "inside" and "outside" belt speed, and velocity (lower frequency) and high frequency enveloped (HFE) data are all very helpful.

Finally, the vibration analyst may need to conduct further investigation in the field to identify other possible sources of vibration – and may find good detective work to reveal unexpected suspects!

GAS TURBINE VIBRATION CASE STUDY – ROTOR SHAFT CRACK EXAMPLES

Author: Timothy S Irwin, P.E.

VIATC | TRAINING CONFERENCE & EXPO
SAVANNAH, GEORGIA · 2022

GAS TURBINE VIBRATION CASE STUDY – ROTOR SHAFT CRACK EXAMPLES

Timothy S Irwin, P.E.
Mitsubishi Power Americas, Inc.
Gas Turbine Service Engineering Principal Engineer
2287 Premier Row
Orlando FL, 32809
Timothy.Irwin@amermhi.com

Mr. Irwin is a Gas Turbine Service Engineering Principal Engineer for Mitsubishi Power Americas and resides in Lakeland FL. He holds a BSME from the University of Illinois and holds Professional Engineer Licenses in both Florida and South Carolina. During his 35+ year career, he has worked for large equipment OEMS, equipment owners and as a consultant providing rotating machinery technical support, vibration analysis, and diagnostic support.

ABSTRACT:
The work that this case study describes was due to changing vibration signatures to high vibration levels in several large frame gas turbines. The vibration signatures exhibited increasing amplitudes during routine operation and changes across unit shutdowns and restarts. The vibration analysis for the various examples was performed based on data provided by the owner operators. The behavior was indicative of shaft cracks and subsequently the shaft cracks were confirmed.

KEY WORDS: Gas turbines; diagnostics; shaft crack; vibration analysis

INTRODUCTION:
There are several thousand large frame gas turbines in operation in the United States. Technological advancements in recent decades have allowed the gas turbine to become the preferred land-based fossil fueled power generation method for base load and load following duties today.

Given the advancements made in these machines, they are still large turbomachinery that can incur the typical problems associated with all mechanical rotating machinery. As the machines acquire operating hours and starts, low cycle fatigue and/or high cycle fatigue stress at levels high enough to result in component degradation can occur. These large complex rotating machines require the designers to balance many considerations to develop a commercially viable and reliable machine.

The machines discussed in this case study had many years of service with varying levels of operating hours and starts. Prior to the change in the vibration signatures there were no signs of a changing or deteriorating mechanical condition.

GAS TURBINE OVERVIEW AND COMPONENTS:

Figure 1 below shows a general cutaway of a large frame gas turbine. Figure 1 should be referenced for the discussion of the issues that occurred and resolution of those issues. The figure referenced shows the air inlet to the left and the exhaust to the right.

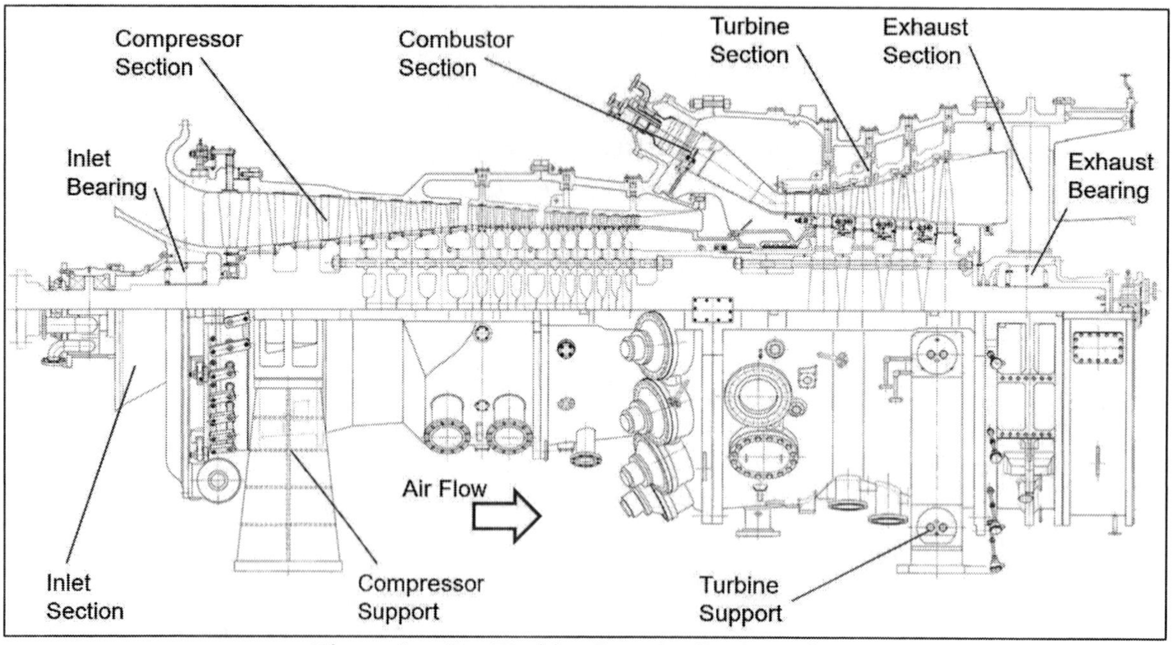

Figure 1 – Gas Turbine Longitudinal Section

There was no specific outage immediately prior to the events that will be presented. The units were between outages during routine operation when the vibration signature changes were observed.

Figure 2 below shows the frame of reference for the vibration analysis. Documenting the frame of reference for vibration analysis efforts is critical for future reference and when communicating with different groups to ensure everyone understands the frame of reference of the analyst.

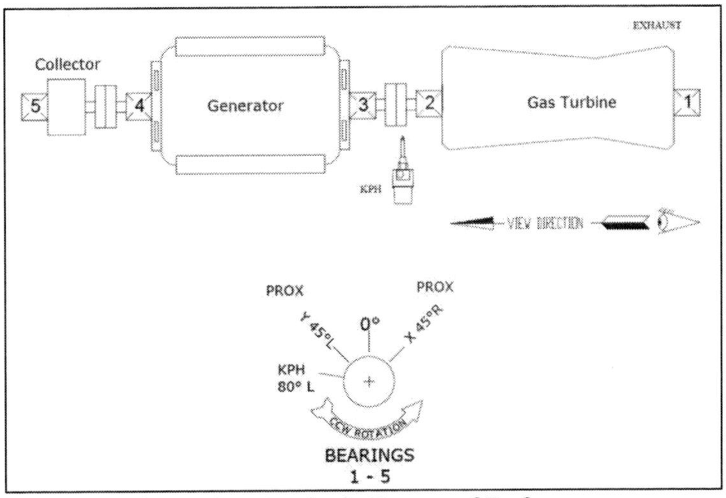

Figure 2 – Analysis Frame of Reference

GAS TURBINE SHAFT CRACK AND VIBRATION RESPONSE:

The focus of this paper will be on shaft cracks and how a shaft crack can affect the vibration signature. Any specific situation needs careful and thorough analysis due to numerous variables that cannot be discussed at length at this time as responses from machine to machine and event to event may have specific differences in response and behavior.

As examples:

- A circumferential crack in a hollow portion of the shaft may respond differently than a through crack in a solid disk
- A radial circumferential crack may respond differently than a torsional circumferential crack event
- A crack on the same rotor in two different locations may respond differently.

This section will provide examples of radial circumferential cracks for consideration during any evaluations that may be performed.

A shaft crack reduces the shaft bending stiffness and the vibration response of a machine is the ratio of the applied Force to the Dynamic Stiffness, see Equation (1) Reference [1]:

$$Vibration = \frac{Force}{Dynamic\ Stiffness} \quad [1]$$

Different crack positions and types can result in different changes in shaft stiffness. For a more in-depth discussion of the mathematical description of the vibration response equation, please refer to the Reference and Bibliography for a few options on books and papers written that provide thorough discussion and development of this equation and its concepts.

VIBRATION DIAGNOSIS:

To simplify the discussion in this paper we will focus on the dominant characteristics of the vibration signature changes. The dominant signature changes were in the overall (unfiltered) and the 1X (synchronous) parameters. The list below is a general list that can be referenced for the most likely causes of synchronous vibration signature changes:

Table I – Synchronous Frequency Possible Causes

Predominant Frequency	Possible Causes
Synchronous 1X (Running Frequency)	Unbalance Lost Rotor Parts (Unbalance) Rotor Bow Casing Distortion Foundation Distortion Misalignment Journal & Bearing Eccentricity Bearing Damage Rotor Critical Speed Rotor-bearing System Critical Coupling Critical Structural Resonances Thrust Bearing Damage Structural Looseness

The above list should be considered during vibration data analysis when the dominant frequency or frequency change is synchronous to the rotor operating speed.

UNIT A

Figure 3 below shows the initial single channel trend (exhaust bearing Y channel) of the overall vibration amplitude and load provided by the machine owner.

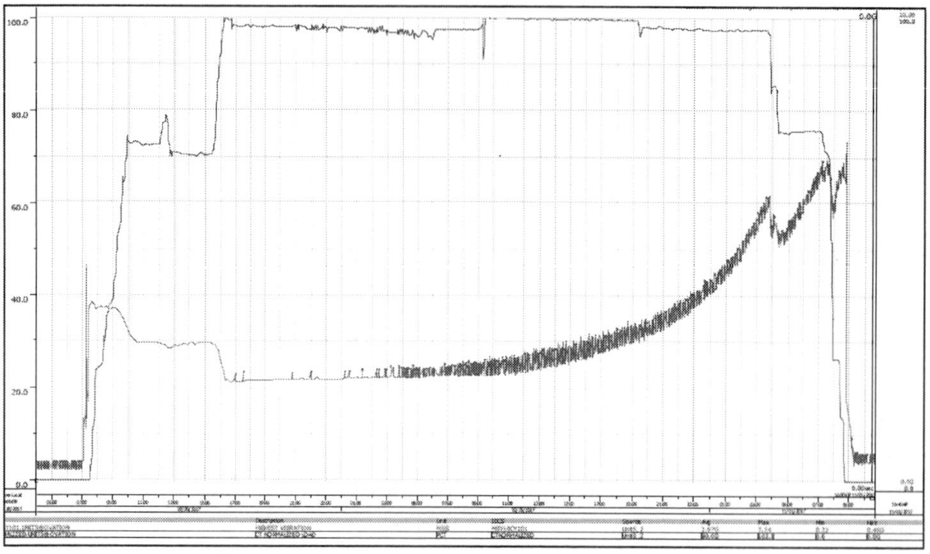

Figure 3 – Overall Amplitude Trend Values

The green trace is the Bearing 1Y overall amplitude and the blue trace is the unit load. Speed was steady at 3600 rpm with typical operational load changes that had never previously resulted in the observed increasing amplitude.

The owner then provided a historical vibration database for review. The amplitude and phase Trend plots below in Figure 4 show the last three operational runs of the unit.

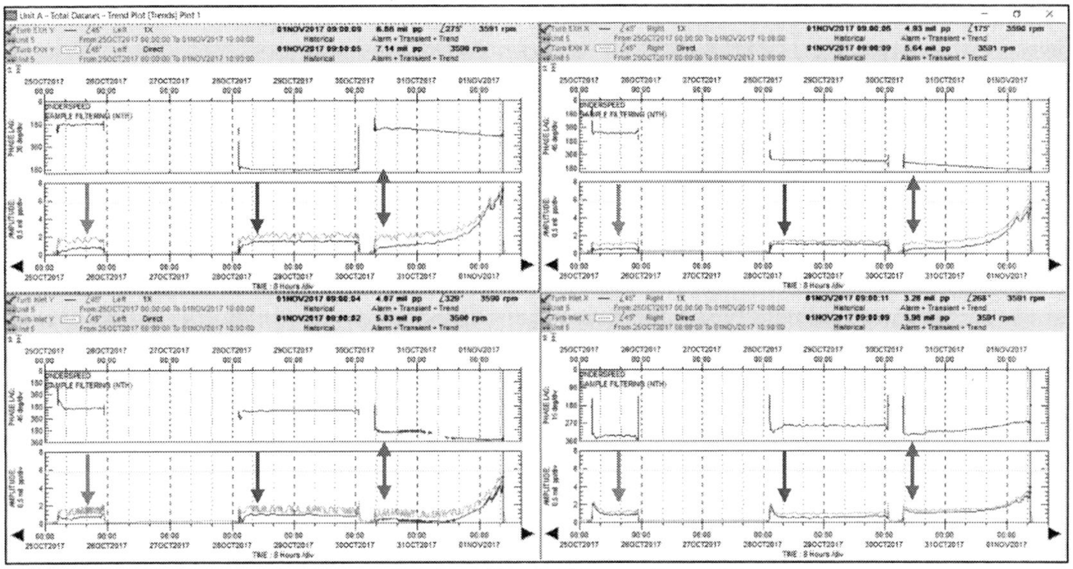

Figure 4 – Amplitude/Phase Trend Plots

The four channels shown in Figure 4 - Bearing 1 (Exhaust) X & Y and Bearing 2 (Inlet) X & Y all indicated the following:

- The exhaust bearing was exhibiting higher response to the changing machine condition than the inlet bearing.
- The blue traces show the filtered 1X amplitude and phase; the orange traces show the Direct amplitude.
- Left side of each plot shows the previous typical Direct amplitudes (below 2 mils) and steady 1X amplitude and phase (green arrows).
- The 2nd operational run indicated both amplitude and phase shifts at both channels on both bearings primarily for the synchronous or 1X parameter (blue arrows).
- At the start of the 3rd run the Direct amplitudes on the #1 bearing appeared to return to near previous levels, but 1X phase shifts were observed (red arrows).
- The dominant amplitude increase occurred over the last 24 hours with smaller amplitude and phase variations during the previous 16 hours.
- The total operational run time during the last run leading to the shutdown was ~50 hours.
- Given the behavior change included all four gas turbine vibration channels with increasing amplitudes and changing signatures over the last 24 hours we did not consider this a potential instrumentation or monitoring system issue.

The slow roll data was reviewed for both the pre-startup and the post operation coast down conditions (See Table II below). Changes in the pre-startup slow roll conditions at ~235rpm were less than 0.1 mils and 5 degrees. The slow roll changes were not indicating a change in condition prior to the operational run where the vibration amplitudes increased.

Table II – Slow Roll Conditions

Data Sample	Bearing 1X		Bearing 1Y		Bearing 2X		Bearing 2Y	
	Startup	Shutdown	Startup	Shutdown	Startup	Shutdown	Startup	Shutdown
1st Run	0.46@142	0.56@132	0.47@230	0.58@220	0.95@188	1.05@181	1.01@272	1.10@267
2nd Run	0.46@142	0.57@132	0.47@229	0.59@223	0.94@188	1.04@180	1.01@272	1.10@266
3rd Run	0.46@141	0.98@165	0.48@228	1.08@245	1.03@189	1.28@182	1.06@273	1.31@268

The shutdown data (at 230rpm) on the 3rd run indicated changes of ~0.25 to 0.5 mils with phase changes of 33 degrees or less (red font). The shutdown slow roll change was indicative of an actual change in rotor condition/bow, further data and operational analysis was necessary to determine the most likely reason(s) for the change.

The amplitude and phase Trend plots below in Figure 5 focuses on the final operational run before the shutdown and subsequent disassembly and inspection.

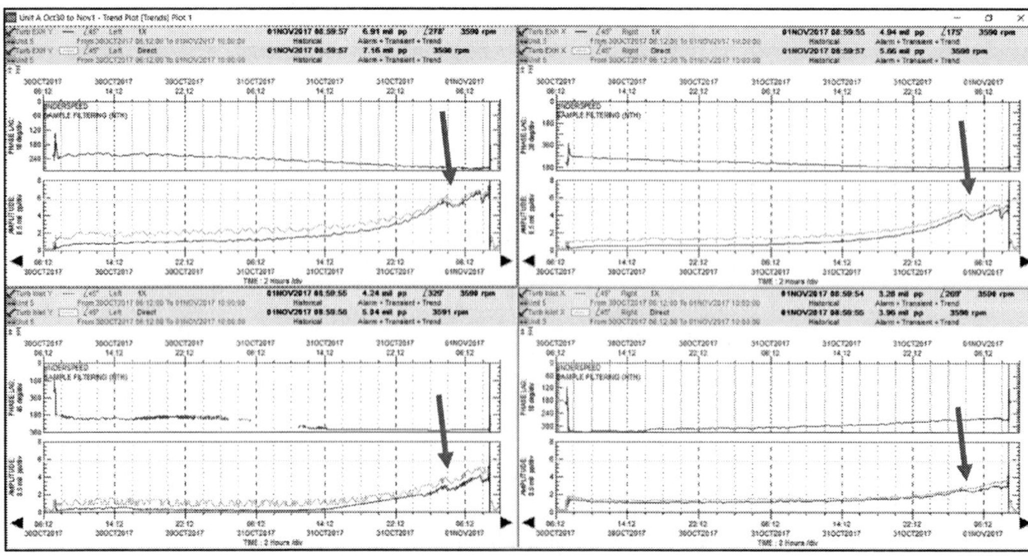

Figure 5 – Amplitude and Phase Trend Plots

The blue traces are 1X amplitude and phase trends; the orange traces are the Direct/Overall trends. The top two plots are the turbine exhaust bearing plots; the lower two plots are the turbine inlet bearing plots. This data indicated that the dominant frequency during the event operation was synchronous or filtered 1X frequency. The Bearing 1 and 2 Y channels indicated more non-synchronous energy than the bearing 2X channel, but the 1X frequency was the dominant factor in the amplitude change.

The amplitude drops towards the end of the run (blue arrows) occurred when load changes were made to determine their effect on the observed vibration amplitude increase. The amplitudes initially dropped from the load changes and then continued to increase.

Spectral content was reviewed indicating synchronous and synchronous multiples were the primary content of the increasing amplitudes. See Figure 6 below for the #1 bearing data.

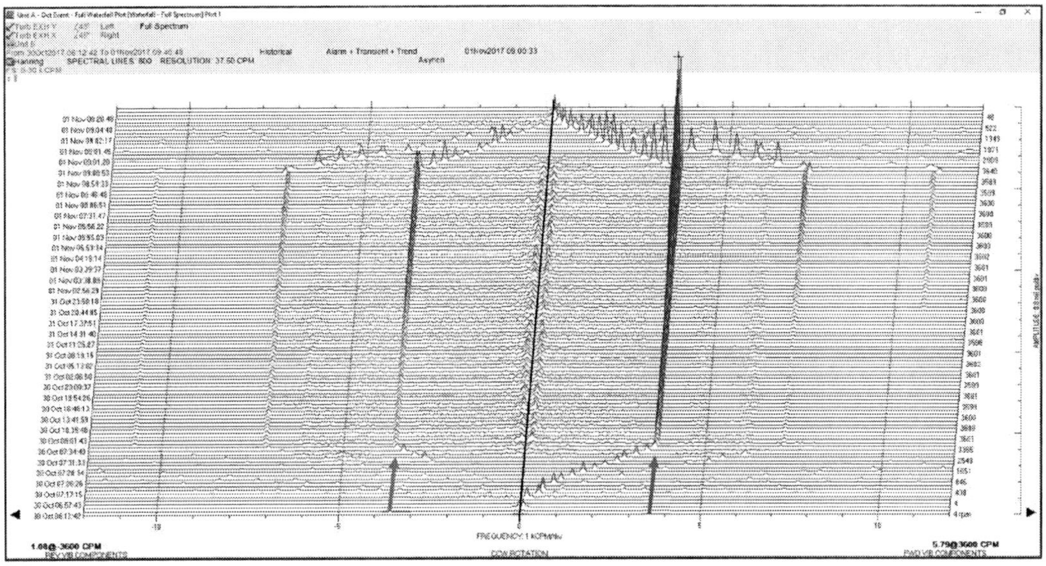

Figure 6 – Full Spectrum Waterfall Plot

The plot shows that the synchronous (1X) frequency (red arrows) was the dominant amplitude as the machine began its loaded operation. As the operating time progressed the forward precessed 1X started climbing first, then the reverse precessed 1X started climbing. Towards the end of the operating period the forward and reverse precessed 2X frequencies started increasing in amplitude. Forward precession energy was dominant, but reverse precession energy was also exhibiting increasing amplitudes.

1X amplitude and phase vector data was reviewed. The baseline October 25 operation indicated steady amplitudes with phase shifts that appeared to be due to thermal stabilization. The filtered and compensated 1X vector during the final operational period indicated a nontypical change, please see Figure 7 below.

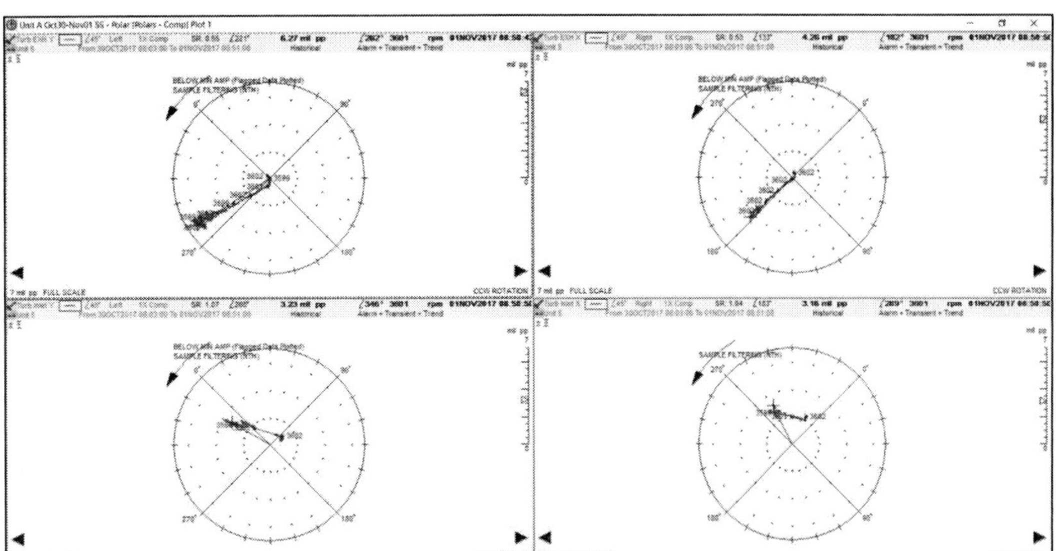

Figure 7 – 1X Filtered and Compensated Bode Plots

After initial thermal stabilization at amplitudes below 1.5 mils the amplitude started to climb. During the amplitude climb the phase shift indicated a 'straight line' vector change pattern from the starting position. This behavior would not be considered normal or typical.

Bearing housing seismic sensors were installed on this unit and overall patterns were reviewed in Figure 8 below.

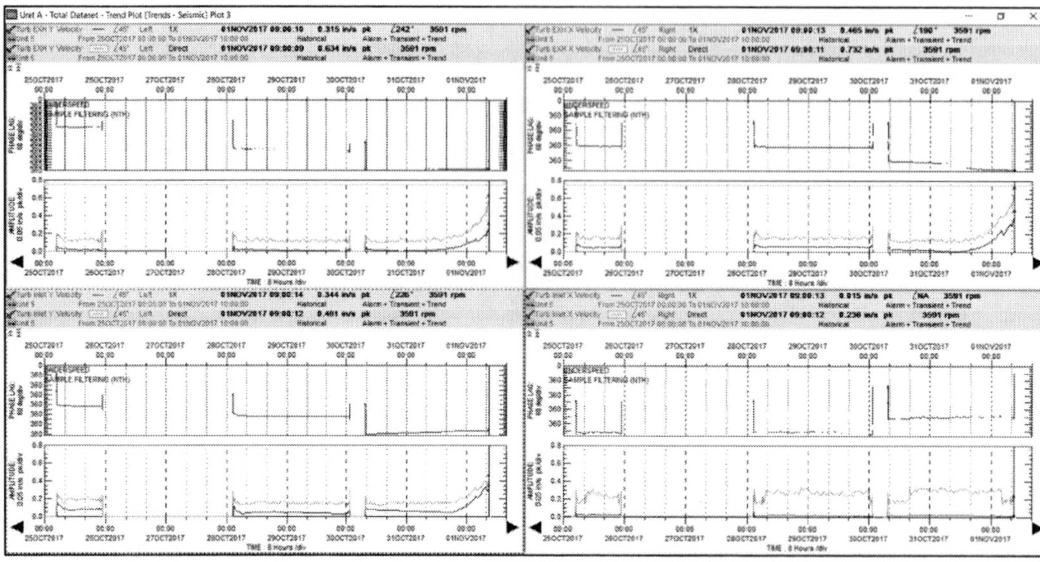

Figure 8 – Bearing Housing Seismic Trends

The blue traces are 1X amplitude and phase trends; the orange traces are the Direct/Overall trends. The top two plots are the turbine exhaust bearing plots; the lower two plots are the turbine inlet bearing plots. The seismic velocity amplitude increases followed the shaft relative amplitude increases. Synchronous amplitudes when converted to displacement were about 25% of the shaft relative amplitudes.

Spectral seismic data was reviewed and is shown in the Figure 9 plot below.

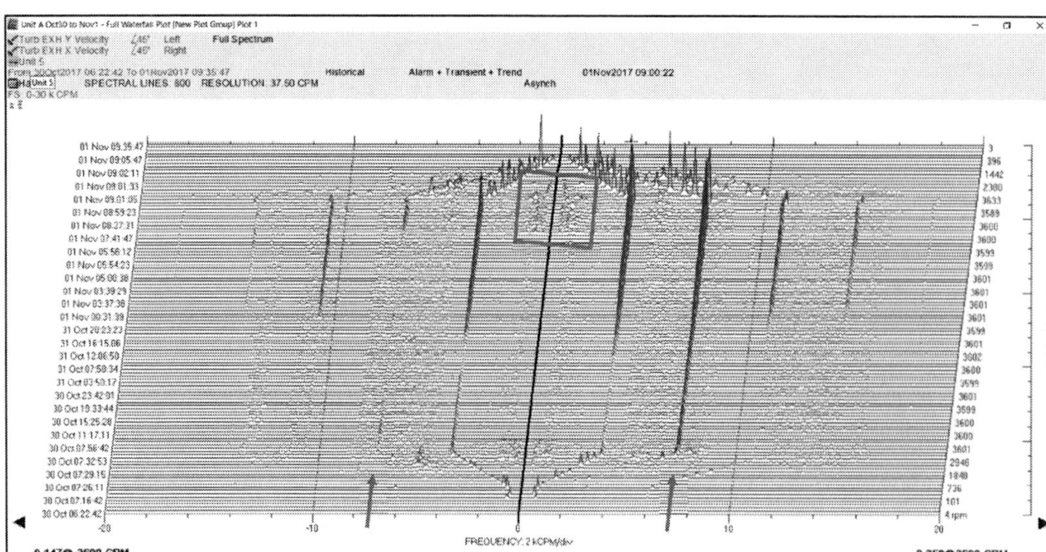

Figure 9 – Bearing #1 Full Spectrum Waterfall - Seismic

Initial operation indicated that forward precession energy and a synchronous 2X multiple (red arrows) was the dominant frequency. Through the operating period the synchronous 1X and multiples all increased in amplitude. By the time of the unit shutdown, the 1X frequency was higher in amplitude than the 2X frequency. During the last hour some sub-synchronous energy was starting to exhibit and was likely due to changing load conditions including the final load drop (red box).

The Direct Orbit patterns were reviewed. The #1 bearing data is shown in Figure 10 below.

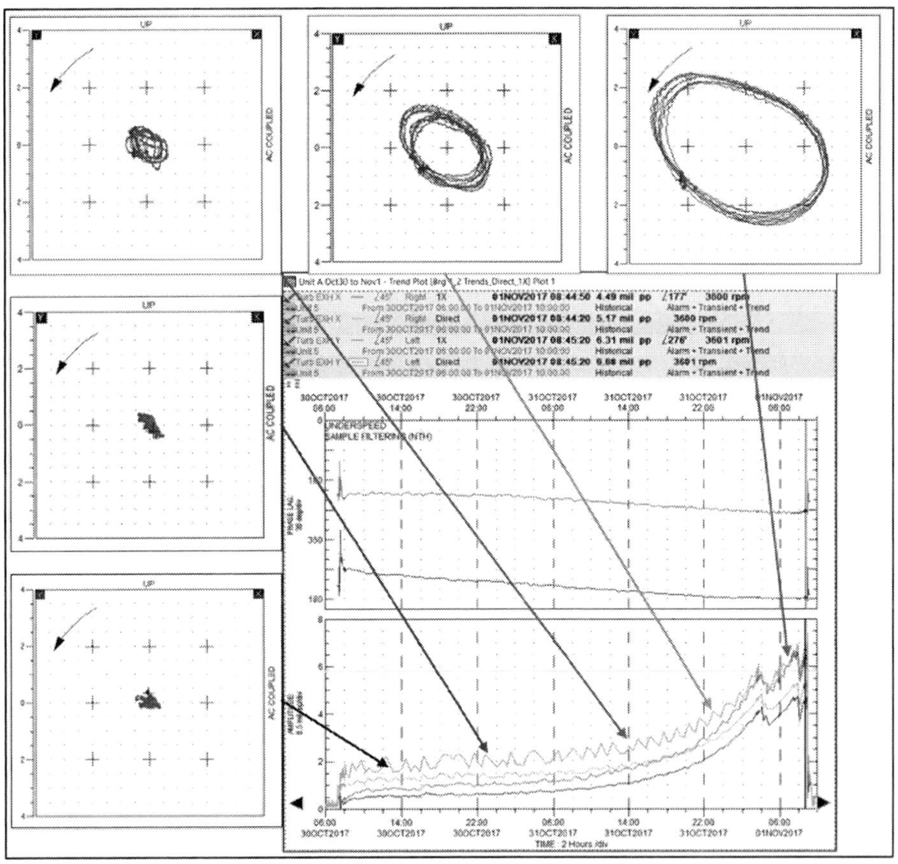

Figure 10 – Direct Orbit Plus Plots

The compensated Orbit plots above show the progression of the amplitude increase. The Trend plot shows the difference in amplitudes between the compensated and uncompensated data. The arrows from the Orbit plots to the trend plot show the approximate timing of each Orbit sample.

- The initial Orbit pattern (black arrow) was low amplitude and without a clear circular pattern.
- The second Orbit pattern (purple arrow) was starting to show an increase in amplitude and a pattern shape was beginning to develop.
- The third Orbit pattern (blue arrow) continued indicating the amplitude increase and a more open Orbit shape was starting to develop while variation rotation to rotation was exhibited.
- The fourth Orbit pattern (green arrow) was dominated by an increasing amplitude and the Orbit was now open with a variation rotation to rotation exhibited.
- The fifth Orbit pattern (red arrow) was dominated by the synchronous 1X amplitude increase with an open Orbit shape.

Once the Direct Orbit pattern began to stabilize, the Orbit shape and phase appeared to remain similar as the amplitudes increased.

The overall analysis summary for this signature change is shown below in Table III. Note that this review was specific to experience on large frame gas turbine behavior.

Table III – Analysis Summary Table

Predominant Frequency	Possible Causes	Analysis Results
Synchronous 1X (Running Frequency)	Unbalance	Not a typical unbalance/changing balance response
	Lost Rotor Parts (Unbalance)	Not a typical lost parts unbalance response
	Rotor Bow	Could be a changing rotor bow – both bearings exhibited final coast down slow roll changes.
	Casing Distortion	Not a typical casing distortion response
	Foundation Distortion	Not a typical foundation distortion response
	Misalignment	Not a typical changing alignment condition response
	Journal & Bearing Eccentricity	Not a typical response for journal or bearing eccentricity
	Bearing Damage	Could be a possible bearing damage response
	Rotor Critical Speed	Rotor critical speeds are not near operating speed, but is critical speed changing? – unknown
	Rotor-bearing System Critical	Rotor-bearing system critical speeds are not near operating speed, but is critical speed changing? – unknown
	Coupling Critical	Not a typical response for a coupling critical
	Structural Resonances	Structural (seismic) response did not indicate a typical structural issue response.
	Thrust Bearing Damage	Not a typical response for thrust bearing damage.
	Structural Looseness	Could be a possible structural looseness condition

Based on the analysis results several actions were recommended:

- External inspection of the structural support components
- Borescope inspection of the gas flow path
- Verify proximity sensor calibration and gap voltages

The inspections did not identify any deficiencies. Based on those results the most likely cause for the condition was a changing shaft stiffness. It was recommended to remove the rotor from service for disassembly and inspection.

Subsequent disassembly and rotor inspection identified a circumferential through wall crack in the hollow compressor to turbine torque tube. This location is not visible for physical inspection until the rotor is disassembled and the air separator is removed. Initial confirmation of the crack was determined using phased array ultrasonic testing prior to rotor disassembly.

See Figure 11 below for approximate location of the crack in the rotor structure:

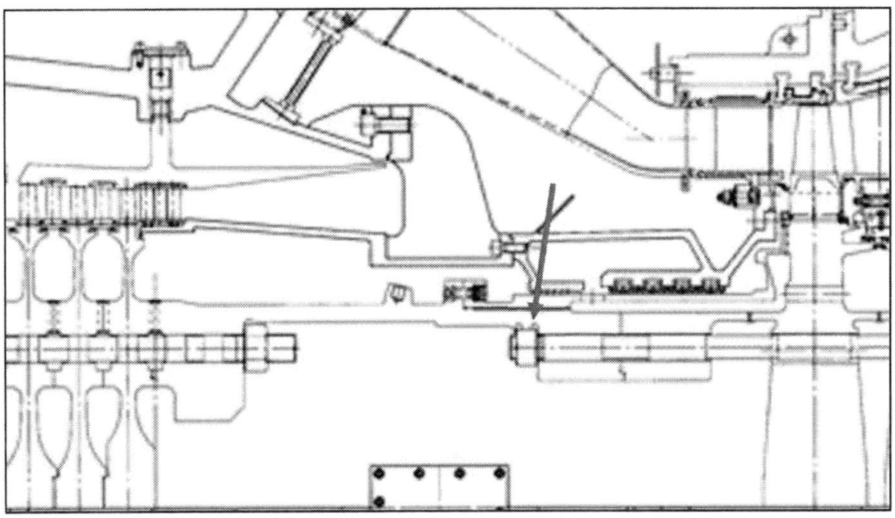

Figure 11 – Rotor Crack Location in the Torque Tube

Additional Unit Shaft Crack Examples:

UNIT B

Figure 12 below shows the initial trend data of the overall vibration amplitude and load provided by the machine owner.

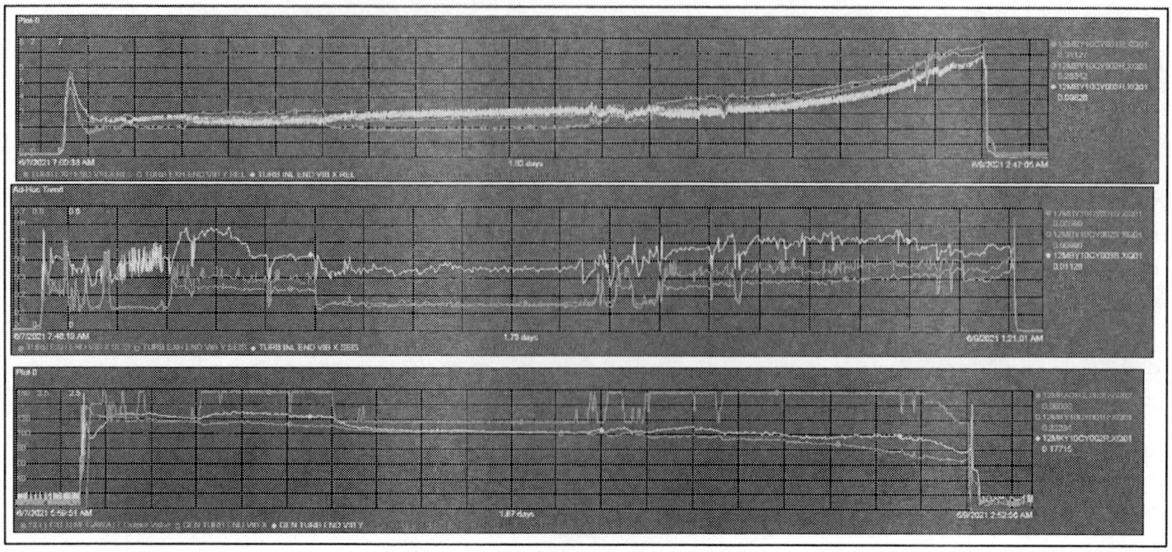

Figure 12 – Unit B Trend Data

The top Trend plot was of the exhaust and inlet bearing shaft relative vibration channels. The middle Trend plot was of the exhaust and inlet bearing seismic vibration channels. The bottom Trend plot was of the generator shaft relative vibration channels.
The Trend data above exhibited very similar behavior to the Unit A example. Initial operation appeared to be typical for the unit and during the operational run shaft relative vibration levels climbed until the unit was shut down. Changing load levels did not result in a measurable vibration response. Bearing housing seismic indicated some response to the condition change towards the end of the operating run. Generator vibration amplitudes had varied responses to the changing Turbine conditions.

Dynamic vibration data plots were also provided and indicated similar behaviors to the Unit A 1X (synchronous) amplitude and phase.

Subsequent disassembly and inspection revealed a similar through wall crack indication at the same location.

UNIT C
Figure 13 below shows the initial trend data of the overall turbine inlet and exhaust bearing shaft relative vibration amplitude and load provided by the machine owner.

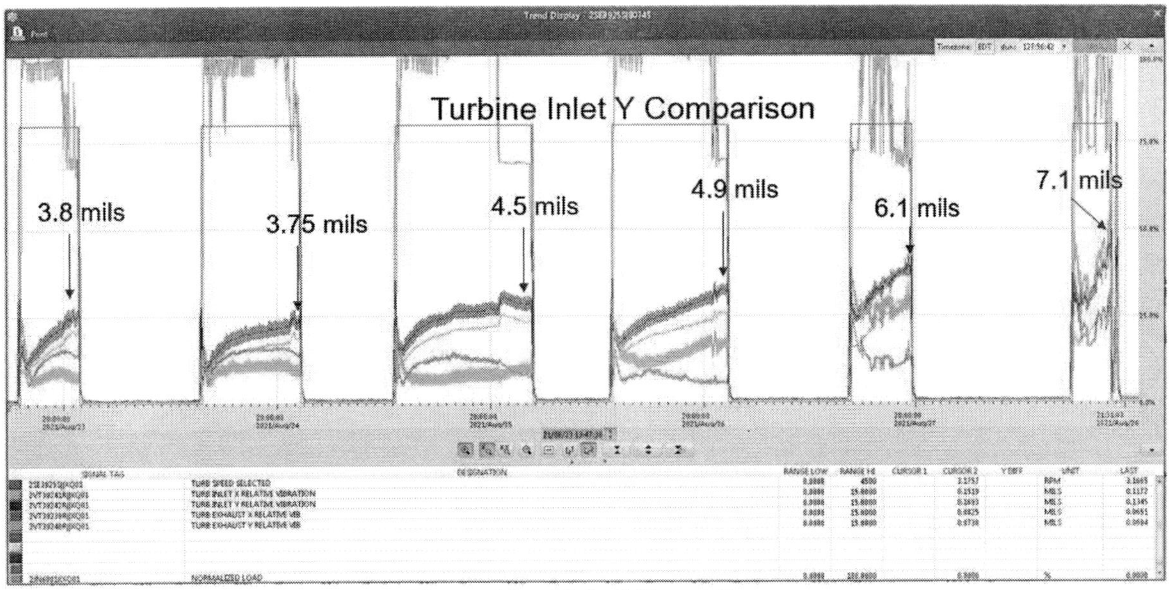

Figure 13 – Unit C Trend Data

This unit's operation is different in that the typical operational runs were shorter during the crack initiation and growth. The longest operational run was ~16 hours. The final run prior to shutdown was only ~4 hours.

By the third operational run on this plot, it appears that the inlet bearing amplitudes (black and blue traces) are responding more (increasing amplitude) to the changing rotor condition than the exhaust bearing. During the final two operational runs, the exhaust and inlet bearing y channels indicated the highest response and were responding similarly during the amplitude increases.

Dynamic vibration data was not available during this event.

Figure 14 below shows a comparison between the first operational run to the last operational run.

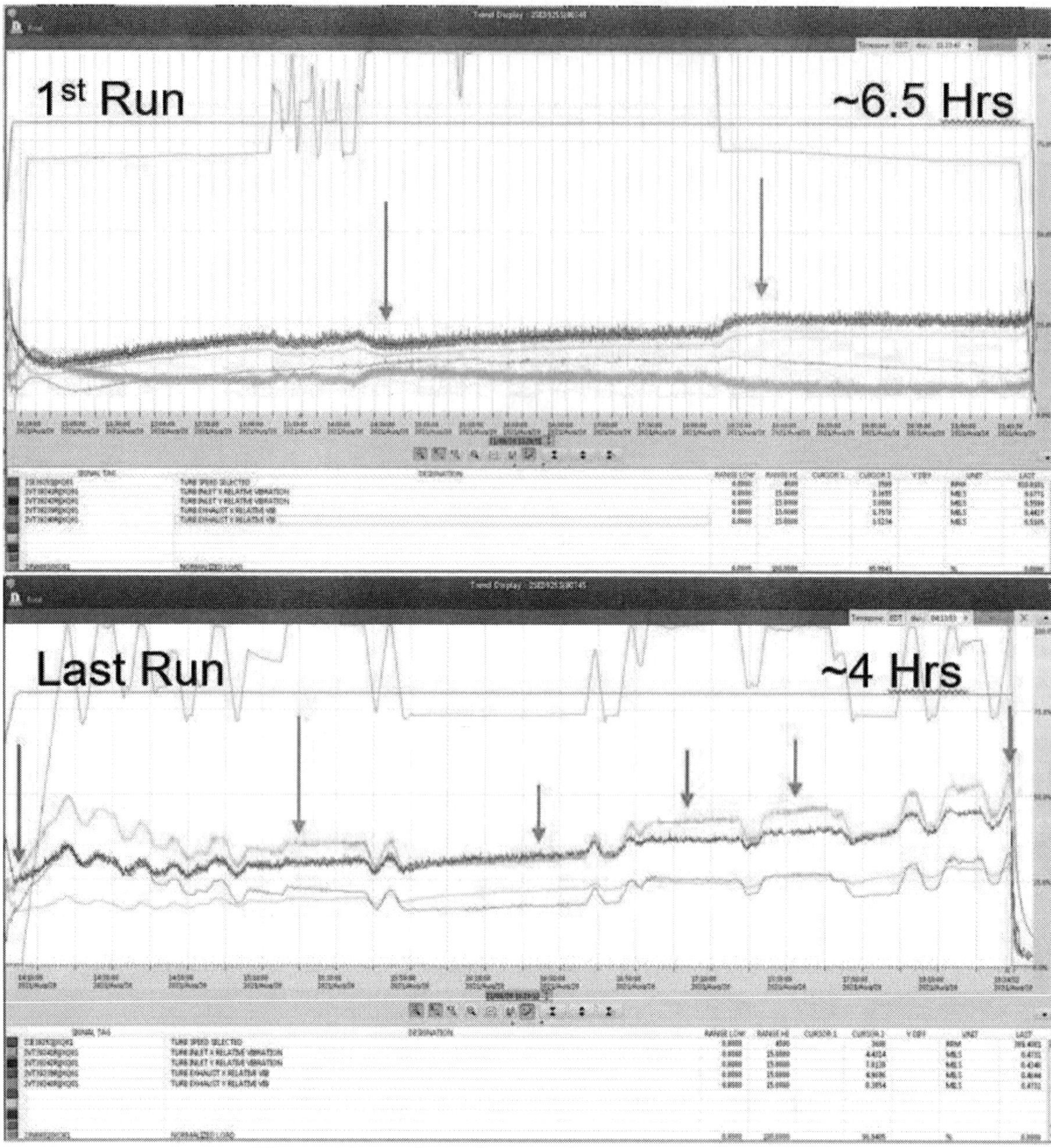

Figure 14 – Unit C First Run/Last Run Comparison

During the first run the unit had low amplitude responses (less than 0.8mils) with the inlet bearing Y channel indicating the highest changes to load swings that were more than 35MW. The green arrows identify various vibration amplitude changes observed.

During the last run the behavior if the unit had changed when compared to the 1st run:

- Overall amplitudes were higher during initial baseload operation.
- The unit vibration amplitude responses to load swings were higher at all bearing channels.
- The exhaust bearing Y channel now had the higher amplitude responses (more than 1.2mils) to load swings.
- The vibration amplitudes continued to climb during the duration of the operational run.

Figure 15 below shows the 1st run to Last run startup comparison.

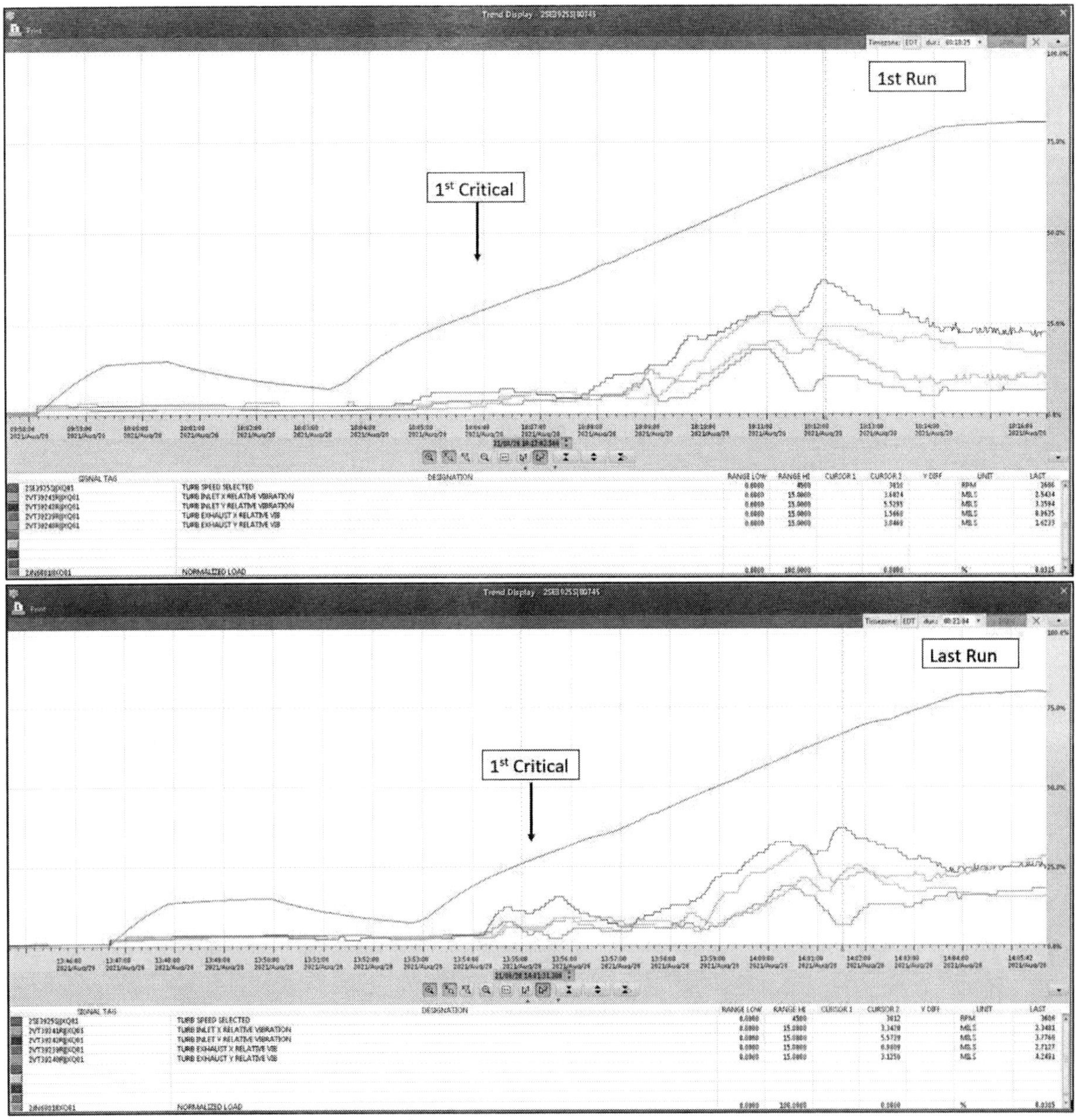

Figure 15 – 1st Run to Last Run Startup Comparison

The most significant change in the startup vibration response signature is the increased amplitudes at the 1st critical. Previous history of the machine indicated a well-balanced rotor with minimal 1st critical response. The last run startup indicated an increasing amplitude and response at the 1st critical.

This rotor indicated a first critical amplitude increase during transient conditions.

Subsequent disassembly and inspection revealed a similar through wall crack indication at the same location.

UNIT D

Figure 16 below shows the initial trend data of the overall turbine inlet and exhaust bearing shaft relative vibration amplitude and load provided by the machine owner.

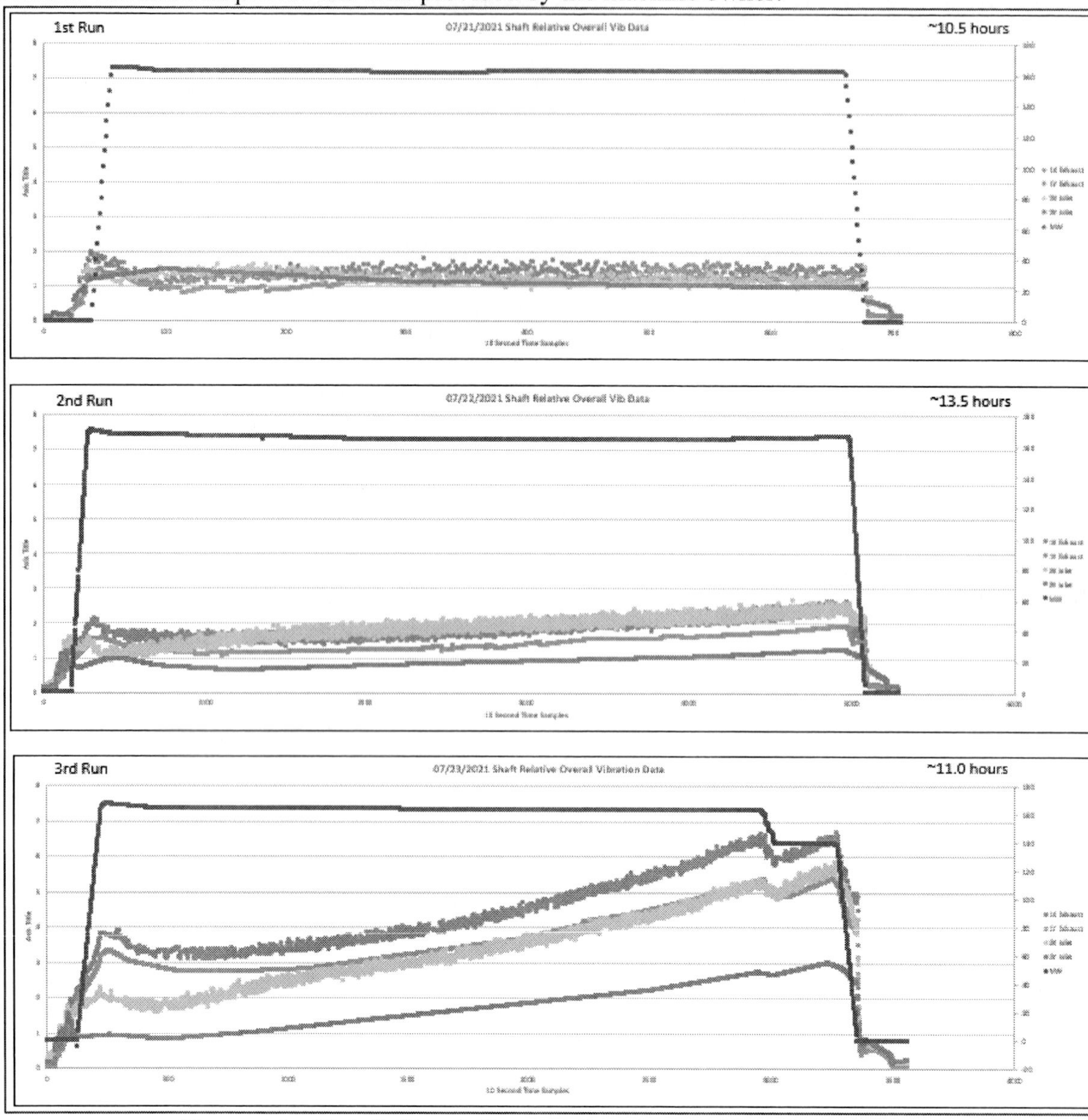

Figure 16 – Unit D Trend Changes Over Last 3 Operational Runs

During the last 3 operational runs the overall amplitude vibration signature of the machine changed. The dark blue trace shows the MW loading during the operational runs.

The 1st run indicated steady overall amplitudes below 2 mils. The 2nd run indicated an increased overall amplitude signature to ~2.5 mils. The 3rd run indicated increasing amplitudes until the unit shutdown with peak amplitudes over 6.5 mils and increasing towards 7 mils.

Figure 17 below shows the exhaust bearing amplitude and phase behavior during the 30 days until the unit shutdown.

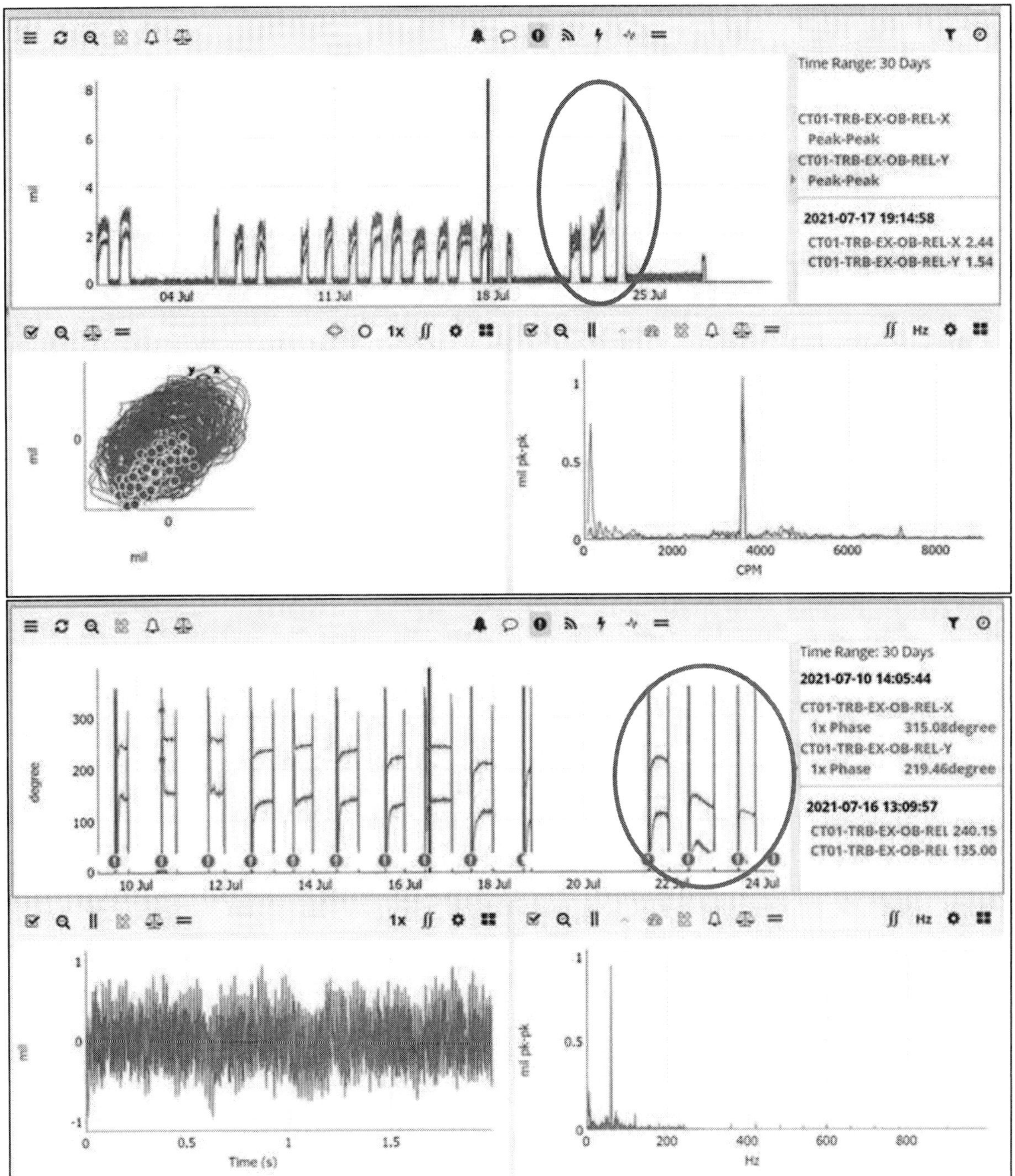

Figure 17 – Exhaust Bearing Synchronous (1X) Amplitude and Phase Trends

General amplitudes prior to the failure were below 3 mils at the exhaust (#1) bearing. The unit exhibited both amplitude and phase variations run to run prior to the event. The amplitude increase and phase shift occurred during the last two runs prior to the inspection.

The Spectral data indicated that the dominant frequency was synchronous (1X) prior to the event.

CONCLUSION:
The examples in this case study indicated two primary themes:

First – Trend monitoring and analysis are critical to machinery protection.

Second – An unexplained change in 1X amplitude and phase was confirmed to be development of a shaft crack. Fortunately, the operators of these units had reliable vibration monitoring and protection systems and performed shutdowns of the units before catastrophic failures occurred.

Additional observations noted depending on data availability:

A change in Hot/Coast Down slow roll signature could be an indicator of a changing rotor condition. A change in the Hot slow roll signature could indicate a change in the rotor bow; a crack growing in a shaft will change the rotor bow.

An increase in the 2X amplitude and phase during operation with the 1X changes could be an indicator of a changing rotor condition. Development of a shaft crack results in an asymmetric stiffness of the rotor that would present itself as an increase in the 2X amplitude.

Thorough analysis of all available data is recommended to ensure other potential contributing factors to a vibration signature change are evaluated prior to making the decision to start disassembly of a large frame gas turbine.

REFERENCES:
[1] C.T. Hatch & D.E. Bently, "Fundamentals of Rotating Machinery Diagnostics", ASME Press (American Society of Mechanical Engineers), New York, 2003.

BIBLIOGRAPHY:
1. D.E. Bently & A. Muszynska (1986). Detection of rotor cracks. Texas: Proceedings of the Fifteenth Turbomachinery Symposium.
2. G. Meng & E.J. Hahn (1997). Dynamic response of a cracked rotor with some comments on crack detection. Journal of Engineering for Gas Turbines and Power, April 1997, Vol. 119. New York: New York: ASME Press (American Society of Mechanical Engineers).
3. M.P. Boyce (2006). Gas turbine engineering handbook 3rd edition. Texas: Gulf Professional Publishing.

TRANSIENT VIBRATION CONTROL FOR CONDENSING STEAM TURBINE USING FORCE REDUCTION & INCREASING DAMPING TECHNIQUES

Author: Mina Shenouda

VIATC VIBRATION INSTITUTE ANNUAL TRAINING CONFERENCE & EXPO 2022 | MFPT ANNUAL CONFERENCE

2022 CONFERENCE PROCEEDINGS

VIATC | TRAINING CONFERENCE & EXPO
SAVANNAH, GEORGIA · 2022

TRANSIENT VIBRATION CONTROL FOR CONDENSING STEAM TURBINE USING FORCE REDUCTION & INCREASING DAMPING TECHNIQUES

Mina Shenouda
MINA@PPTOTAL.COM
PPTOTAL Company
Nasr City-Cairo-Egypt

Mr. Mina Shenouda is Cat IV vibration analyst since 2015, previous member at VI certification committee and works as a Predictive maintenance Manager at Helwan Fertilizers Company
He is a Shareholder for PPTOTAL Company for industrial services, in addition he is a Vibration consultant at El-Salam workshop for Balancing of high-speed rotors, and he is a well-known vibration analysis courses instructor in Egypt, Middle East & Europe, He is a third party for Oil& Gas, petrochemicals and Power generation Projects

ABSTRACT:
This case study declares how important the vibration analyst must be aware that analyzing a machine vibration will not be accomplished successfully unless the analyst shall first understand the machine nature, drawings, operating conditions and its history very well.

Turbomachinery vibration analysis requires high skills and expert for such kind of machines but defining the root cause of the vibration needs a well understanding to the machine in concern.

Rubbing between rotors and stators represents only 2% of rotating machine faults. When rubbing is suspected to be the cause of the machine vibration, the analyst will need to check other vibration analysis techniques beside the ordinary FFT spectrum.

Bode, Polar, Orbit, shaft centerline Plots and observing 1X phase values during the high vibration events are important in such cases.

Rubbing forces in a rotating machine is considered one from the causes of Non-Linear Response, that's why the analyst must be aware of kind of rubs and how the machinery system responds to it. In most cases Rubbing is an effect (i.e., Not the primary cause of high vibration), that's why it is important to define the causes of the rubbing.

Rubbing types are many as annular rubs, partial rubs and bouncing rubs and rubbing area may vary (seals area, bearings area…etc). Some kinds of rubs can lead to catastrophic failures, so vibration control by reducing rubbing forces or eliminating the rubbing action should take place.

KEY WORDS: Damping; Force reduction; Oil coking; Preload; Rubbing; Shaft centerline; Transient vibration; Vibration control.

INTRODUCTION:

In this case study, A steam turbine which is in service since 2006 till 2021 in an Egyptian Petrochemicals Company had experienced high transient vibration which usually take place for a short period (from 20 to 30 minutes), after then the vibration returns back to normal values.

The vibration Consultant here worked on Vibration data, Site inspection, Machine drawings review and Operating Conditions check to diagnose and solve the transient high vibration problem.

It will be shown how the site inspection, operational parameters review beside the vibration analysis plays an important role to apply the appropriate vibration control techniques to solve that high transient vibration problem.

PLANT OVERVIEW:

Helwan Fertilizers company (HFC) is one of the biggest fertilizers companies in Egypt, it produces UREA granules (about 2000 Tons/day), the company was constructed in 2005 and started production in March 2007. The company area is divided to five main sectors (Ammonia- Urea- Utilities- Compressor house - Bagging).

In the Compressor House sector - which is the heart of the plant- there are five steam condensing turbines which are the drivers of four centrifugal compressors and one generator. All the five turbines are manufactured by Siemens company.

The turbine which showed a high transient vibration was the CO2 compressor driver, it had been in service for fifteen continuous years without any internals inspection. The driven CO2 compressor has 4 stages, Stages 1& 2 in LP casing are running with same turbine speed, but stages 3 & 4 in HP casing are running with nearly double speed (as the gearbox increase the speed ratio by factor of 1.95), The machine train is shown in Figure 1.

Figure 1 Machine Train

HIGH TRANSIENT VIBRATION PROBLEM:
During normal operation for the CO2 compressor turbine (320MT02), the vibration values on the drive end side bearing (Bearing no 913) increase and then returns back to its normal values in a short period (usually from 20 to 30 minutes), then the vibration values stabilize few days before the same phenomenon is repeated.

The normal vibration values for the turbine drive side bearing are 22µm P-P, when the vibration increase it may reach 100 µm P-P. Also, it was noted that sometimes the vibration gets lower than the normal Value and may reach 10 µm. This phenomenon of the elevated high transient vibration started from March 2019 till 2022. A strip chart for turbine drive side bearing vibration is illustrated in Figure 2.

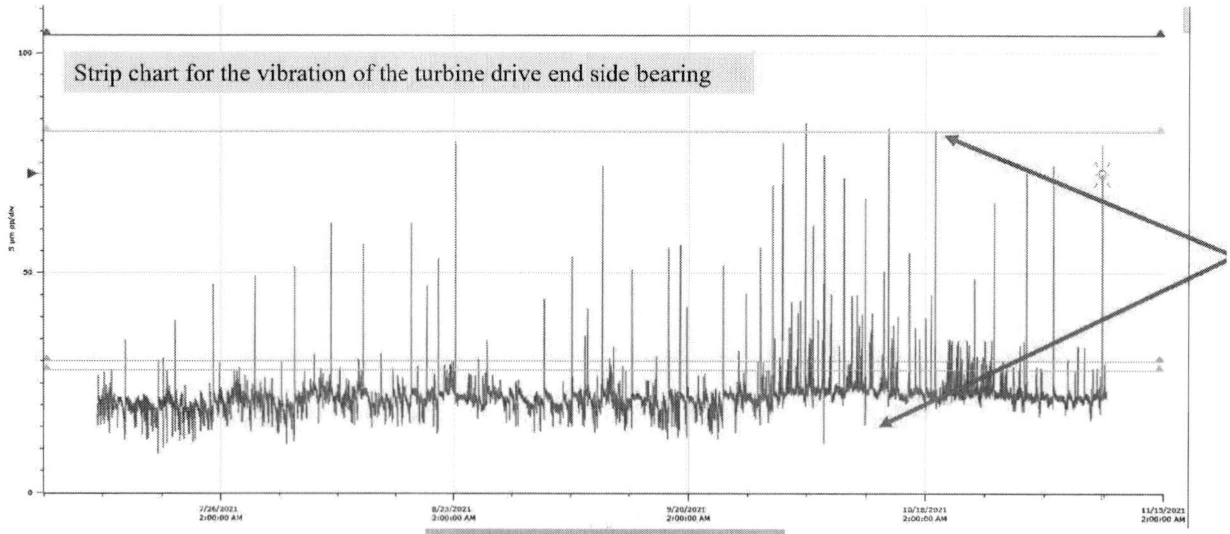

Figure 2 Vibration Strip Chart

These transient vibration values exceed the DCS first alarm (50 µm P-P). This enforces the operation team to lower the turbine speed and consequently the plant load decrease from 105% to 80%. This action is very costly for the plant.

The Operators return the turbine to its rated speed when the vibration reaches down to its normal value. This is illustrated in Figure 3.

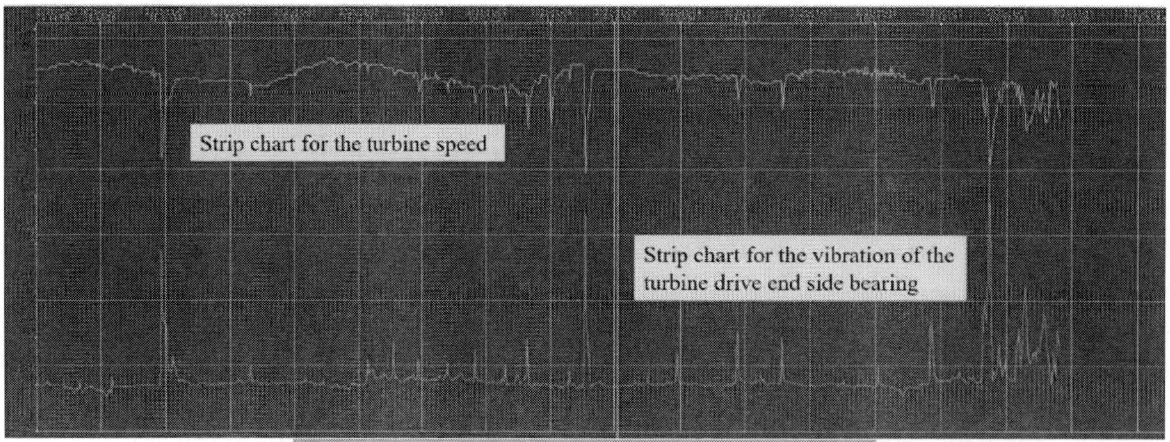

Figure 3 Vibration VS Turbine Speed Strip Chart

There were no trip values upon radial vibration, the trip values are only set for axial displacement, The radial vibrations only have 2 alarm levels (50 µm, 103.8 µm). This is shown in Figure 4.

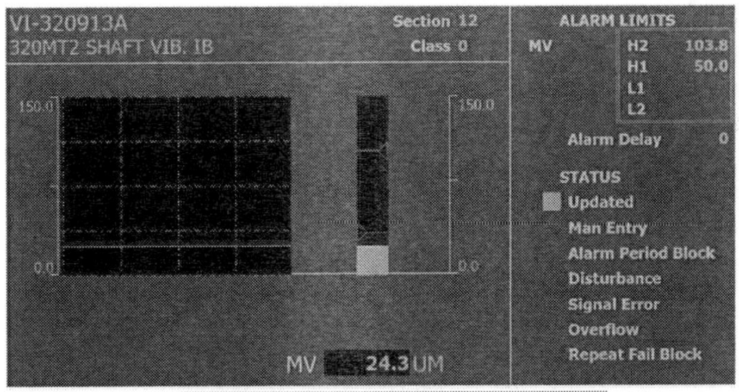

Figure 4 Vibration Alarm Levels

The average vibration values for turbine drive side bearing are 22 um P-P, so the vibration increasing to 100 um represents 5 times greater than its normal values. This phenomenon appears obviously in summer more than winter as well as it happens most often between midnight and early morning. Axial displacement and bearing temperature are not affected. The vibration values of the turbine non drive side bearing and the compressor LP stage DE bearing is not affected, so it is thought that the problem is localized at turbine drive end side as illustrated in Figure 5.

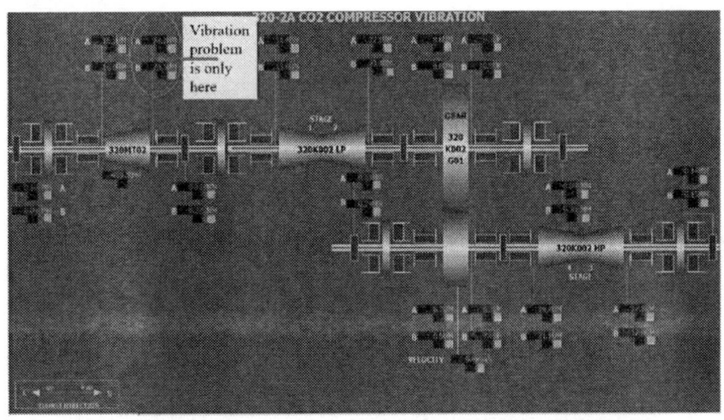

Figure 5 Vibration Localized for drive side side

LUBRICATION OIL:

The inlet lubrication oil temperature was 46^{0}C while it is stated in OEM manuals, that the lubrication oil temp shall be adjusted to 40^{0}C. Oil temperature at site versus oil temperature stated in OEM is illustrated in Figure 6.

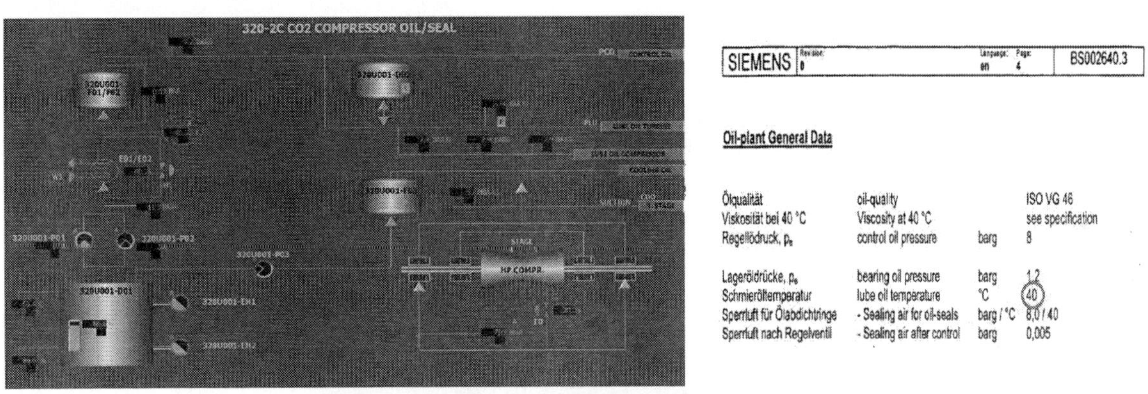

Figure 6 Inlet Oil Temperature (OEM vs Actual)

From Oil (Temperature -Viscosity) curve, the vibration analyst extracted that the viscosity will decrease from 46 cSt to 35 cSt in case the lubrication oil is 46^0 C instead of 40^0 C. That means that oil viscosity is decreased by 25%, oil curve is shown in Figure 7

MACHINE DATA/DETAILS:
The Vibration analysis Consultant started to study the machine details and all information about the Turbine Configuration, Construction, design, rotor bearing system & lubrication oil system.

Figure 8 shows the machine Configuration, while Figure 9 & Table 1 shows the machine construction.

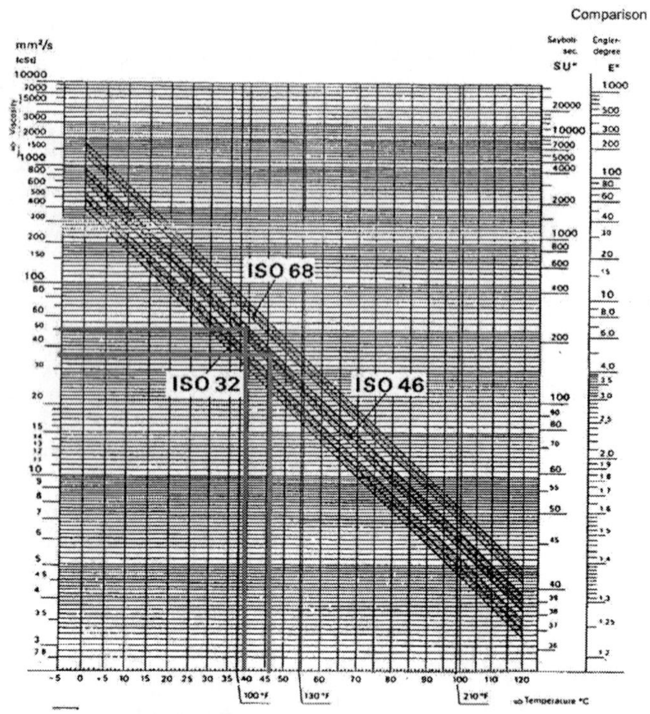

Figure 7 Oil viscosity change Vs Temperature

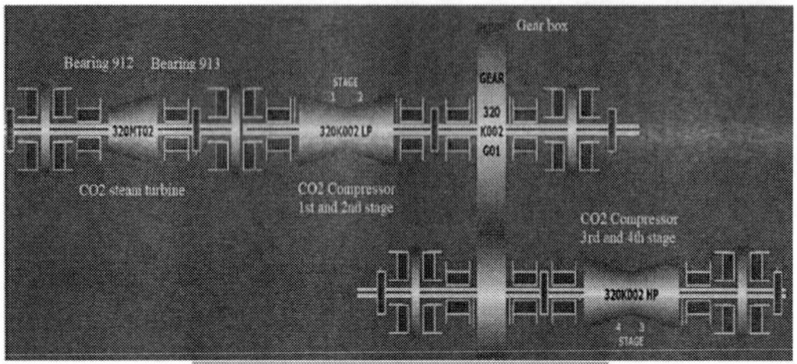

Figure 8 Machine Configuration

Figure 9 Machine Bearings at Site

Table 1 Machine Construction

Machine Name	CO2 Compressor steam turbine
Driven Machine	CO2 compressor (4 stages)
Turbine Manufacture	Siemens
Turbine Model Design	VE25
Turbine Type	Ext-Injection-Condensing
Turbine Radial Bearing Tag No	(Front end, non-Drive side bearing 912)
	(Rear end, Drive side bearing 913)

Beside Rotor and Bearing details, all operating /Design parameters are listed from the OEM manuals to compare it with the real operating parameters as shown in Table 2.

Table 2 Turbine Operating/Rotor Design Parameters

Turbine Design Details	Operating Parameters Details	Rotor Details	Bearing Details
Turbine Power: 10.2 MW	Inlet steam pressure: 107 bara	Rotor Total mass: 1134.2 Kg	Bearing Type: Tilting pad bearings
Turbine Rated Speed: 6655 RPM	Inlet Temperature: 505^0 C	Total shaft length: 2568 mm	No of pads: 4 pads LOP
Critical Speed: 4300 RPM	Extraction Steam Pressure: 23 bara	Bearing Distance: 1968 mm	Bearing diameter: 110 mm
Operating Speed Range: 5324 to 6988 RPM	Outlet steam pressure: 0.134 bara	Average Shaft diameter: unknown	Bearing axial length: 55 mm
Turbine Stages: 17 stages	Outlet Steam temperature: 50.8^0 C	Shaft diameter at bearing: unknown	Diametral clearance: 220 µm
Lubrication Oil Type: ISO VG 46	Lubrication oil Temperature: 40^0 C	Balancing Grade: G 2.5	Bearing Preload and offset: unknown

The machine cross section drawing is shown in Figure 10 shows that the transient vibration which happened in Rear side (Drive End Bearing 913) is located at the LP steam side.

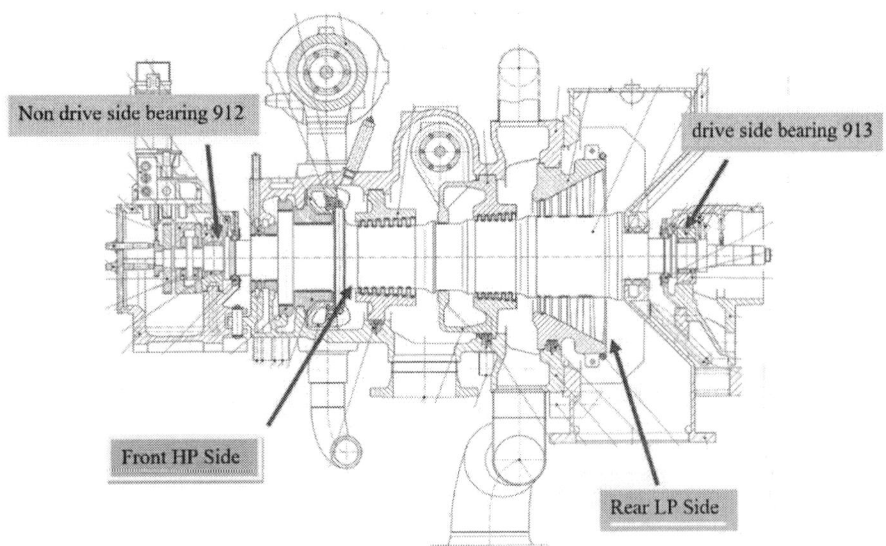

Figure 10 Turbine Cross Section Drawing

VIBRATION TRANSDUCERS:

The machine is only equipped with 4 Proximity probes 3300 XL with sensitivity 7.87 V/mm (2 orthogonal probes at each bearing) and 1 accelerometer on the gear box, all the Transducers are connected to Bentley Nevada 3500 Rack configuration.

No online System was installed, that's why vibration analysis team utilize a portable 4-channel analyzer which is capable to record vibration data for 48 hours, The vibration analyzer was left in Bentley Nevada Rack room aiming to catch a high transient vibration to start analyzing it.

A high transient Vibration values were recorded on 18 August 2021 to catch the vibration peak as to analyze it using Time, FFT spectrum, Phase, Direct orbit, Bode and Polar & shaft centerline Plots., The vibration increased at 00:15 AM (Early morning), the turbine speed was 6648 RPM. Probe 913 vibration data was analyzed carefully.

VIBRATION ANALYSIS PLOTS:

<u>Trend (Strip chart):</u> The overall vibration reached 63 μm as it is illustrated in the strip chart. The time until the vibration reaches its normal value again is 20 minutes as shown in Figure 11.

<u>Time Plot:</u> Time plot during maximum vibration shows pure sinusoidal wave, phase between the 2 probes is nearly 90 degrees as illustrated below in Figure 12.

<u>FFT Spectrum:</u> FFT spectrum plots during maximum vibration show dominant 1X vibration (110.8 HZ coming from machine speed). This is clearly shown in Figure 13.

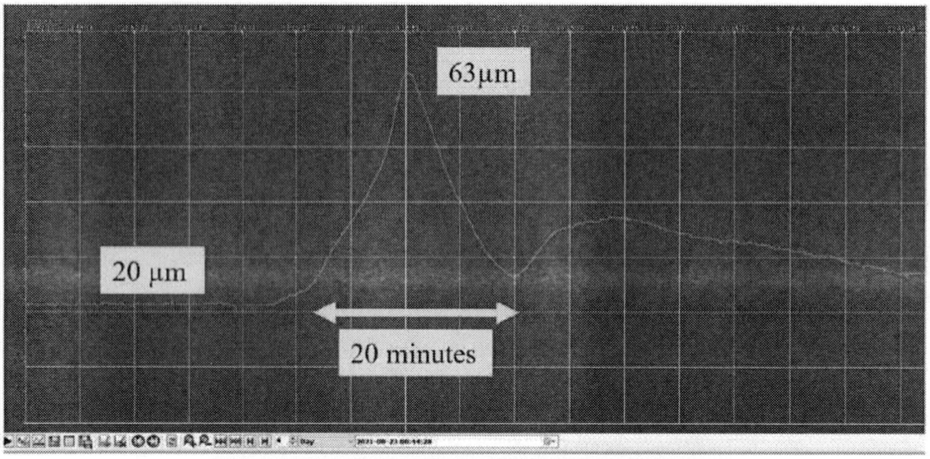

Figure 11 High Transient Vibration Event

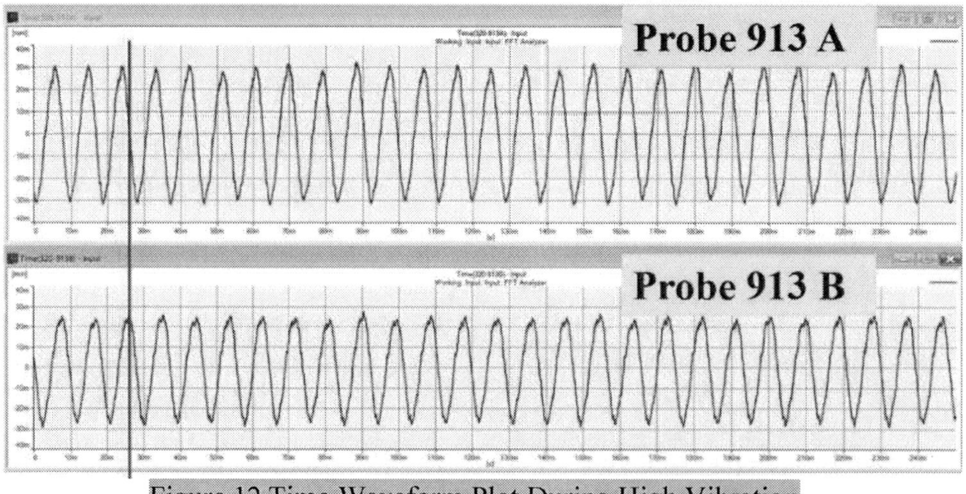

Figure 12 Time Waveform Plot During High Vibration

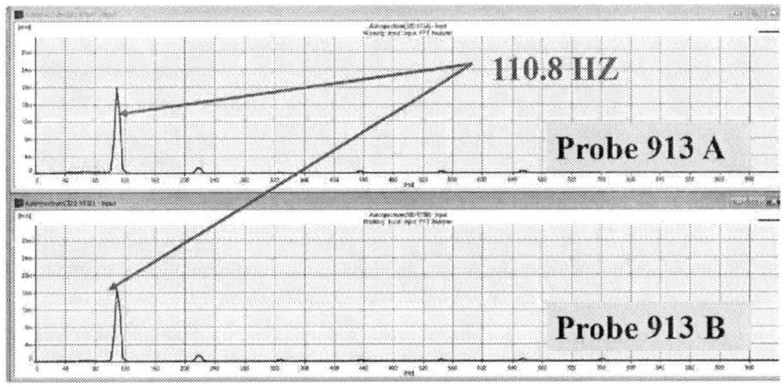

Figure 13 FFT Spectrum Plot During High Vibration

Orbit Plot: Orbit plot during maximum vibration shows high vibration Smax with minor preload as illustrated, Preload Source appears as coming from the bottom right side as shown in Figure 14.

Bode Plot: The 1 X magnitude increase while its phase shows a high variation reaches 60 degrees at maximum vibration. The speed during the high vibration event was constant (6648 RPM) as shown in Figure 15.

Polar Plot: Same result as Bode plot, The polar shown plotted in Figure 16 is at constant speed (6648 RPM), It shows amplitude maximum peak and phase shift about 60 degrees.
Shaft Centerline (SCL): The Shaft Centerline (SCL) shows minor change, less than 1 μm, the shaft location slightly moved to the left bottom then returns back again to its normal position as illustrated in Figure 17.

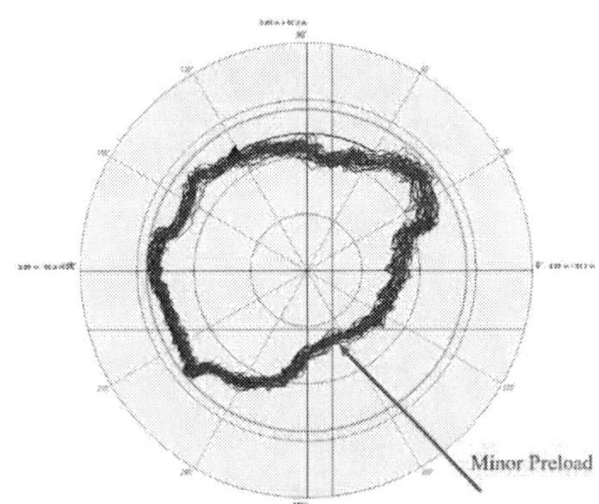

Figure 14 Orbit Plot During High Vibration shows preload

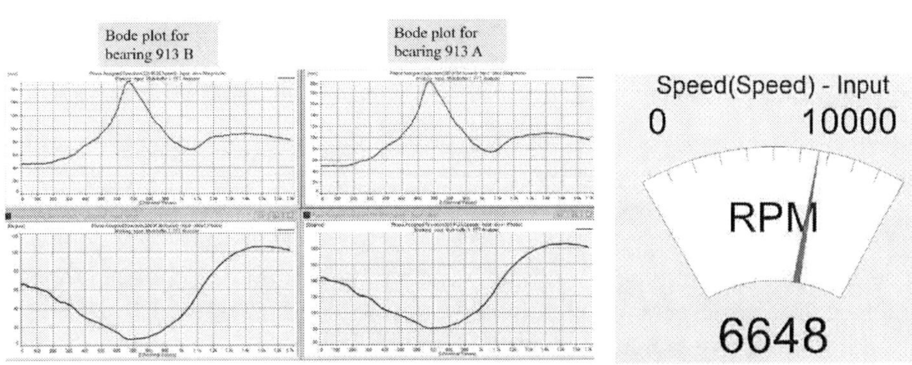

Figure 15 Bode Plot During High Vibration

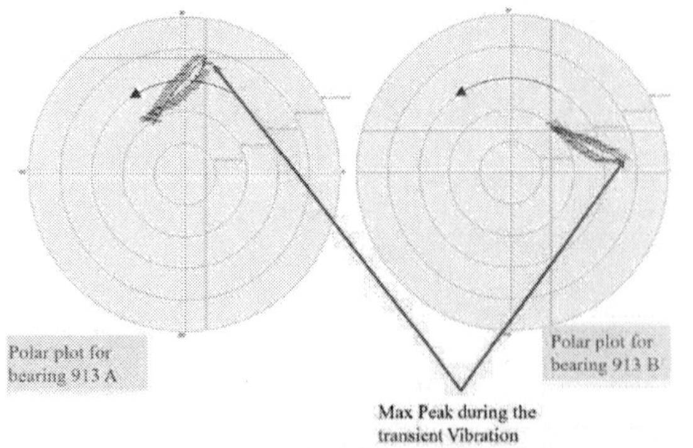

Figure 16 Polar Plot During High Vibration

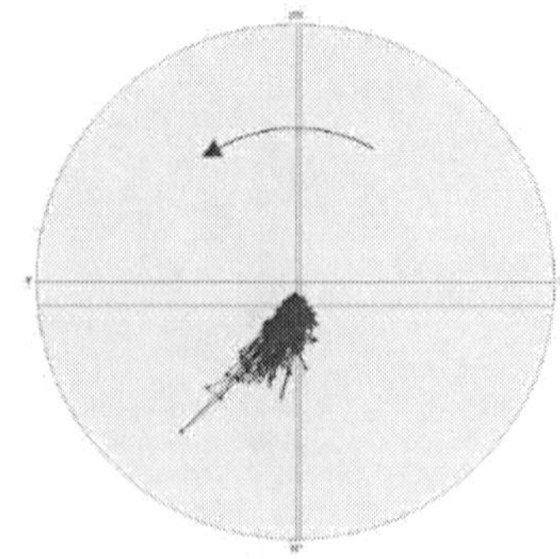

Figure 17 Shaft Centerline

Beside the vibration plots, the vibration analysis specialist then started to review all operating parameters values and trends during the high vibration event, everything was normal as shown in Figure 18.

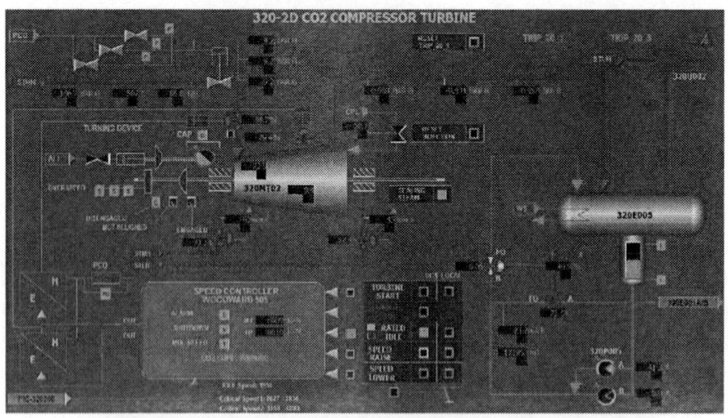

Figure 18 Operating Parameters Review at Site

1X VIBRATION:

The 1X amplitude and phase for Vertical and Horizontal probes at normal vibration and also at maximum peak of the transient vibration are listed in table 3, It was noticed that a phase difference change between the 2 probes from 80^0 to 73^0 which confirms the minor preload during high vibration.

Table 3 1X Vibration Amp & Phase variation During High Vibration

1X value	Probe 913 A	Probe 913 B	Phase Difference Between A and B Probes
Amplitude and phase at Normal vibration	14 µm (P-P) 145^0	14 µm (P-P) 65^0	80^0
Amplitude and phase at Maximum peak of transient vibration	56 µm (P-P) 80^0	50 µm (P-P) 7^0	73^0
The values above are at same turbine speed (6648 RPM)			

Due to the presence of high 1X Component in the FFT spectrum, this may lead the analyst to suspect mass unbalance problem especially that the vibration is localized at the LP side (condenser side) and it is well known that in steam turbines the blades of LP stages are suspected to be eroded due to high moisture in this area.

Misalignment/Coupling Lockup also may be suspected as a minor Preload is found and vibration problem is near the coupling, but if the vibration is coming from mass unbalance or misalignment, that will cause a high steady vibration, not transient vibration as in that case.

VIBRATION ANALYSIS RESULT:

The vibration specialist suspect Rubbing problem due to: (1) Preload found in orbit plot, (2) Phase angle difference between probe A &B changed from 80^0 at normal vibration to 73^0 at maximum vibration, (3) Bode & Polar plots confirm a phase change during rubbing action, it is thought that phase angles return back to its normal values after rubbing action finish, (4) Minor change in shaft centerline during maximum vibration.

MAINTENANCE TEAM QUESTIONS:

After the analysis report was delivered to maintenance team, some questions were raised by them as follows:

Q1. Why the rubbing phenomenon appears suddenly and disappears after a period from 20 to 30 minutes and why it appears from time to time?

Q2. Why in some cases the vibration is transient decreasing instead of transient increasing?

Q3. Why the rubbing action is high only at turbine drive side?

Q4. Where the rubbing action take place? At the bearing? Labyrinth or inside the turbine itself?

Q5. Is it dangerous to continue operating the machine in this condition?

Q6. Do we need to stop the machine and change the bearings?

So, to answer all these questions, the vibration analysis specialist asks for a site inspection and all machine drawings and operational data.

SITE INSPECTION:

During the site inspection, the machine was running in a steady condition, the vibration specialist notice visually the presence of oil coking and shiny part on shaft which confirm that a rubbing action take place between rotating shaft and hard oil coked layer as shown in Figure 19, The presence of oil coked layer in this area is a result of oil leak.

Also, during the site inspection, it was noticed that oil leakage indications is found on the turbine condenser, The condenser is 5 meters lower the turbine drive side bearing (LP Steam side). This is shown in Figure 20.

Figure 19 Rubbing Sign under Coked oil

Figure 20 Oil Leakage indication

Machine Drawing:
By revising the machine drawing, it was found that the Oil Coking layers are found under oil retaining ring labyrinth as shown in Figure 21.

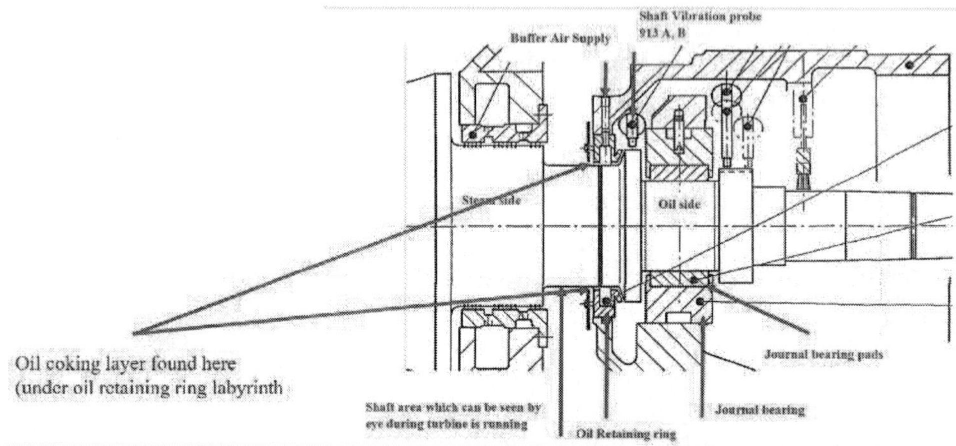

Figure 21 Oil Coking Layer Location Shown on Turbine Cross-Sectional Drawing

High Transient vibration Sequence:
After the vibration analysis and site inspection took place, the analyst confirmed that an oil leakage from the bearing area cause some oil coked layer under the labyrinth of oil retaining ring.

When this coked area thickness increase by time, the shaft which is vibrating in its orbit path will touch this hardened coked layer (rubbing action), This will cause a hot thermal spot at a certain shaft area and the shaft is thermally bowed, this will sound a high 1X vibration and the vibration will go up.

Phase is changing as the hot thermal spot-on shaft acts as a new heavy spot (kind of mass unbalance), this will create a new resultant heavy spot and consequently a higher different high spot, After the coked layer broke down to pieces and the shaft cools to its normal temperature, vibration will go down to its normal values as the shaft is no longer thermally bowed. Figure 22 shows the high transient vibration sequence.

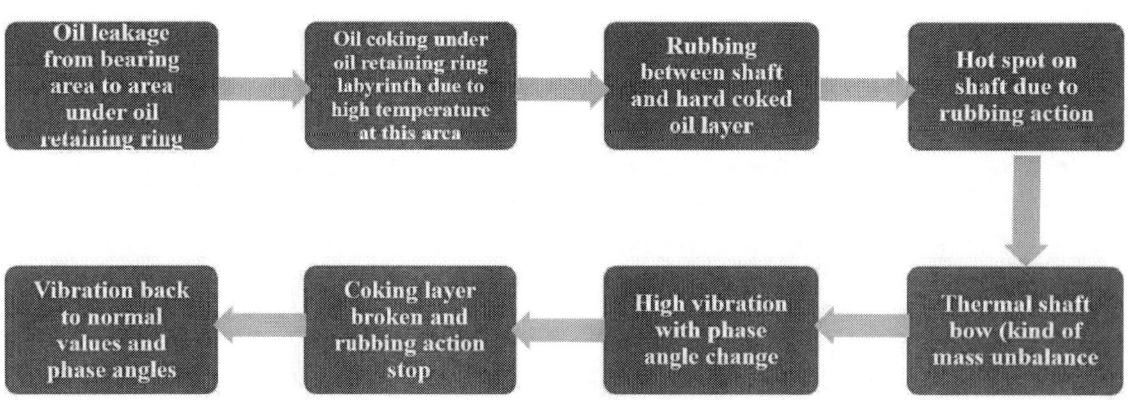

Figure 22 High Transient Vibration Sequence

Answering The Maintenance Team Questions:
We can now answer the 6 raised questions by maintenance team

Q1. Why the rubbing phenomenon appears suddenly and disappears after a period from 20 to 30 minutes and why it appears from time to time?
A1. This is because the coked layer is formed in a random way, when it fills the clearance around the shaft, rubbing action takes place which will cause thermal bow (kind of mass unbalance) and all of this will vanish after these hardened coked layers broke and the shaft cools down so the vibration decreased again

Q2. Why in some cases the vibration is transient decreasing instead of transient increasing?
A2. the resultant new heavy spot is random depending upon on which part of the shaft circumference the rubbing action will take place, so the resultant heavy spot may be less than the one already exists in the machine and consequently the vibration may decrease.

Q3. Why the rubbing action is high only at turbine drive side?
A3. Because the oil is leaking only from the oil retaining ring at that side and the hardened coked oil is visually seen at this area.

Q4. Where the rubbing action take place? At the bearing? Labyrinth or inside the turbine its self?
A4. Now it is obvious the rubbing is at the circumference under the oil retaining ring labyrinth, one may estimate where is the coking at every high vibration event from orbit shape preload side reference to X, Y vibration probes as shown in Figure 23.

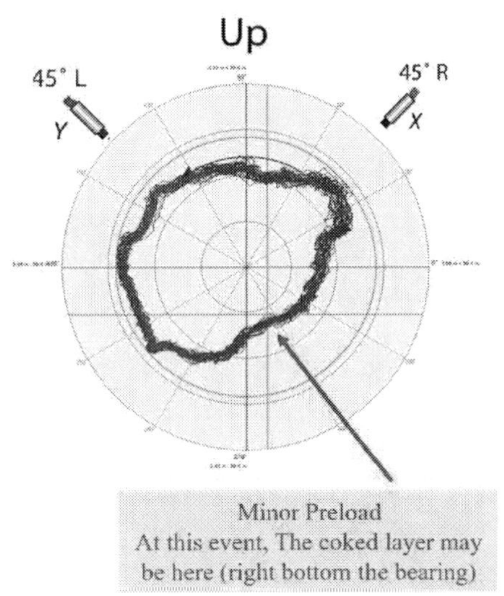

Figure 23 Orbit shows High Preload may indicate Location of Coked Layer

Q5. Is it dangerous to continue operating the machine in this condition?
A5. Running the compressor in this condition is not dangerous but with Monitoring the vibration values and plots daily and applying some recommendations to reduce the rubbing action.

Q6. Do we need to stop the machine and change the bearings?
A6. The bearings are good because the vibration when lowered, it decreases to its normal value (i.e., the bearings are acting well) plus that the shaft centerline measured at previous startup shows that the eccentricity ratio = 0.15 + knowing the diametral clearance (220 μm) confirm that even at high vibration (100 μm P-P), the shaft still far from bearing pads.

Additional Site Inspection:
A power failure had stopped the plant in late September 2021, So the maintenance team decided to open the bearing cover to have a look on the bearing pads and to measure the bearing/oil retaining ring clearances and also to inspect the shaft area under the oil retaining ring.

The vibration specialist was invited to attend the maintenance actions and to do extra site inspection on the machine. The turbine had to operate at turning gear speed for 48 hours before inspection as to avoid any thermal bowing in turbine rotor. The oil retaining ring show coked oil layers which is assigned to be the cause of high vibrations as shown in Figure 24. No spare oil retaining ring was available so it was not changed but just only cleaned.

Deep Scoring signs at the shaft area under the labyrinth of oil retaining ring was found as shown in Figure 25, this may be a reason for enlarging the area under the labyrinth and may be a possible cause of the oil leakage.

Maintenance team checked the bearing clearance and it was OK and the pads were in a good condition as shown in Figure 26 but although everything in bearings appear normal, they decided to change the 4 bearing pads as they believe that this may enhance the situation. After running the machine again, the phenomenon still the same as it is.

Figure 24 Coking Layers under Oil Retaining Ring

Figure 25 Deep Scoring Signs on Shaft

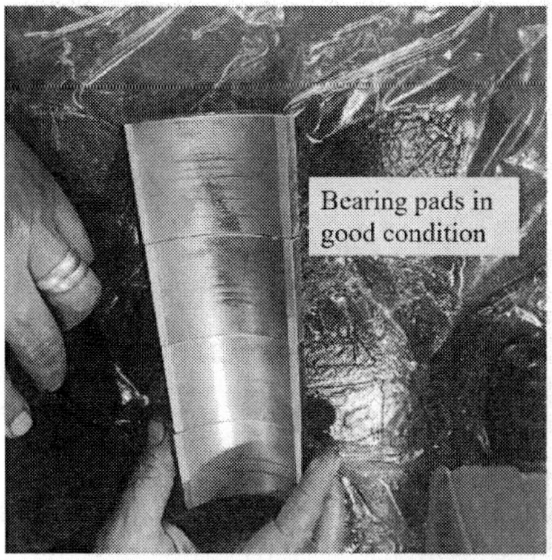

Figure 26 Bearing Pads in Good Condition

Vibration Control:
The vibration analyst was asked if he can advise some recommendations to reduce / eliminate this transient vibration especially that the plant operators reduce the turbine speed during these events and consequently the load of the plant decrease, this situation was costly for the plant.

Regarding that Controlling vibration techniques, they are only five which are: Isolation- Mass Addition-Tuning - Damping and Force reduction (Figure 27)

The analyst has to think what technique he shall use in this case, The vibration analyst decided to apply the last two methods (Damping and Force Reduction) to control this high transient vibration, Reducing the opportunity of oil coking formation plus increasing the damping provided by bearing oil is the way to control this high transient vibration.

Recommendation:
The vibration analysis and Site inspection Confirmed that the machine is suffering from light rubbing due to oil coking so the analyst recommendation was as an attempt to control this rubbing phenomenon. As an attempt to reduce/eliminate the oil coking, a nitrogen purge line was installed between the oil seal labyrinth area and turbine side, this may lower the temperature at this area, unfortunately this is only available at the upper half but not for the lower as shown in Figure 28.

The- turbine drive side bearing -buffer air pressure had to be checked to reduce the possibility of oil leakage, oil leakage is not only cause for high vibration but may cause fire due to oil spill on hot turbine components.

It was also recommended to operate the turbine at inlet lubrication oil = 40^0 Celsius as to reduce the opportunity of oil coking formation and increase bearing damping (i.e., increasing the viscosity will accordingly increase bearing oil damping).

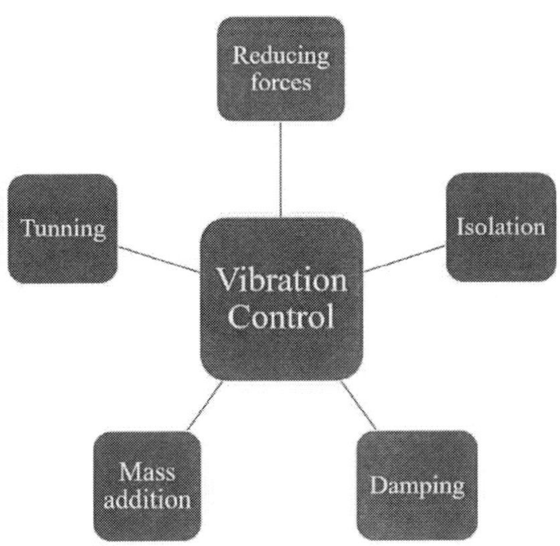

Figure 27 Five Methods of Vibration Control

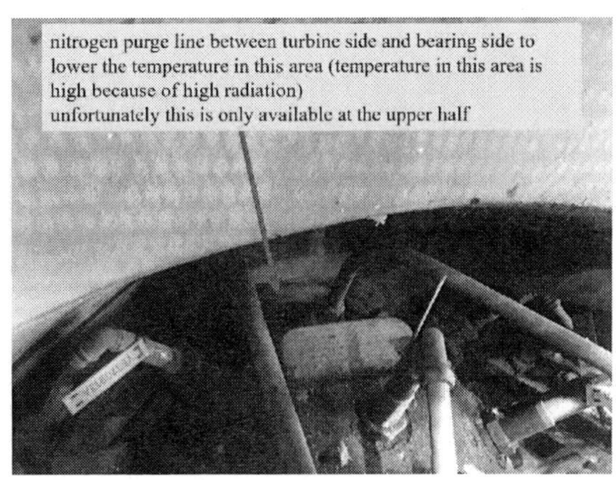

Figure 28 Nitrogen Purge on Bearing

Result:
After applying these recommendations, the phenomenon of this high transient vibration did not vanish but its repetition and maximum values were reduced.

The customer will change the rotor in the next plant Shutdown (June 2022) as the scoring on the shafts shall be repaired at vendor (Siemens) facility in Germany.

Also, during the next shutdown, the customer will change all internals (Bearing pads-oil retaining ring, labyrinth, ...etc.)

Root Cause:
Oil leakage was responsible for this vibration problem as this high transient vibration is coming from rubbing action between the shaft and hardened oil coking layers.

The root cause of this oil leakage may be: (1) Deep scoring signs on shaft under oil retaining ring labyrinth, (2) Turbine bearing to bearing misalignment which should be checked in next turnaround, (3) Faulty oil retaining ring (may have some ovality).

Conclusion:
In this case study, A high transient vibration caused an unstable operating condition for a big Petrochemical company. It is obvious that in the case of transient vibration presence, vibration analyst should focus on Phase angles variation, Trends (DC gab + Direct vibration), Plots (Time waveform-FFT spectrum- SCL & Bode/Polar).

Site inspection-Machine drawings & Operational parameters play a very important role with the vibration data, beside requiring high skills and experience, Controlling the vibration needs high knowledge by all machine parameters and operating conditions and all possible ways to control the vibration.

From Orbit plot, analyst may be capable of knowing the physical location of the hardened coked layer which cause the vibration at certain events.

WIRELESS VIBRATION MONITORING AS A COMPLIMENT TO MONTHLY VIBRATION ROUNDS

Author: Chad Dyson

VIATC | TRAINING CONFERENCE & EXPO
SAVANNAH, GEORGIA · 2022

WIRELESS VIBRATION MONITORING AS A COMPLIMENT TO MONTHLY VIBRATION ROUNDS

Chad Dyson,
Application Specialist - MHM
Everactive
2986 Oakmead Village Ct
Santa Clara, CA 95051

Vibration analyst Cat. III. Over 18 years in the vibration industry in varying roles from team analyst, lead analyst, Sales consultant, and online systems SME.

Chad.Dyson@everactive.com

ABSTRACT:
With the increase in availability of wireless vibration sensors and programs becoming more cost-effective, the question of how this fits with a monthly vibration round program.

- Most vibration programs tend to collect data monthly. The monthly rounds are performed on roughly 80% of the equipment, the balance of plant equipment. These machines typically are smaller in size and are not critical enough to warrant an online machine protection system.
- Wireless vibration monitoring is becoming more available and more cost effective to deploy.

KEY WORDS: Analysis, data collection, facilities, frequencies, vibration monitoring, wireless

INTRODUCTION:
Monthly vibration rounds typically provide one snapshot of all running equipment. The readings typically are fine tuned to each machine and make the most of the capabilities of the data collector. The ability to focus on exact areas of the frequencies in question can provide a more in-depth analysis of the monitored equipment. Typically fault frequencies are built into the program to provide accurate analysis of faults detected in the monthly readings. Alarming is typically performed to get an analyst attention to changes over the previous readings and help dictate what machines should get the focus of a deeper analysis.

Wireless vibration monitoring usually consists of vibration monitoring of the balance of plant, roughly the 80% of all the equipment in a facility. Wireless vibration monitoring typically is collected to provide a general overall health of the monitored equipment either up to the minute, hour, or daily with more detailed data collected less frequently, once daily or monthly. The vibration data is usually displayed in some form of dashboard providing multiple different levels of data depending on who is reviewing the data.

If only using monthly vibration rounds to collect data, there can be large gaps in between data collection. 29 days, 23 hours and 50 minutes in some cases. If wireless vibration monitoring is only being performed, there can be a less prescriptive approach for machine repairs. High vibration levels and in some cases low resolution spectrums leading to general repair scopes lacking more detailed direction for the repair.
These gaps can lead to high repair cost, unexpected failures, downtime, and even environmental fines in some facilities. The workflow for the analyst should change with the addition of a wireless monitoring program.

WIRELESS MONITORING CONSIDERATIONS:

- The first check in the morning should be of the wireless vibration system dashboard.
- Review any alarms from the previous day/night.
- Document the machines in alarm and form a plan for analysis.
- Does the analyst need to visit the machine to collect more robust data, or are there enough capabilities in the wireless program to be able to make an accurate diagnosis? Determine if the machine has exhibited these patterns and or alarm states previously.
- Is the machine currently running? Can a collection be triggered at that moment, trigger a collection if possible?
- Perform analysis of machine in question using the wireless software or handheld collected data.

DOCUMENTATION:

Some things to consider part of the documentation process are: Number of alarms each day, Is the alarm status and level, correct? Should the alarm level be adjusted and if so, why the change is being made? Keeping this record may be required for your organization. Think MOC (Management of change). Who reviewed the data? It is recommended that a failure report be generated for each finding. This report should be sent along with any recommendations for a repair.

USING WIRELESS VIBRATION MONITORS:

Elevating the wireless data for other software and applications for use should also be considered. With advancements of machine learning and AI, a wireless vibration solution can be a great source of data for these types of applications. Don't keep this data siloed. Moving data out of the program and into others will provide and extra set of "eyes" looking after the health of the equipment. Consider sending some of the more important machines vibration levels to operations and their DCS or PLC. Especially the machines that have the likelihood of elevated vibration caused by changing operating conditions.

A benefit of having a wireless vibration monitoring system is that there should be no more overdue vibration readings due to the analyst not being present when the machine was last operated. A wireless vibration system should collect data when the machine is run automatically. In some plants, this is KPI used to score how well a vibration program is performing. We can use this data to elevate concern that a machine has not run and should be exercised. If the spare has high vibrations, we need to address this and perform a corrective action.

DEALING WITH CHANGE:

Culture change is a major part of merging the two systems. At some point making monthly rounds will be eliminated. There will be concerns over not needing someone to collect the data and not needing a vibration analyst. The goal should be to have the analyst performing a high value task of analyzing the collected and not having to collect it.

We will free up the analyst to perform more analysis. There will be more data collected, more frequently. There is also very high value in visiting the machine. Defects like oil leaks, seal issues, high temperatures on machines, water leaks, and other outside factors will still need to be addressed. Performing weekly walk-throughs are highly valuable. This will typically be done in the morning, a handheld vibration meter and or temperature gun can be brought along for spot checks of equipment in question.

Things to consider when adding a wireless vibration monitoring solution to your site:

- **Equipment Choices**

Consider the machine speeds and component types inside of the equipment. Make sure they match with the proposed solution. Is the solution capable of collecting valuable data for the machine in question?

- **Infrastructure**

Is there IT support for the proposed solution? Where is the data stored? Onsite or offsite? If onsite, work with the local IT team early in the process. Ensure the proposed solution will align with the local site policies. Discuss the possible systems that could use the collected data and coordinate what level the data reside. (Think Perdue model). If the data is stored offsite, what is the method for getting the collected data offsite and back onsite? Some solutions can send the data offsite without local IT infrastructure. These solutions are great for remote locations. If this is the proposed solution, consider what IT requirements are needed to keep the local network safe.

- **System Maintenance**

Consider how the system will need to be maintained and who will be providing that support. Most wireless solutions require batteries. Consider up to 10% of the installed sensors to have batteries in the warehouse. Some of the time savings seen in not having to collect the data could be used to provide upkeep and maintenance on the wireless vibration system. Some wireless solutions are now not requiring batteries to power the sensor.

- **System Access**

The local vibration analyst is a likely user of the system daily. Consider having access for outside contractors and consultants. Managers, engineers, and key stakeholders are another likely user of the data. Consider using different levels of login access to restrict certain users from making changes. Managers are likely to use the dashboard feature to get an update of overall machine health in the facility. Engineers would use the access to direct the vibration analyst to questionable equipment.

- **Battery life/collection expectations**

What is the battery life expectation of the solution and what is the collection rate? What is the ideal data collection for your program? Ensure the solution fits the local vision. Is an overall reading once a day ok, or once a minute? Consider that upping the collection rate will have an effect on battery life of the solution. How often will the solution be asked to collect an reading outside of the normal collection schedule? This task is most likely to have a big effect on the battery life.

- **Installation methods**

Consider how the sensor is best mounted for your site. Drilling and tapping the sensor gives the best level of data transmission over all frequency ranges. This is not always possible given the size and type of machine. This solution is the most labor intensive and intrusive of all the mounting solutions. Epoxy is another method to consider. Easier to install than drilling and tapping. It also has great data transmission characteristics the frequency ranges expected. Using magnets will be the quickest and easiest mounting method. Magnets are less permanent and can easily be knocked off by personnel performing maintenance on the machine. This does make moving the sensor to a new location easier and gives flexibility to the solution.

THE FUTURE OF VIBRATION ANALYSIS – UNDERSTANDING MEMS (MICRO-ELECTROMECHANICAL SYSTEMS) TECHNOLOGY

Author: Dr. M. David Howard, CRE, CMRP, CSSBB

VIATC TRAINING CONFERENCE & EXPO
SAVANNAH, GEORGIA · 2022

THE FUTURE OF VIBRATION ANALYSIS – UNDERSTANDING MEMS (MICRO-ELECTROMECHANICAL SYSTEMS) TECHNOLOGY

Dr. M. David Howard, CRE, CMRP, CSSBB
CEO – ERBESSD INSTRUMENTS
19 Exchange Street, Glens Falls, NY 12801

Michael Howard, D.Sc is an American entrepreneur, a combat veteran of the United States Air Force, and respected leader in the predictive maintenance industry. Michael is an avid CrossFit® athlete, CrossFit® CF-L1 Trainer and passionate advocate of revolutionary concepts in the wireless instrumentation and the IIoT communities for the maintenance & reliability industries.

Dr. Howard is a native of South Glens Falls, New York and a graduate of Excelsior College, Capella University, & New Charter University with degrees in Electro-Mechanical Engineering, Leadership, & Organizational Management, & Engineering Management. Mike is a Certified Reliability Engineer, Six-Sigma Black Belt & Certified Maintenance & Reliability Professional. Dr. Howard is the CEO of Erbessd Instruments and is responsible for Strategic Direction, Distribution, Sales, Marketing and Operations throughout the English-speaking markets.

ERBESSD INSTRUMENTS is a leading manufacturer of Vibration Analysis Equipment and Dynamic Balancing Machines with facilities in Mexico, Canada, Europe, and the United States

ABSTRACT:
Vibration Analysis has long been established as one of the pedigree technologies utilized to diagnose and correct a variety of rotating equipment machine faults. Traditionally ICP accelerometers have been the natural choice for this type of analysis, however recent breakthroughs in industrial MEMS technology have significantly reduced the cost of entry for online machine health monitoring, thus challenging the status quo within the vibration analysis community.

It is essential that users, system integrators and other stakeholders clearly understand the benefits and challenges associated with deploying MEMS based technology for online condition monitoring.

We will explore the unique nature of MEMS technology and how it compares with existing ICP technology so that informed deployment and analysis decisions can be made based on a solid understanding of the fundamental differences between the two technologies.

KEY WORDS: Accelerometers; ICP accelerometers, Machine health monitoring; MEMS, Vibration analysis

MACHINE HEALTH:

According to (Dictionary, n.d.), machines are defined to be a mechanically, electrically, or electronically operated device for performing a task or as an assemblage of parts that transmit forces, motion, and energy one to another in a predetermined manner as depicted in fig 1. Thus, machines are critical to diverse fields ranging from the automobile industry, the food industry, clothing industry, chemicals industry and many others. Majority of machines consists of synchronized movable parts; however, there are machines which might not have movable parts in the true essence of their operation mode such as electronic machines. Electromechanical machines are however more commonplace and have seen applications in many fields from aviation, automobile, robotics, space exploration, industrial machinery etc.

Figure 1: Typical Machines

The health of machines are critical, and thus it is vital that the state of operational health of all parts that constitutes a particular machine are determined to ensure longevity and optimal performance. Machine health inspection and evaluation is mostly paramount for machines operating in critical spheres of life such as aviation, nuclear power systems, military equipment etc. According to (Monk, 1973), Vibration analysis has long been regarded as an indicator of the condition of rotating machinery. The study went further to examine applications of vibration monitoring to industries employing large numbers of process machines, which are engaged in expensive processes.

Vibrations in machinery can take various forms, and most of the time these vibrations are unintended and undesirable. Therefore, monitoring the vibration characteristics of a machine can provide the information of its health condition, with this piece of information being used to detect problems that might be incipient or developing (Arturo, et al., 2016). Thus, machine condition monitoring systems are integral for the monitoring of machine health, in order to detect faults at the incipient stage or at any other stage, before it leads to the overall failure of a system. Machines on operation will give out vibration signatures which can be detected and measured with a machine condition monitoring system (Arturo, et al., 2016), the signatures can thus serve as input for processing in order to extract features that relate to a fault or a diagnosis of the machine's condition even if the machine is currently functional.

(Nabby, et al., 2018) Also identified Vibration analysis as an effective tool for identifying and predicting failure in machines, using condition monitoring signal processing. The study also examined three types of non-stationary signals to determine the health of a paper machine forming roll. The forming roll consisted of frequency drive motor, gears and a bearing supported symmetric rotor. Further examinations using vibration analysis resulted in their submission that Impact demodulation is recommended to specify faulty bearing source location and severity, while time waveform or spectrum analysis techniques can detect fault progress but couldn't distinguish severity due to bearings similarity as well as advising on continuous machine monitoring for 100% health.

Similar vibration analysis studies of machine health were carried out by (Brown, et al., n.d.) On paper mills, (Mohamadi Monavar, et al., 2008) on electro-motors at rpm levels of 500, 1000 and 1500 respectively, (Manish, et al., 2017) presented a review of some vibration feature extraction methods applied to different types of rotating machines.

The highlighted studies above indicate the efficacy of vibration analysis as an integral part of a machine condition monitoring system. (TWI, n.d.) defines vibration analysis as a process that monitors vibration levels and investigates the patterns in vibration signals. It is commonly conducted both on the time waveforms of the vibration signal directly as depicted in fig 2, as well as on the frequency spectrum as depicted in fig 3, which is obtained by applying Fourier Transform on the time waveform.

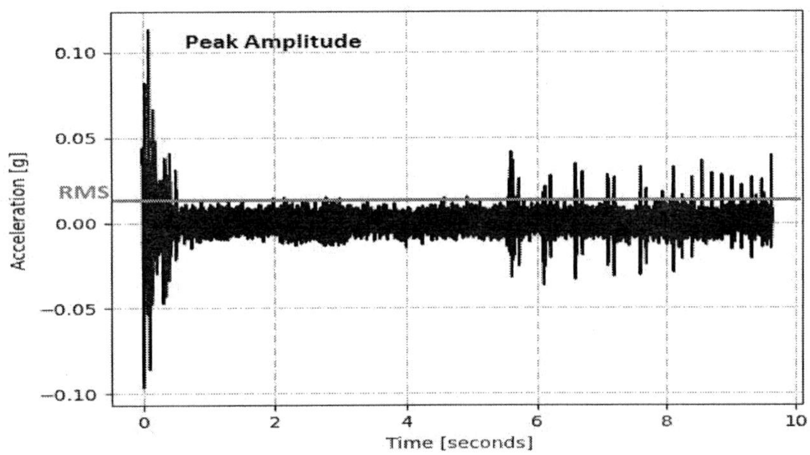

Figure 2: Vibration Time Waveform captured of a Moving Mechanism (TWI, n.d.)

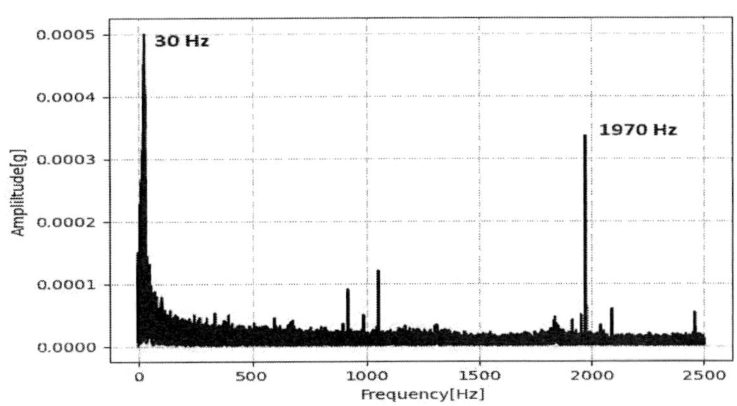

Figure 3: Vibration Frequency Spectrum from a moving mechanism (TWI, n.d.)

A vibration monitoring system is thus a complete system that is capable of acquiring vibration signals according to pre-determined parameters such as sampling frequency, vibration level, recording length, recording intervals and frequency bandwidths. But there must be some means by which vibration data can be measured directly from the machine in order for vibration analysis to take place, thus where accelerometers come into the picture.

ACCELEROMETERS:
According to (Tinder, n.d.), an accelerometer, which is depicted in fig 4, is a tool that measures proper acceleration, where Proper acceleration is the acceleration (the rate of change of velocity) of a body in its own instantaneous rest frame. There are two general classes of accelerometers: the AC-response accelerometer and the DC-response Accelerometers.

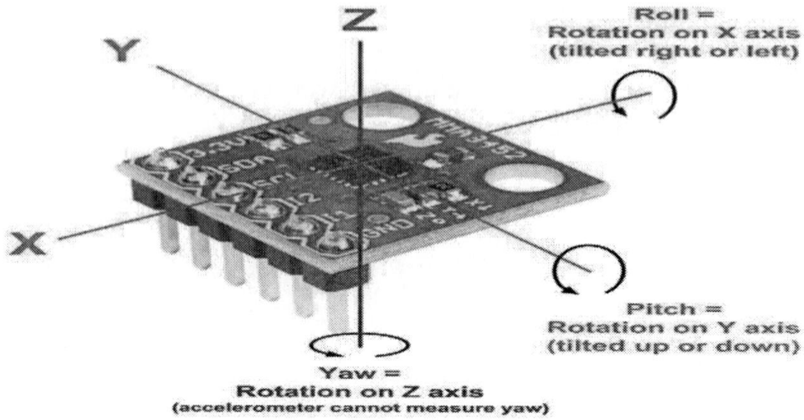

Figure 4: An Accelerometer

According to (TE Connectivity , n.d.)In AC-response accelerometer as depicted in fig 5, as the name implies, the output is AC coupled. An AC coupled device cannot be used to measure static acceleration such as gravity and constant centrifugal. In a DC-response accelerometer as depicted in fig 6, is dc coupled, and can respond down to zero Hertz. It therefore can be used to measure static, as well as dynamic acceleration. Measuring static acceleration is not the only reason a DC-response accelerometer should be selected acceleration

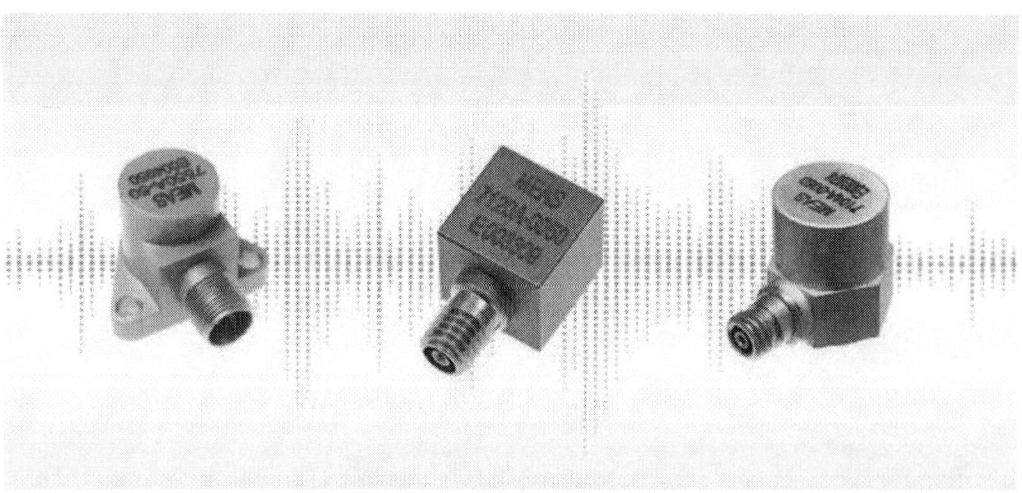

Figure 5: AC Accelerometers

Figure 6: DC Accelerometers

Two popular sensing technologies are used in making DC-accelerometers, namely the Capacitive and Piezoresistive technologies. While the most common AC-response accelerometers use piezoelectric elements for their sensing mechanism.

There are two types of piezoelectric accelerometers available on the market, namely the charge output type and voltage output type. Most piezoelectric sensors are based on lead zirconate titanate ceramics (PZT) (TE Connectivity , n.d.) Which offer very wide temperature range, broad dynamic range, and wide bandwidth (usable to >10kHz).

MEMS ACCELEROMETERS:

MEMS stands for micro electromechanical system and applies to any sensor manufactured using microelectronic fabrication technique (PCB Piezoelectronics, 2020). According to (Matej, 2008), MEMS accelerometers are one of the simplest but also most applicable micro-electromechanical systems. They find application many areas of application such as aviation, automobiles, computer, and audio-video tech industries. Typical MEMS accelerometers as depicted in fig 7, are composed of movable proof mass with plates that are attached through a mechanical suspension system to a reference frame.

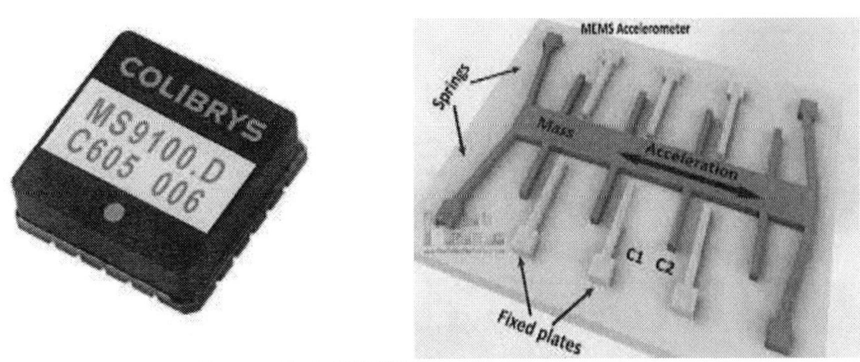

Figure 7: MEMS Accelerometer

According to (Alhussein, et al., 2008) performances of three of the MEMS accelerometers from different manufacturers are investigated in this paper and compared to a well calibrated commercial accelerometer used as a reference for MEMS sensors performance evaluation

Tests were performed on a real CNC machine in a typical industrial environmental workshop and the achieved results indicated that MEMS accelerometers has high sensitivities at low and high frequencies respectively, the MEMS accelerometer showed a linearity of $\pm1\%$ per 1Khz and for 5Khz.

MEMS are great for machinery monitoring with frequency ranges now available between 1 Hz and 10 kHz, but typical MEMS accelerometers suffer from floor noise. According to the study by (Lainé & Mougenot, 2014), They posited that the current generation of MEMS has a noise floor (40 ng/√Hz from 10 to 200 Hz) that is lower than the ambient noise , and that this noise increases toward low frequencies, particularly below 5 Hz, where it can exceed ambient noise. This is highly undesirable, however there has been active research to minimize the floor noise as described in (Mohd-Yasin, et al., 2009) who provided a comprehensive survey of noise research in MEMS and suggested areas for the reduction of floor noise, (Izudin, 2011) who reiterated the relationship between noise floor and sensor resolution, where he said that, the noise floor of the sensor determines the resolution, which is dependent on bandwidth. The study also stated that Thermo-mechanical noise (or Brownian noise) derives from the fact that MEMS accelerometers consist of small moving parts. These small parts are susceptible to the mechanical noise that results from molecular agitation.

The benefits of MEMS accelerator for machine health monitoring are immense and not limited to the fact that they are, According to (Spence, 2016), economical and highly integrated solutions, which help reduce the cost of deployment and ownership, they are also important in weight restrictive applications , such as the aviation industry, where conventional sensors add a lot of undesired weight to the aircrafts, MEMS have brought a small size, small weight solution to their doorsteps.

The topology of a MEMS signal-conditioning circuit with both analog and digital output variations is common and opens options to adapt the sensor to a wider variety of situations. This allows for a transition to digital interfaces commonly found in industrial settings.

MEMS ACCERLERATOR – PHANTOM:

The PHANTOM, as depicted in fig 8 is a perfect example of a MEMS accelerometer. It is a tri-axis +/-8g, +/-16g or +/-32g silicon micromachined accelerometer with integrated 2048 byte buffer, orientation, tap/double tap, activity detecting, and free fall algorithms.

Figure 8: PHANTOM

It has high shock survivability, excellent temperature performance, user configurable wake up function, digital high pass filter outputs, small footprint, signal bandwidth (-3db) up to 8KHz and user selectable data rate of up to 25600Hz.

Acceleration sensing is achieved based on the principle of a differential capacitance arising from acceleration-induced motion of the sense element, which further utilizes common mode cancellation to decrease errors from process variation, temperature, and environmental stress. Voltage regulators are used to maintain constant internal operating voltages over the range of input supply voltage and this results in stable operating characteristics over the range of input supply voltages.

ICP ACCELEROMETERS:

According to (PCB Piezoelectronics, 2020), an ICP accelerometer as depicted in fig 9 is a sensor that generates an electrical output proportional to applied acceleration. ICP accelerometers are designed to measure vibration and shock for a wide variety of applications. They are simple to use and accurate over a wide frequency range which makes them the recommended choice for many testing situations.

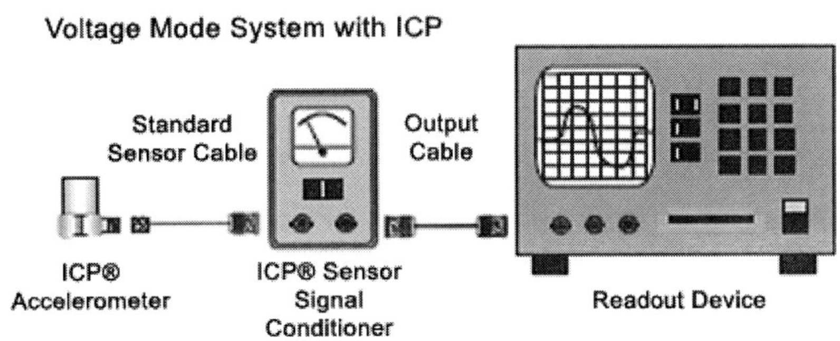

Figure 9: ICP Accelerometer Connection

ICP means Integrated Circuit Piezoelectric" and identifies sensors that incorporate built-in microelectronics. The electronics convert a high-impedance charge signal generated by a piezoelectric sensing element into a usable low-impedance voltage signal that can be readily transmitted, over ordinary two-wire or coaxial cables to any data acquisition system or readout device.

ICP accelerometers has the following benefits for use, such as; the use of standard coaxial cable or ribbon wire for connection as depicted in fig 9, it operates through slip rings, offers better system reliability, it connects directly to many read out instruments, it drives long cables without noise increase or loss of resolution, it has a very simplified mode of operation requiring less training or expertise, Bias monitor detects cable faults - shorts or open circuits, Low impedance systems have greater resistance to contamination and electrical interference, operates from a low cost constant current source, Has very low floor noise and unlimited sample rates

The drawbacks to using ICP accelerometers are that Limited temperature range (-320 °F to + 325 °F), Electrical characteristics, sensitivity, range, and discharge time constant are fixed within the sensor, May not be interchangeable in system if power requirement is not the same, low frequency roll off, require higher power to power them 5V+ etc.

AC140:
The AC140 as depicted in fig 10, features a welded stainless-steel construction for survival in harsh factory environments. A dual case design shields the sensing element from RFI & EMI interference are utilized to provide the highest signal to noise ratio available. Two pin MIL spec connectors are used to carry the signal output from the accelerometer, protecting the shielding and hermetic sealing of the accelerometer it uses PZT (lead zirconate titanate) over quartz sensing elements due to its high charge output and sensitivity over lower vibration levels.

Figure 10: AC 140

The accuracy of high frequency response is directly affected by the mounting technique that is selected for the sensor, while Low frequency response may be accurately obtained by all the mounting techniques available. AC 140 will function well with a constant current power unit, which provides 2-10 mA with a DC voltage level between 18 and 30 VDC. The 100 ohm, ¼ watt resistor and 1N4148 diodes as depicted in fig 11 are used to suppress electrostatic discharge. The CR220 current regulator diode provides the necessary bias current for the sensor. The power supply can be virtually any regulated supply that provides a clean 24-volt DC output Pin "A" is utilized for the Power/ Signal (+), and Pin "B" is utilized for Signal/Common (-). The case of the sensor is electrically grounded to the machinery that is mounted to and electrically isolated from Pin "A" and Pin "B" as depicted in fig 11.

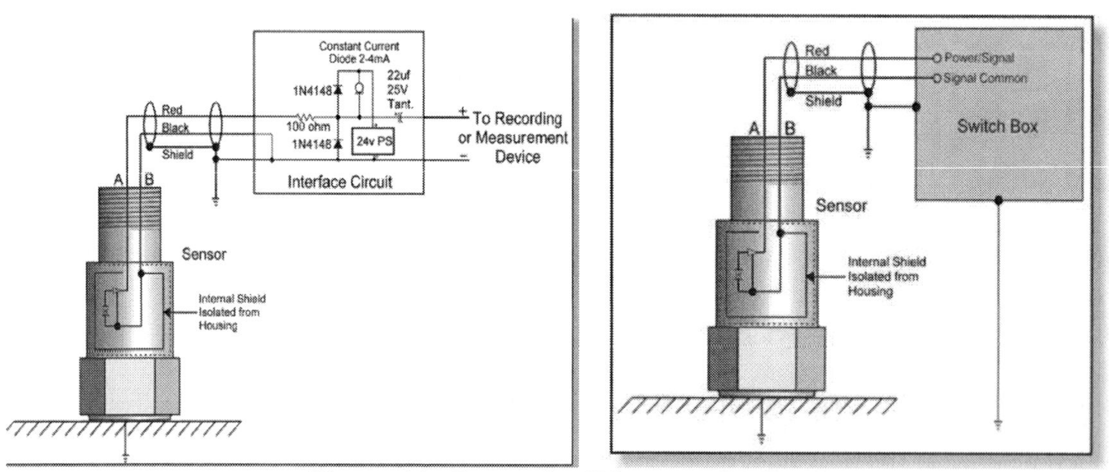

Figure 11: Connection Diagram of AC 140

DISCUSSION AND CONCLUSION:
Machine health monitoring cannot be over emphasized because of the key roles it plays in ensuring the longevity of machine life. Vibration analysis is a key technique in estimating machine health and thus sensors that measure vibrations are key

Accelerometers are sensors that measure acceleration and can be used for vibration detection and measurement. MEMS accelerometers and integrated circuits piezoelectric accelerometers are two key families of accelerometers that can be used for diverse application. Table 1 below specifies their overall features and specifications.

Table 1: Specification Comparison between MEMS and ICP Accelerometers

Features /Specifications	PHANTOM	ICP Accelerometers
Acceleration sensing	differential capacitance	Piezo velocity sensing, PZT ceramic
Sensitivity	1024 – 4096 counts/g	100 mV/g
Operating Temperature	-40°C to +85°C	-50°C to 121°C
Package Dimensions	2 x 2 x 0.9 mm LGA Plastic Package	20g weight, Stainless Steel, welded hermetic
Supply Voltage (VDD)	1.71V to 3.6V DC	18V to 30V DC
Serial Peripheral Interface	4 -wire	2-Wire
Frequency Range	.5Hz – 10000Hz	0.6Hz – 15000Hz
Low Power Mode/High Resolution Mode	Yes	No
Tilt Angle (mounting)	45°	90°
Internal Voltage Regulator	Yes	No
Temperature Performance	Good	121°C
Shock Tolerance	Excellent	5,000 g, peak
Non-Linearity	0.6% of FS	
Noise	630µg/√Hz (Density)	30µg/√Hz (Density)

Capacitive technologies (MEMS) have an edge over piezo technologies-resistive, velocity, electric (ICP), as piezo technologies have poor bias stability after years of usage, aren't optimal for static measurements, are sensitive to humidity with very poor non-linearity. The floor noise issue with MEMS accelerometers is a major drawback as can be evidenced in a noise density of 630µg/√Hz as opposed to 30µg/√Hz of the ICP variant, but the power requirement of 1.71V to 3.6V DC for the MEMS accelerometer is a great advantage over the high-power requirements of the ICP accelerometer standing at between 18V to 30V DC. The small dimension of the MEMS with a lightweight plastic packaging, compared to the stainless steel 20g weight ICP that the MEMS accelerometer would find more application in processes where light weight is of utmost importance in the design considerations. Despite the slight drawback of high floor noise, MEMS accelerometers, such as the Phantom can tolerate much higher levels of shock levels than the ICP accelerometers, with no impact on its sensitivity specifications. MEMS also offer great stability in temperature, good stability over years of usage, good repeatability, and ease of compensation than the ICP accelerometer. This thus makes it the optimal choice for use in machine condition monitoring.

REFERENCES:

Alhussein, A., Samir, M., Andrew, S. & Robert, P., 2008. Suitability of MEMS Accelerometers for Condition Monitoring: An experimental study. *sensors ISSN 1424-8220,* Volume 8, pp. 784-799.

Arturo, G.-P., Juan, P. A.-S., Daniel, M.-S. & Konstantinos, N. G., 2016. Vibration Analysis as a Diagnosis Tool for Health Monitoring of Industrial Machines. *Shock and Vibration, Article ID 1235139, https://doi.org/10.1155/2016/1235139,* Volume vol., p. 2 pages.

Brown, D., Jensen, T., Briiel & Kjaer, n.d.. *The use of Spectrum Comparison for Bearing Fault Detection A Case Study from Alma Paper Mill, Quebec, Canada ,* s.l.: Briiel&Kjaer, https://www.bksv.com/media/doc/bo0253.pdf, Retrieved 10/23/2020.

Dictionary, M.-W., n.d. *"Machine".* s.l.: Merriam-Webster, https://www.merriam-webster.com/dictionary/machine. Accessed 23 Oct. 2020..

Izudin, C., 2011. *Noise Measurement,* s.l.: Fierce Electronics, https://www.fierceelectronics.com/embedded/noise-measurement, Retrieved 10/24/2020.

Lainé, J. & Mougenot, D., 2014. A high-sensitivity MEMS-based accelerometer. *Sensor technology and nanotechnology,* pp. 1234-1242.

Manish, V., Rajesh, P., Harshlata, V. & Rajput, P., 2017. Vibration Analysis & Condition Monitoring for Rotating Machines: A Review. *Materials Today: Proceedings,* 4(2), pp. 2659-2664.

Matej, A., 2008. *MEMS Accelerometers,* s.l.: University of Ljubljana, Faculty for mathematics and physics, Department of physics.

Mohamadi Monavar, .., Ahmadi, H., Mohtasebi, S. & Hasani, S., 2008. Vibration Condition Monitoring Techniques for Fault Diagnosis of Electromotor with 1.5 Kw Power. *Journal of Applied Sciences,* Volume 8, pp. 1268-1273.

Mohd-Yasin, F., Nagel, D. & Korman, C., 2009. Noise in MEMS. *Measurement science and Technology,* 21(1), pp. 1-9.

Monk, G. R., 1973. Machine-health monitoring — the application and payback. *Non-Destructive Testing,* 9(3), pp. 127-130.

Nabby, E. A., Belal, M. & Gomaa, F., 2018. *Machine Health Monitoring Using Vibration Analysis.* Military Technical College Kobry El-Kobbah, Cairo, Egypt., Proceedings of the 18th International Conference on Applied Mechanics and Mechanical Engineering..

PCB Piezoelectronics, 2020. *Introduction to MEMS Accelerometers,* s.l.: PCB Piezoelectronics, https://www.pcb.com/resources/technical-information/mems-accelerometers#:~:text=MEMS%20stands%20for%20micro%20electro,microscopic%20size%2C%20typically%20on%20silicon, Retrieved 10/24/2020.

Spence, E., 2016. *Bring the Benefits of MEMS Accelerometers to Condition Monitoring ,* s.l.: Electronics Design, https://www.electronicdesign.com/technologies/test-measurement/article/21801738/bring-the-benefits-of-mems-accelerometers-to-condition-monitoring, Retrieved 10/24/2020.

TE Connectivity , n.d. *Accelerometer Types,* s.l.: TE Connectivity , https://www.te.com/global-en/industries/sensor-solutions/insights/types-of-accelerometers.html#:~:text=There%20are%20two%20classes%20of,constant%20centrifugal%20acceleration%2C%20for%20example, Retrieved 10/24/2020.

Tinder, R. F., n.d. *Relativistic Flight Mechanics and Space Travel: A Primer for Students, Engineers and Scientists..* s.l.:Morgan & Claypool Publishers. p. 33. ISBN 978-1-59829-130-8. Extract of page 33.

TWI, n.d.. *What is Vibration Analysis and What is it used for,* s.l.: TWI, https://www.twi-global.com/technical-knowledge/faqs/vibration-analysis#:~:text=Vibration%20analysis%20is%20a%20process,condition%20of%20the%20test%20object, Retrieved 10/23/2020.

IMPLEMENTING INDUSTRY 4.0 WIRELESS VIBRATION TO IDENTIFY ROTATING EQUIPMENT ISSUES, IMPROVE RELIABILITY AND REDUCE UNPLANNED OUTAGES

Author: Frank Mignano, CMRP

VIATC TRAINING CONFERENCE & EXPO
SAVANNAH, GEORGIA · 2022

Implementing Industry 4.0 Wireless Vibration to Identify Rotating Equipment Issues, Improve Reliability and Reduce Unplanned Outages

Frank Mignano, CMRP
Manager – Condition Monitoring Americas

Industry 4.0 Service Solutions

308 Springhill Farm Rd

Fort Mill, SC 29715

Tel: 803-548-8955   Cell: 803-659-6737

mignafan@schaeffler.com - www.schaeffler.com

Drawing upon 30+ years of experience with reliability and vibration systems, Frank Mignano leads the condition monitoring team for Schaeffler in the Americas from the company's regional headquarters in Fort Mill, South Carolina. A Level 2 Vibration Analyst as well as a Certified Maintenance and Reliability Professional, Frank holds a Bachelor of Science in mechanical engineering.

ABSTRACT:
Utilizing the experience gained at one of Schaeffler's production facilities in Fort Mill, South Carolina, this paper will provide an overview of how a large global manufacturing organization can benefit from the implementation of IIoT technologies in its own manufacturing plants around the world.

Real-life case histories will be presented based on an award-winning, Industry 4.0-based condition monitoring solution that consists of wireless, battery-powered vibration sensors, cellular gateways and associated digital services.

As a leading global supplier to the industrial and automotive sectors, Schaeffler has been pioneering groundbreaking inventions and developments in the fields of motion and mobility for over 75 years. With a worldwide network of manufacturing plants, R&D facilities and sales companies at approximately 170 locations in over 50 countries, Schaeffler is one of the world's largest technology firms in family ownership.

KEY WORDS:
Battery powered; Digital Services; Industry 4.0; Machine Learning; Vibration; Wireless

INTRODUCTION:
First, a little background on Schaeffler: With approximately 75 manufacturing facilities and 20 R&D centers throughout the world, the Schaeffler Group is a leading global supplier to the automotive and industrial sectors. This includes significant bearing business with OEMs and the aftermarket (MRO). With approximately 83,000 employees worldwide (many of whom are based in our manufacturing plants), Schaeffler is extremely knowledgeable about industrial maintenance. This is due in large measure to the fact that we develop many of our processes and tools internally.

As part of our Schaeffler Lifetime Solutions group, and in perfect complement to our aftermarket and OEM business units, Schaeffler develops and offers condition monitoring tools as well as related technology and supporting services to support the longevity of bearings. While it may sound somewhat counterproductive when you think about it – a bearing manufacturer helping end users make their bearings last longer – at the end of the day, everything we do at Schaeffler is about helping our customers operate more efficiently as part of an enduring partnership.

This paper will discuss how a large global manufacturer can and is benefitting from deploying its own technology at its own facilities. The focus will be on our Fort Mill, SC, manufacturing plant, known as FM1 (shown below). In this facility, we make many components that end up in motors and transmissions via a stamping and heat treatment process. On the heat treatment side, we have many critical motors/fans and numerous twin-screw compressors supplying the furnaces. Each heat treatment line consists of two furnaces, an annealing and tempering furnace, to develop the proper characteristics in the metals. Any failure on any of the six motors/blowers and two compressors per line results in two-and-a-half days of unplanned downtime – because it takes a day to cool off the furnace and a day to heat it back up, setting aside time to perform the maintenance.

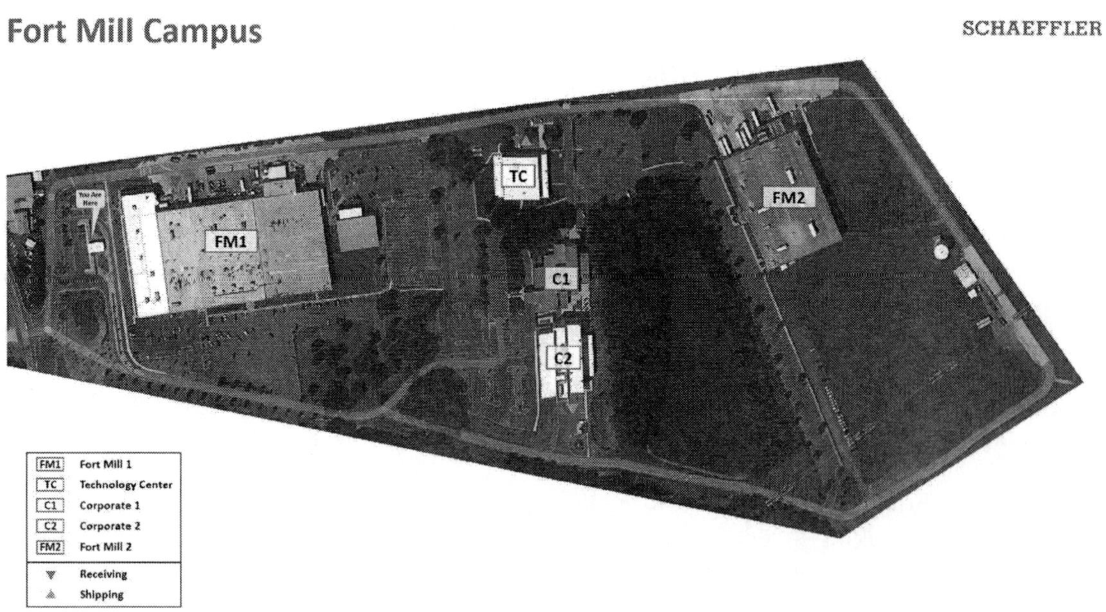

Figure 1: Schaeffler's corporate campus in Fort Mill, SC

Limited traditional predictive maintenance (PdM) activities, such as a monthly route-based program or lubricant analysis, were in place at the time it was decided to implement IIoT/Industry 4.0/Machine Learning/AI-based battery-powered vibration technology. As with all manufacturing facilities, unplanned downtime comes at a tremendous cost. Moreover, with technology and signal processing techniques (enveloping/demodulation), there is absolutely no reason a piece of rotating equipment should fail without warning today – and for at least the past 30 years or so. The technology has been there for decades, although the processes may not have always been executed properly. The implementation of this technology allows costs to be removed from our processes while at the same time limiting and even eliminating unexpected machine failures – all in a concerted effort to ensure that we can meet market demand for our products, much like every manufacturer attempts to do. The goal is to not let machines dictate the maintenance cycle; instead, we use condition monitoring technologies to determine the equipment's condition and plan the maintenance – preferably during the day shift when parts and personnel are available to maintain the asset in the most cost-effective manner.

We accomplish this using condition monitoring as well as lubrication delivery technology. The focus of this discussion will be the work and success in our Fort Mill, SC, facility which mainly supplies parts to automotive manufacturers. We produce over 20 million parts per year in this plant, and each one is important to our end users. Many components developed here – such as spacers, needle roller & thrust bearings, to name just a few – end up in engines and transmissions. These components are a vital part of the supply chain for building cars.

SECTION TITLE:

The Current Situation:

Like many factories in industry today, Schaeffler's FM1 manufacturing plant in Fort Mill faces the daunting daily task of keeping production rates up while effectively managing the health of the facility's manufacturing assets. As we all know, healthy assets feed good production yields. Nonetheless, sometimes machine reliability takes a back seat to other seemingly more important tasks. You need to keep your assets functional to keep the process moving with the ultimate goal of delivering saleable, quality products to your existing customers at a profit. In our FM1 facility, manufacturing parts involves a lot of stamping and heat treatment activities; not surprisingly, many rotating fans play a critical role in this process.

Figure 2: Example of a furnace and one of the three critical fan assemblies

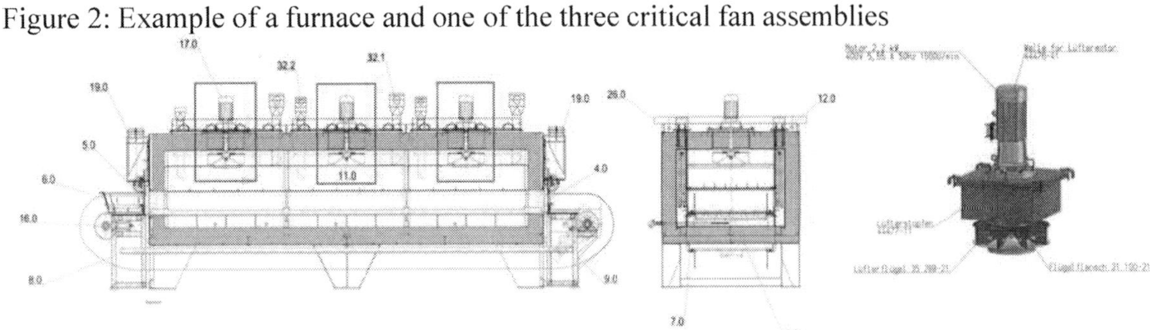

Components enter from the left, move through the furnace on conveyors, and exit on the right. Three motor/fan assemblies circulate the air within the furnace to ensure a homogeneous temperature so that the components are imbued with the correct metallurgical properties.

CONCLUSION:

The Results:

In 2017, Schaeffler initially implemented then-current technology on its GA4 line. This included 10 Schaeffler SmartCheck wired monitoring devices, along with an 8-port automatic lubricator to supply grease to the lube ports on the three critical fan assemblies (Fig. 3). Each line has an annealing and tempering furnace featuring three fan assemblies. The systems shown in Fig. 3 are still in use today and have provided tremendous value with an ROI period of less than one year.

A victim of personnel changes, adjusted priorities and shifts in focus and funding, this system received diminished attention for a time. However, coinciding with the development and deployment of newer Industry 4.0-compliant technologies, the project regained momentum. In 2020, Schaeffler added more than 100 wireless sensors with cellular gateways to capture data on the other nine heat treatment lines (see Fig. 4) as well as assorted exhaust fans, compressors, and plant air and water systems. In addition, many single point lubricators were also added at a greatly reduced installation cost compared to prior generations.

Figure 3

Figure 4

The IIoT-based system was installed and started providing data on the very first day. Consisting of wireless, battery-powered, industrial-grade meshing sensors and cellular gateways to transmit the data to the cloud, the system collects seven overall data sets (six vibration measurements plus temperature) several times per day. A time waveform/FFT is stored each day. Two 15-day learning modes automatically set the alarm levels for each location, after which the collected data is run through proprietary algorithms that convert the data into actionable information (such as a suspected balance/alignment issue, lubrication issue, or identified bearing damage). Machine status, which is a culmination of all sensors on a machine train, can range from "Normal" to "Suspect" to "Warning" to "Severe."

Success #1

PIFO 2 – Oil Circ Pump

Oil circulating pump on PIFCO Line #2. While many of the plant's 10 lines involve tempering and annealing furnaces moving air, the PIFCO 2 line use an oil quenching process. The wireless sensor system identified and suggested a "Coupling/alignment" issue on the pump shown in Fig. 5 below. This occurred early in the program, right after the sensors had been installed. Upon alarm notification, an inspection revealed that the coupling was indeed worn and needed to be replaced. A few months later, the same alarm was issued by the system's automated diagnostics.

Following an RCFA review of the system, the spare pump was found to be spinning in reverse due to a faulty check valve. Whenever power was applied to the spare pump, tremendous torque was transmitted through the coupling, which was causing the damage. A new check valve was installed (shown at right on Fig. 6), and the coupling was replaced again. The confidence level is high that the issue has been resolved for good. Total potential savings as a result of this fault diagnosis were incalculable – had both pumps failed, the oil could have heated up to its flash point and caused a fire. Documented savings were $7,500 per incident.

Figure 5

Figure 6

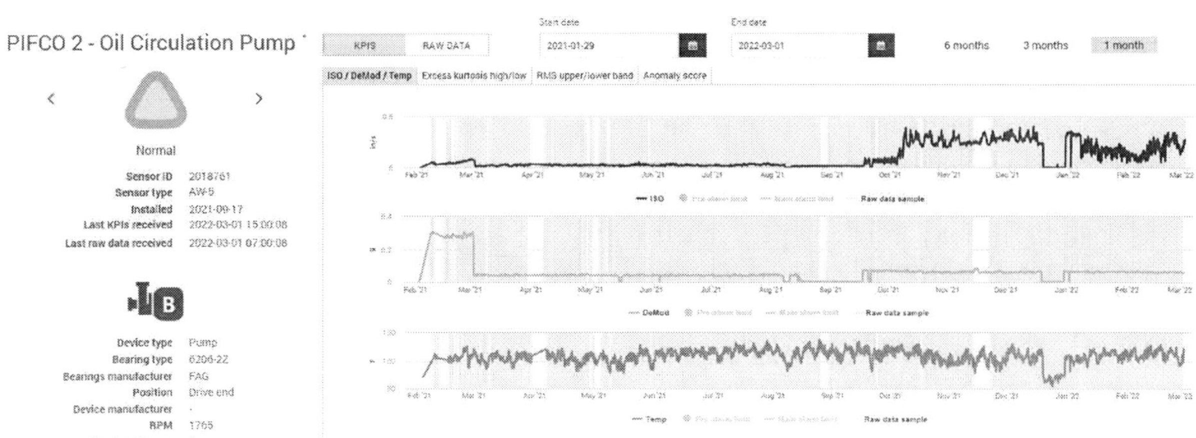

Figure 7

Success #2

GA 3 – Compressor 2

A twin-screw compressor suddenly advanced to a "Severe" alarm notification and remained there for several months while maintenance ordered a replacement and found the time to schedule the work. This is an example where the automated diagnostics suggested issues such as balance and bearing damage. The machine was ultimately replaced, and readings returned to acceptable levels. After the repair/replacement, the learning mode was activated again to set the alarm levels to the characteristics of the newly replaced machine. Lost production savings alone were $9,460.80

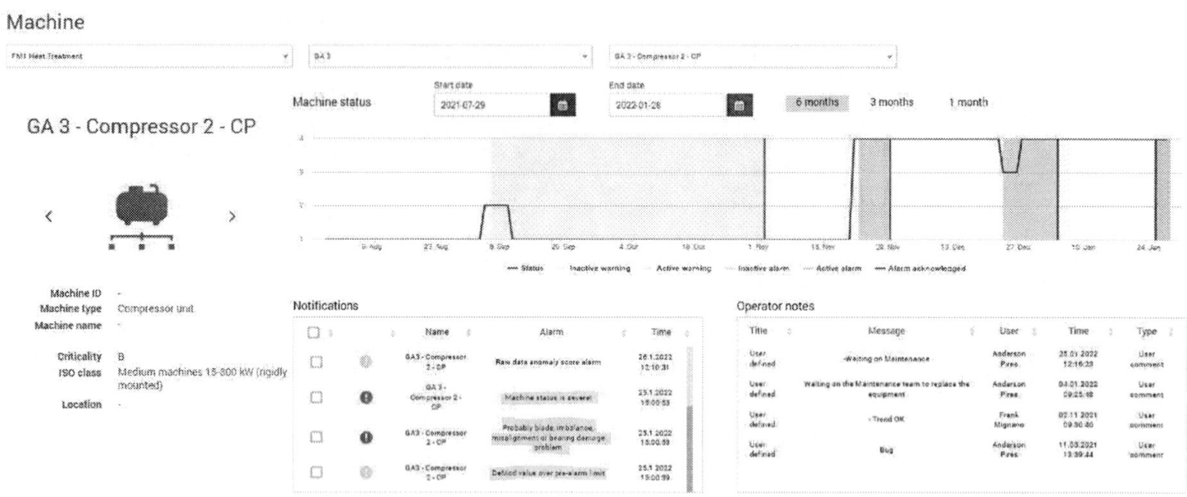

Figure 8

Success #3

PIFCO 1 – Washer Fan - DE

While this example is not a typical "save," it shows the advantages of having a system that automatically captures data, sets alarm levels and generates 180 trend points each month on a facility's rotating assets. On this particular washer fan, based on very high vibration velocity readings, the maintenance team performed a visual inspection that identified several loose and/or missing bolts. Tightening the loose bolts led to a 30% drop in vibration velocity, while the maintenance team planned for a full balance of the fan as the predominant peak was at running speed.

Figure 9

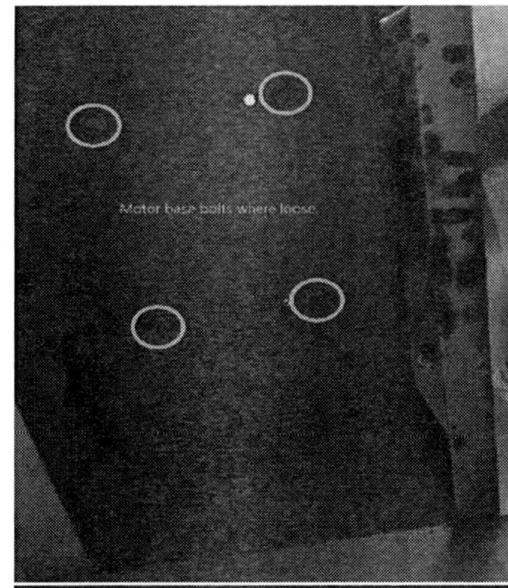

Figure 10

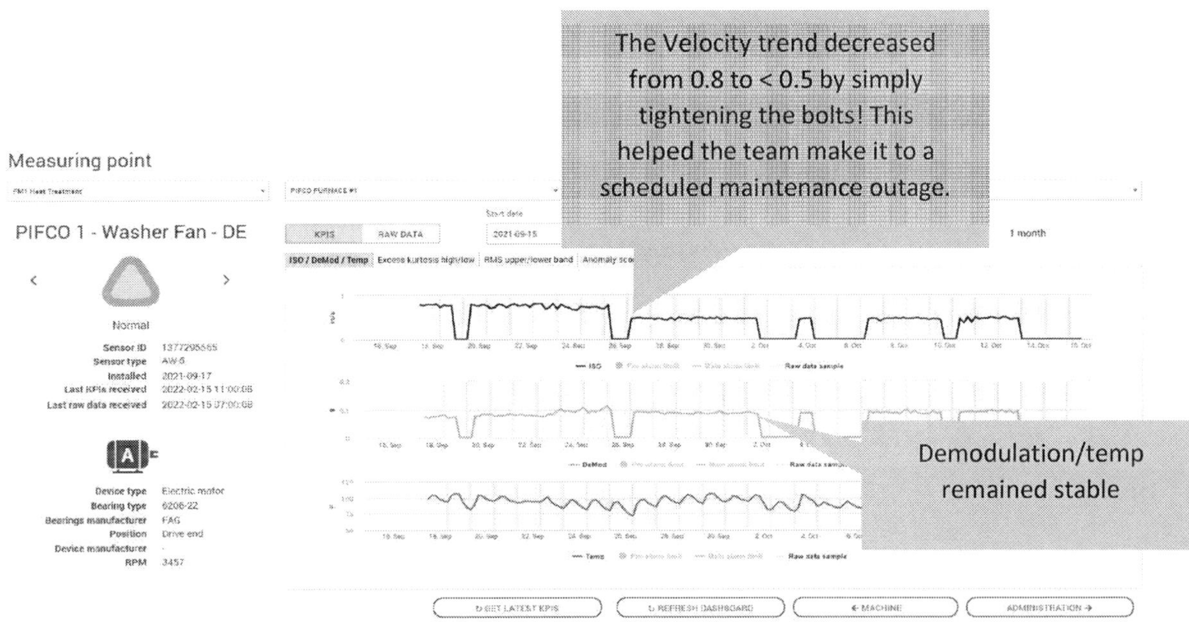

Fig 11

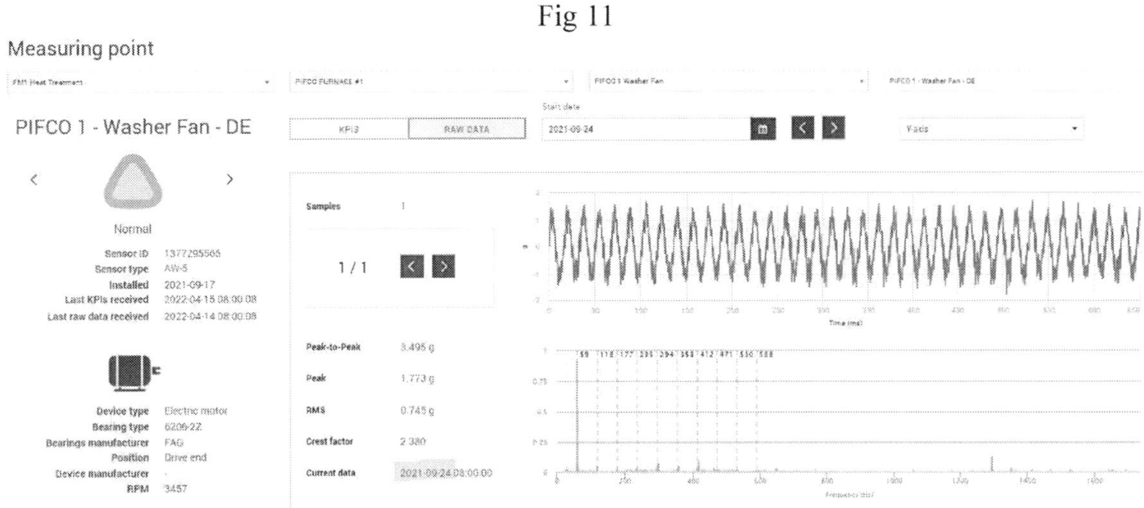

Fig 12

WHAT DO YOU DO WHEN YOUR ONCE PER TURN PHASE MARKER DOESN'T WORK?

Authors: Everett (Rett) Jesse
and John P. Edwards

VIATC TRAINING CONFERENCE & EXPO
SAVANNAH, GEORGIA · 2022

WHAT DO YOU DO WHEN YOUR ONCE PER TURN PHASE MARKER DOESN'T WORK?

Everett (Rett) Jesse
Vice President of Sales & Marketing, Metrix Vibration
8824 Fallbrook Dr. Houston TX 77064

Rett Jesse is a registered professional Mechanical Engineer, and has an MBA from Regis University in Denver, Colorado. His undergraduate work was in Civil Engineering from the University of Nevada. After graduation he became a Naval Nuclear Submarine Officer for United States Navy. His participation in two shipyard overhauls provided him an excellent background in machinery and processes. He became the Training Manager for Bently Nevada Corporation where he documented the Machinery Diagnostic Course, Advanced Machinery Dynamics Course and Applied Diagnostic Courses. He and Don Bently traveled the globe teaching employees and customers about Machinery Diagnostics. At Bently Nevada he became the Service Director overseeing the Project, Product Service, Machinery Diagnostic and Training Groups. After the GE acquisition of Bently, Rett became the Quality Leader for Services, earning his Lean Six Sigma Master Blackbelt certification. He became the Managing Director of Nevada Controls, and currently serves as the Vice President of Sales and Marketing at Metrix Vibration.

John P. Edwards
Technical Product Specialist, Metrix Vibration
8824 Fallbrook Dr. Houston TX 77064

John Edwards is recent addition to Metrix Vibration and to the machine vibration industry. John works as a Technical Product Specialist at Metrix headquarters in Houston, TX. In this role, John supports the Applications Support Group to ensure customer satisfaction by providing application support to customers pre-sale and post-sale. Prior to joining Metrix, John had 20 years Management and Field Service experience in the offshore oil and gas sector working in both Europe and Asia.

ABSTRACT: There are many installations where the vibration monitoring system has been rendered virtually useless because of a faulty once per turn phase marker. The faulty phase trigger causes the 1X amplitude and phase, 2X amplitude and phase, not 1X and nX to not be reported, or even worse, to be reported inaccurately – leading to bad decisions. Up to now the only way to fix a faulty once per turn phase marker is to shut down the machine, open the machine up and determine the problem. Times have changed, the Digital Proximity System (DPS) from Metrix can be used to gain the scale factor to obtain an adequate pulse for the vibration monitoring system to see, to report correct vibration values. This paper will discuss how a faulty once per turn phase marker, caused by improper installation, wrong keyway or hole size, thermal growth, or thermal change, can be corrected for, in both a proximity driver going to a vibration monitoring system and proximity transmitter going to a control system.

KEY WORDS: Proximity; Proximity transmitter; Phase marker; Keyphasor™; Thermal growth; Thermal expansion; Vibration monitoring modal testing.

INTRODUCTION:

<u>What is a phase marker and why are they used?</u>

Typically, vibration data acquisition captures the raw waveform, which can be broken down into multiple sine waves that when added together create the Complex Time Waveform. The Fast Fourier Transform (FFT) separates the individual sine waves according to frequency. This is known as a Frequency Spectrum. Once separated into a Frequency Spectrum, individual amplitudes can be derived for each of the sine waves, and the severity of each vibration component can be understood.

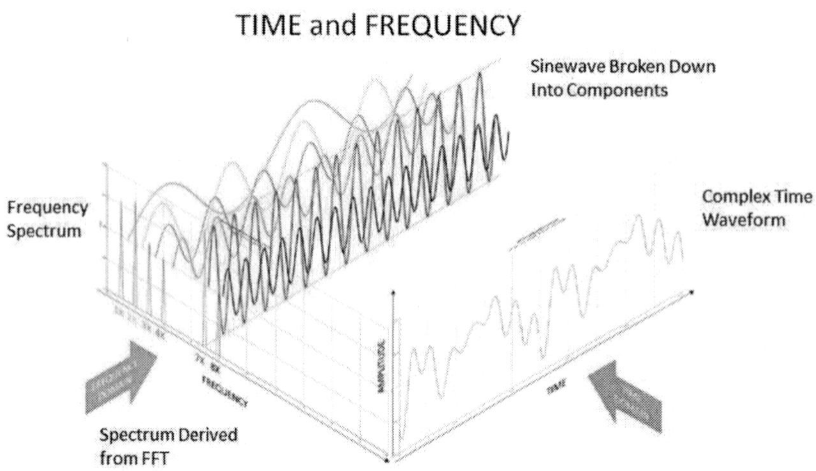

Figure 1. Frequency Spectrum

The Frequency Spectrum supplies frequency and amplitude information for vibrations in a system but does not point out what direction this energy is acting in. This is where the once per turn phase marker is used and supplies a third dimension to the user's vibration analysis, that being phase measurements.

A once per turn phase marker is an absolute timing reference to which all events can be referenced by using degrees and supplying an angle or direction to any filtered vibration signal. The absolute phase measurement is measured from the phase trigger marker to the first positive peak in the filtered vibration signal, measured in degrees.

Figure 2 below uses a setup with a proximity probe and a phase marker on a shaft. The phase marker (Event 1) is triggered and 260° later the proximity probe sees a peak pulse as the shaft moves closest to the probe. As the phase marker serves as a zero-phase reference, it is true to say that Event 2 lags the phase marker by 260°.

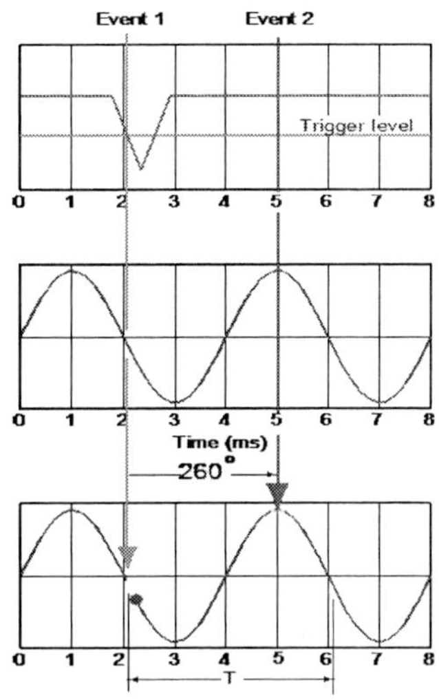

Figure 2. How a Phase Marker Works

With this data it's possible to diagnose defects such as unbalance, misalignment, looseness, bending, rubs, fluid induced instability, shaft cracks and resonance issues.

<u>How To Determine If a Phase Trigger Marker Is Adequate?</u>

A phase trigger signal in a monitoring system usually has to be greater than 6V peak-to-peak, in order to obtain a proper phase reading. This 6V peak-to-peak voltage is chosen to overcome any noise issues in the system. A dead band within the 6V peak-to-peak is used to ensure a clean phase trigger pulse is obtained.

Figure 3 below shows two signals, one which will break the dead band, and another that won't, yellow and grey signals respectively. The peak-to-peak amplitude of the signal must pass the dead band for the monitoring system or the 4-20mA speed transmitter to work. The point where the signal breaks through the dead band and then returns to the initial gap voltage determines the period and the phase trigger output. This diagram depicts positive going phase trigger events, yours may be negative or you may choose to invert them.

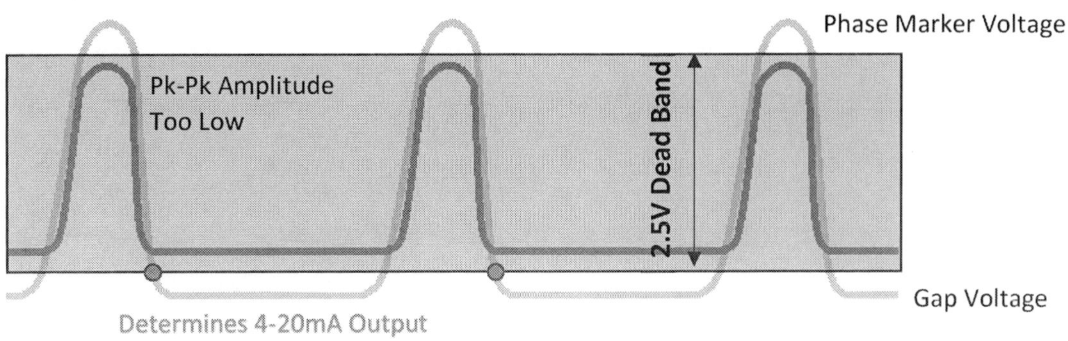

Figure 3. How a Proximity System Works with a Phase Marker

Let's investigate what can be done to make a faulty phase marker adequate without the need to open up the machine and go near the phase trigger probe. Current practice allows one to correct the phase trigger pulse in a matter of minutes.

The following sections illustrate two examples where bad once per turn markers were encountered, namely:

1) Due to thermal growth or thermal change, and
2) When the phase trigger marker notch (e.g., hole, keyway, etc.) is too small to be picked up by the monitoring system.

Thermal Growth or Thermal Change – This phenomenon occurs when the machine changes thermally and it expands or contracts in various directions. In certain circumstances, it is possible for the phase trigger probe and the shaft once per turn notch, no longer line up.

Due to the eddy current found near the probe tip not picking up a portion of the notch (keyway), the once per turn voltage pulse will be weak. This leads to a smaller peak-to-peak amplitude as the notch passes the probe face, and the effect is a reduction in the notch diameter. It is necessary to see at least a 2.5V change in once per turn voltage pulse, in order to obtain clear delineation of a phase trigger event versus noise that could be found in the system.

The following test demonstrates the issue. The probe is gapped to 9V or 45mil from the rotor.

Thermal Growth/Offset Probe Test	
Keyway Dimensions	0.196" x 0.04"
Gap Voltage	9 V
Keyway Voltage	10.05 V
Motor Frequency	60 Hz
Motor RPM	3,600 rpm
Output Loop Current	4 mA (FAULT – below system threshold)

In Figure 4, the position of the probe relative to the phase marker can be seen for the test undertaken. Only around half of the keyway overlaps with the probe as it passes. The signal using an oscilloscope next to this picture shows that the raw signal voltage is only registering a 1.9V peak-to-peak pulse as the probe passes the keyway. This is not enough to break the system dead band threshold of 2.5V, or 6V for certain vibration monitoring systems, and does not provide a phase trigger output.

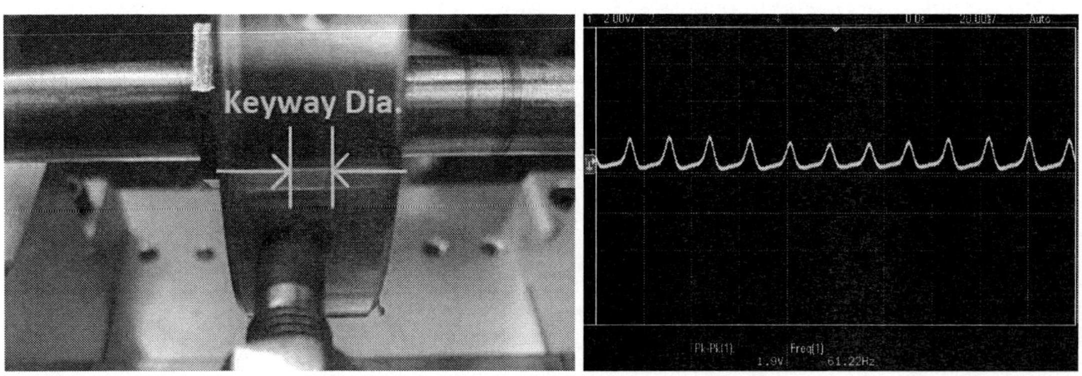

Figure 4. Offset Phase Marker to Probe

Typically, at this point the user would have to make some decisions, either to continue without a phase trigger marker, or shut the machine down to see if adjustments can be made to the phase trigger probe position. Instead of choosing these options, it is possible to make some adjustments to the phase trigger pulse using the Metrix DPS software by utilizing the **Speed Pulse Adjustment Slider** feature in the **Advanced Settings** tab.

By connecting the oscillator demodulator of the DPS, in a phase trigger or speed application, it is possible to gain the scale factor of the proximity system to alter the peak-to-peak voltage of the once per turn marker to obtain a greater pulse. This means for every mil of displacement we can increase the raw output amplitude. So as the probe passes over the notch, the voltage change will be greater than prior to the adjustment. In machine monitoring for vibration and position, it is generally required that the equipment should comply with the API 670 Standard [1] where the Incremental Scale Factor (ISF) is 200mV/mil or 7.87mV/µm when measured at specified increments throughout the linear range. For a phase trigger pulse this is not necessary, an adequate pulse is what is required.

To increase the scale factor and thus the peak-to-peak voltage of the raw once per turn voltage pulse, one must connect the DPS unit to a computer, and then using the DPS software click on the **Advanced Settings** tab and click the **Speed Pulse Adjustment** in the bottom righthand corner. Next with the system setup, and the machine rotating past the probe, we can take a baseline peak-to-peak voltage reading, in our example it's 1.98V. By moving the marker on the **Slider** to the left, as shown in Figure 5, we begin to increase the peak-to-peak voltage. At any point we can click **Get** to capture the peak-to-peak voltage of the raw signal output. By comparing the **Slider** tool before and after the changes, it can now be observed that the new voltage is 4.15V peak-to-peak, enough to surpass the threshold voltage of 2.5V.

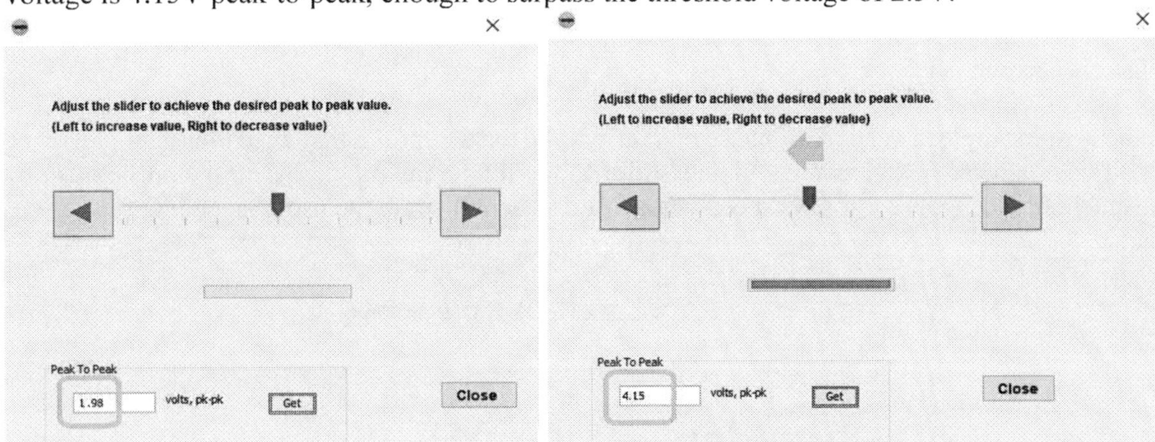

Figure 5. Adjusting the Slider tool – Offset Probe

In Figure 6, the oscilloscope waveforms are investigated before and after the introduction of the **Slider**, it can be seen how the peak-to-peak of the raw signal changes in amplitude.

This process of moving the slider and getting the voltage change can be repeated until one obtains a voltage change of 6V peak-to-peak as required by a vibration monitoring system.

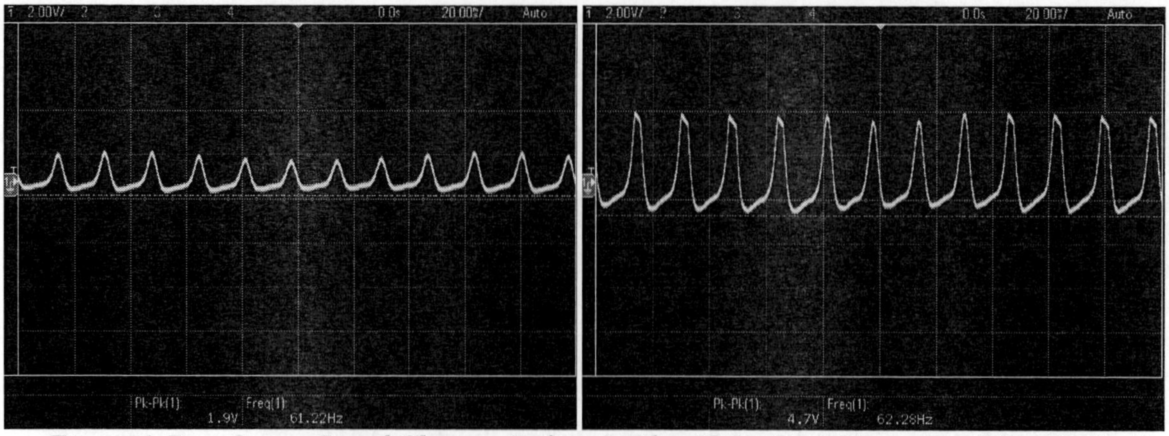

Figure 6. Raw Output Signal Changes Before & After Slider Tool Changes – Offset Probe

<u>Phase Markers with Notches that are Too Small</u> – In high-speed applications it is necessary to remove as little material as possible to use a phase marker so that the risk of unbalance is not introduced. In certain situations, the phase marker notch is small making it difficult for the monitoring system to identify. This could lead to additional time spent at the commissioning stage adjusting the probe to the point where a satisfactory signal from the phase trigger proximity sensor is accomplished.

In the next example, a very small phase marker will be used to prove that a smaller than specified notch can be overcome with the **Slider** tool without having to go near the probe or stop the machine. See Figure 7 for the speed wheel used to demonstrate the use of a smaller phase marker.

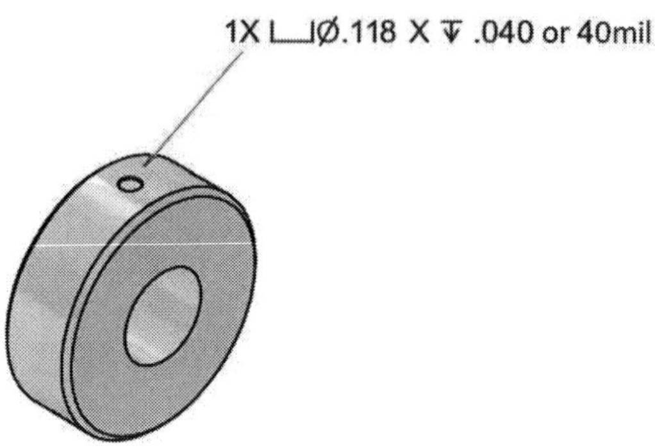

Figure 7. Small Phase Marker Setup

Once again, the probe is gapped to 9V or 45mil from the shaft/target.

Thermal Growth/Offset Probe Test	
Keyway Dimensions	0.118" x 0.04"
Gap Voltage	9 V
Keyway Voltage	10 V
Motor Frequency	60 Hz
Motor RPM	3,600 rpm
Output Loop Current	4 mA (FAULT - below system threshold)

With the system setup in this manner, it can be observed in Figure 8 that although the probe creates an electrical signal as it passes over the phase marker, it is not enough to surpass the dead band. The 1.4V the peak-to-peak signal will not be recognized.

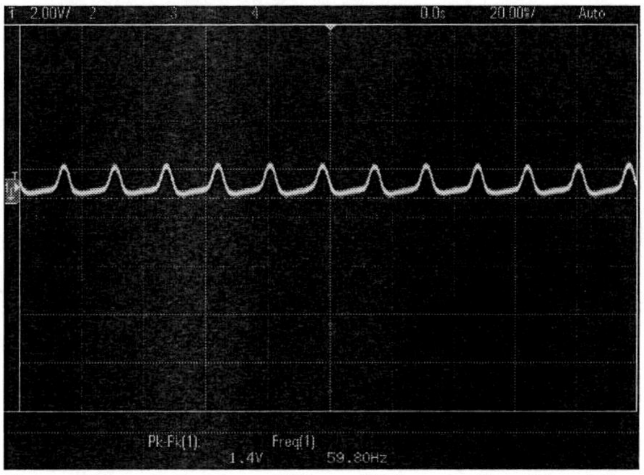

Figure 8. Raw Signal Output When Using a Smaller Phase Marker

Once again in Figure 9, we can connect to the DPS unit to make changes to the phase marker, or speed, scale factor using the DPS software which will increase the peak-to-peak voltage raising it above the 2.5V dead band. By using the **Get Peak-to-Peak** voltage feature, it is possible to gauge whether the changes in scale factor are sufficient to bring the system into a functional state.

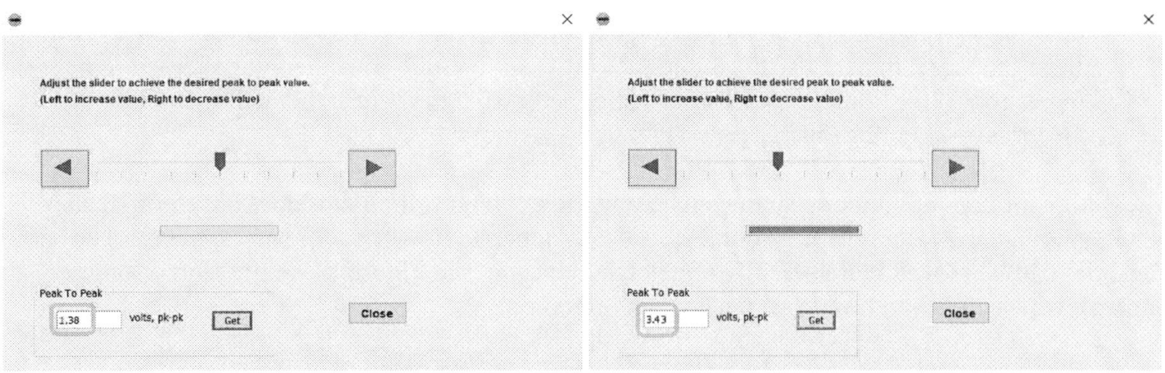

Figure 9. Adjusting the Slider tool – Small Phase Marker

By moving the **Slider** tool, changes in the once per turn pulse can be observed as shown in Figure 10.The graph on the left shows the waveform before the introduction of the **Slider,** and the on the right shows the waveform after the adjustment has been made. The peak-to-peak voltage of the raw signal increases in amplitude so that it surpasses the dead band and an adequate once per turn marker, or speed pulse, can be obtained.

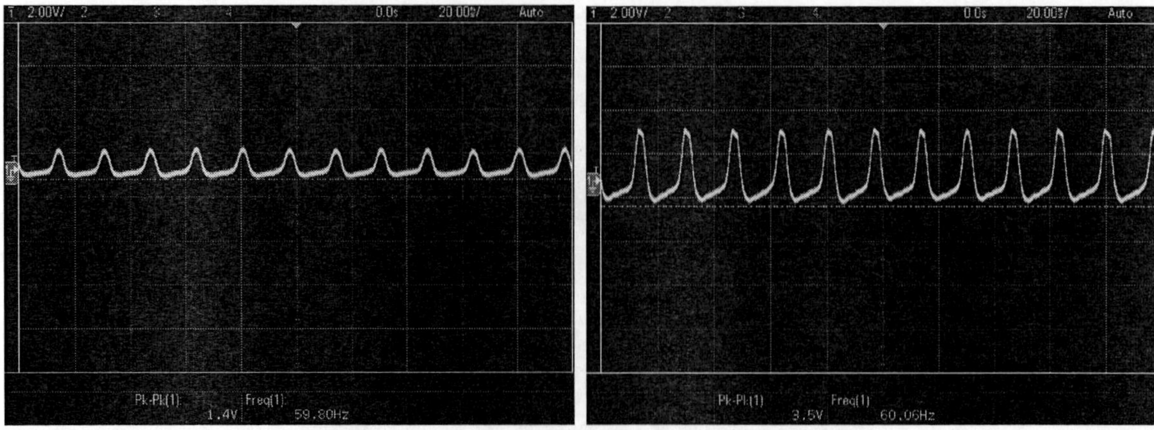

Figure 10. Raw Output Signal Changes Before & After Slider Tool Changes – Small Phase Marker

What Happens When We Introduce the Slider Tool?

The industry standard for the Incremental Scale Factor (ISF) response is 200mV/mil or 7.87mV/µm. When introducing the **Slider** tool for once per turn marker or speed pulse adjustment, it makes changes to the scale factor in the DPS unit. As the system is looking for a voltage pulse, it is acceptable to deviate from the industry standard ISF. By moving the **Slider**, we might change the ISF from 200mV/mil to 300mV/mil, this will have the following outcomes:

1) The phase marker pulse will have a larger peak-to-peak voltage as it passes the probe.
2) The gap voltage will also increase, since increasing the scale factor means that the amount of voltage over a given distance has also increased. If the probe is gapped 45mil from the target material with a scale factor of 200mV/mil, a 9V gap voltage would be measured. If the slider increases the scale factor to 300mV/mil, this will mean that the electrical gap voltage increases to 13.5V.

The **Slider** tool has no numeric settings when making changes, though it does have non-scalable incremental notches along its horizontal axis. To understand how the tool works, the marker on the slider can be moved incrementally one notch at a time, see Figure 11 below, and a static linearity verification can be performed at each increment.

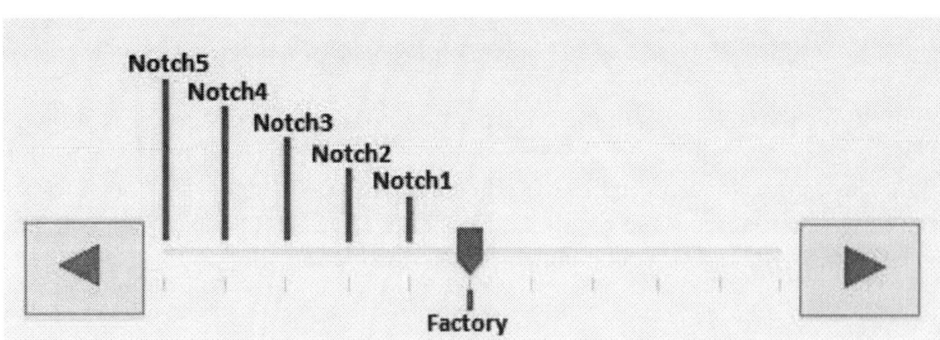

Figure 11. Incremental Scale of the Slider Tool

Figure 12 below details how performing a static linearity verification using a linear proximity calibration device shows changes when the slider is moved to different setpoints along its scale. As the slider is increased, the plot gets steeper resulting in a higher mV/mil reading at each increment. This means the signal amplitude will change resulting in increased pulse height. The increase in pulse height depends on what the gap is set to, the further the sensor is from the target material, the steeper the scale factor becomes.

It should also be noted that at 10 mils (254 μm), the difference between the target and probe face voltage has tripled (between the Factory and Notch 5 setting), from a voltage of just over -1Vdc to just over -3Vdc respectively.

The scale factor for each increment is also noted in Figure 12. This scale factor was taken from 10 mils (254 μm) to where the voltage over-scales, although, it should be noted that this scale factor is not linear along the plot.

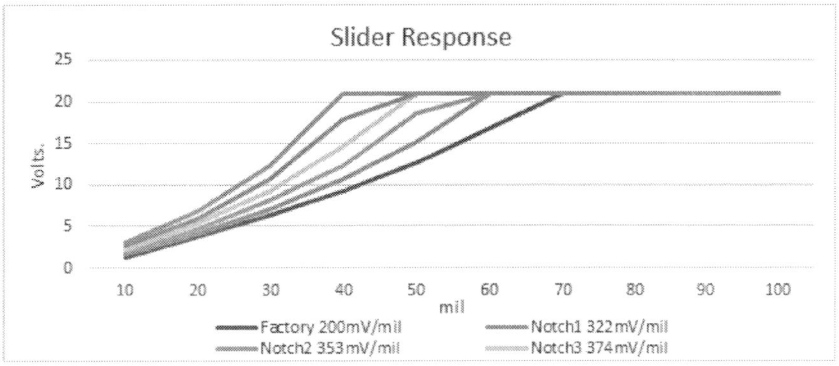

Figure 12. Scale Factor Response to Incremental Changes to the Slider Tool

CONCLUSION:

A once per turn phase marker may exhibit issues due to various scenarios such as improper installation, wrong keyway or hole size, thermal growth, or thermal change. This can lead to loss of time and revenue by not having an adequate phase trigger pulse. By connecting the DPS unit to a computer and using the DPS software, one can make gain adjustments for a once per turn marker or a speed pulse without needing to stop production or to open up the machine.

Users now have security in knowing that phase markers that would not have been useable in the past, can now be implemented consistently and yield great results. Being able to adjust the phase trigger or speed pulse when the machine is operating, means boundaries can be pushed in terms of using smaller phase markers at higher speeds than in the past.

Using the **Slider** tool, in the DPS software, in once per turn marker and speed applications, one can now gather information on voltage pulse peak-to-peak output without the complications of setting up an oscilloscope. Furthermore, in once per turn marker and speed applications, it is now possible to adjust the scale factor and check if those adjustments were able to bring the voltage pulse into specification so the vibration monitoring system or speed transmitter system will work properly.

REFERENCES:

1. API 670 Standard for Machinery Protection Systems, American Petroleum Institute

EXPANDING THE POWER OF VIBRATION DIAGNOSTICS USING MAGNETIC DATA

Author: Robert W. Alexander

VIATC VIBRATION INSTITUTE ANNUAL TRAINING CONFERENCE & EXPO 2022 | MFPT ANNUAL CONFERENCE

2022 CONFERENCE PROCEEDINGS

VIATC TRAINING CONFERENCE & EXPO
SAVANNAH, GEORGIA · 2022

EXPANDING THE POWER OF VIBRATION DIAGNOSTICS USING MAGNETIC DATA

Robert W. Alexander
Augury Inc.
263 West 38th Street, 16th Floor, New York, NY 10018, United States

Rob Alexander has been working in the reliability engineering space since 2015 serving global customers in a wide variety of industries (power, oil & gas, food & beverage, raw materials, etc.). Some of the most common services provided during this time have been vibration analysis, machinery/system performance testing and optimization, and reliability audits performed both locally and remotely. He is currently a Vibration Analyst Team Lead at Augury with a Category III vibration certification and a bachelor's degree in Mechanical Engineering from Temple University.

ABSTRACT:
Vibration analysts rely on multiple sources of information as they seek to decipher and interpret asset conditions in real-time. In some cases, as with assets using Variable Frequency Drives (VFDs), this information is not readily available and may require lost time to acquire it, even if that just means walking into a different room to read a line-voltage. Using modern sensor technologies, acquiring more of this "extra" information becomes more straightforward and accessible. One example of this extra information is capturing magnetic data from an asset in conjunction with traditional vibrational data. Magnetic data becomes the source for calculating a number of factors about motors, which can quickly enable vibration analysts to make better and more accurate assessments of asset condition. Factors such as line frequency, motor rpm, slip, or changes in load may lead to improved decision-making regarding asset condition which traditional methods might miss. This presentation will focus on use cases which demonstrate how the use of magnetic data can augment traditional vibration analysis through understanding in real-time how motor conditions affect asset performance.

KEY WORDS: Condition-based monitoring; line frequency; magnetic field strength; order-normalization; slip.

INTRODUCTION:
As sensor and communication technologies continue to become more sophisticated and economical, more reliability experts seek to leverage the power of as many real-time performance data streams as available to improve performance and uptime of their equipment than ever before. Many condition-based monitoring programs are now able to rely on collecting and analyzing vibration data from a remote location in real time as opposed to the traditional manually collected vibration route data. This provides the ability for vibration analysts to review more machines via real-time or near real-time data streams at a much higher recording interval, from all over the world without physical access the machines after vibration sensors are installed.

As is the case today for both remote and traditional manual vibration data collection, being able to leverage vibration data with any available process data from the facility's Data Collection System (DCS) system (when available) is not a seamless integration. In most cases, it typically involves reviewing the available DCS data points, requesting trend data during the collection period for those parameters, and manually reviewing this data for patterns that would indicate signs of a correlation between the multiple parameters reviewed.

Collecting data remotely has its own additional challenges and one of the most prominent is monitoring vibration on machines that are using VFDs as is common in industries such as pharmaceutical, food and beverage, water treatment, building materials, etc. to change the speed of the motor depending on system demand. This makes analysis much more difficult for the vibration analyst as many machine faults (such as unbalance, resonance, cavitation, etc.) will vary in magnitude depending on the turning speed and/or load conditions. Additionally, vibration analysis relies heavily on understanding the prominent vibration peaks' relationship to the running speed of the component being analyzed. Best practices for traditional vibration routes dictate ensuring the machine is operating under similar conditions each time vibration data is taken to ensure there is a fair comparison of the datasets and if vibration data needs to be collected at multiple operating speeds, the turning speed of the machine is recorded at each collection manually to order-normalize the data.

For Augury's remote condition-monitoring application, confidence in knowing the running speed of each asset is critical to the use of both our algorithms and our vibration analysts. This has been achieved by additionally incorporating a hall sensor to collect magnetic data (as will be discussed later) in addition to the vibration data we collect, and it is the intent of this paper to highlight some use cases of leveraging this magnetic data in conjunction with vibration analysis for more accurate machine health analysis. The following examples are all based off of real-world data collected using Augury's current Halo product offering that incorporates both a tri-axial accelerometer and hall sensor integrated into each sensor mounted non-intrusively with a high-strength adhesive to the best position on the outside of the motor closest to the relevant bearing. All trend data presented in this paper was collected at an hourly interval.

DETERMINING LINE FREQUENCY/ SPEED USING MAGNETIC DATA:

Traditional vibration routes can often simply rely on the person collecting the vibration data recording the speed of the shaft with an IR tachometer or stroboscope simultaneously or near-simultaneously to when the vibration data is collected as a run speed reference. This is an easy and reliable method of determining shaft speed but requires the physical presence of the observer and one of these speed measuring tools on hand. The most common technique for recording speed changes from a remote location such as the control room of a plant when monitoring vibration is permanently installing a laser tachometer that is pointed at the machine of interest's shaft with a demarcation of some sort that the tachometer will read each shaft revolution and convert this into a revolutions per minute value. This is highly accurate, but time consuming to install both at the machine and routing this data back to the DCS system. For remoting monitoring beyond the plant's control room, this would be even more difficult to implement.

An alternative to the above methods is to collect magnetic data that can be used to determine motor turning speed. The sensor used to collect all the magnetic data shown in this paper is a hall sensor. This sensor takes advantage of the Hall Effect, the phenomenon observed when a current-carrying conductor is placed into a magnetic field a voltage will be generated perpendicular to both the current and the field. This "Hall Voltage" can be used to calculate various parameters such as magnetic field strength, speed, proximity, and robotic positioning as relevant to the varied applications these are commonly used for. Due to the nature of this phenomenon, the sensor can operate non-intrusively for many applications, including the ones described in this paper. In Figure 1 below, Augury's Halo sensors are shown on a customer motor and pump that is collecting both vibration data and magnetic data.

Figure 1: Sensors shown in blue include a tri-axial accelerometer for vibration data and hall sensor for magnetic data.

For typical motor-driven applications, with the hall sensor and accelerometer, the rated motor nameplate speed data at a defined line frequency, and some clever data filtering (additional filtering is typically required for DC motors and some atypical drivers), a motor's actual turning speed can be calculated with nearly as much precision as the purchased sensor can attain (see Figure 2 below). This allows all received vibration data to be order-normalized regardless of external factors such as speed change, operational demand, product change, etc. Additionally, with speed and vibration data being collected simultaneously, many notable correlations between speed and the vibration data can be detected that would be much more difficult to detect using a traditional vibration route that only collects data periodically.

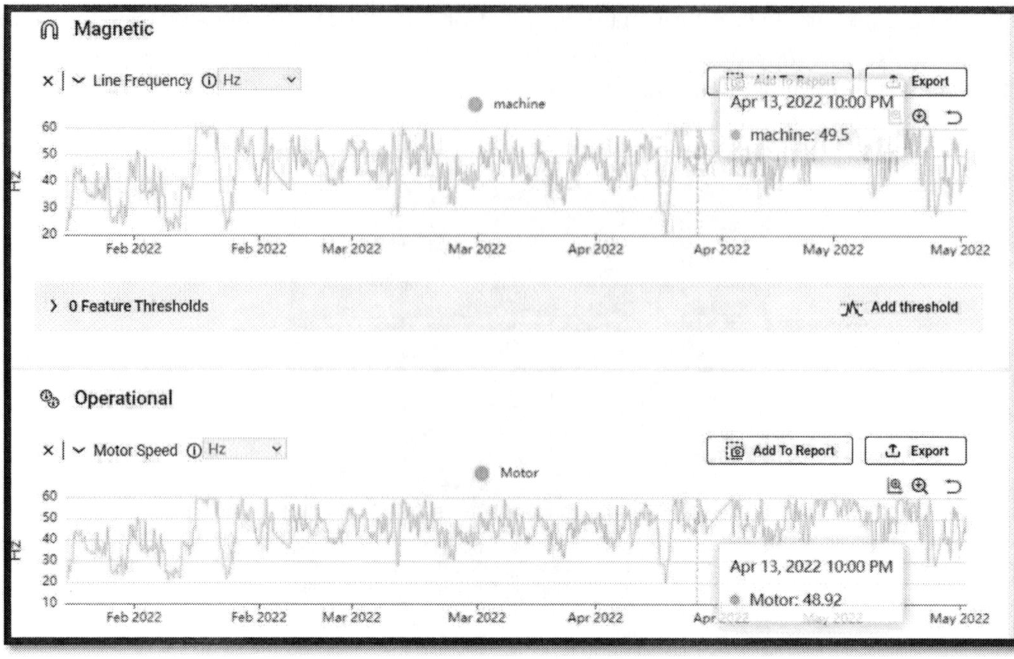

Figure 2: Line Frequency (in Hz – top trend) and Calculated Motor Speed data (in Hz – bottom trend) trend plots

Depending on the magnetism of the turning shaft, speed can also be determined for the non-driven component(s) in a similar fashion, but unless a minimal magnetic field strength is available to employ this accurately, any speed changes downstream of the motor will be more accurately tracked by confirming the speed ratio between the motor and driven component(s).

ORDER TRACKING FOR VFD/FREQUENT SPEED CHANGING MACHINES:
The most common and practical use-case for the magnetic sensor is to be able to accurately order-normalize fault frequencies for machines that can change speed. The data presented in Figure 3 below was collected from the same machine and three-month timeframe shown in Figure 2. In this case, from Figure 3 we can see that the vibration levels at motor sensor 1 (which has a trend for each axis: 1V for vertical, 1H for horizontal, 1A for axial) changed very minimally despite operating speeds varying more than 38 Hz.

This itself is a great insight to consider during analysis, but for spectral analysis, the strength of the magnetic data being used in parallel with vibration data is the ability to easily order-normalize your motor run speed regardless of the operational mode. Figure 4 and Figure 5 shows motor sensor 1 vertical operating data selected randomly at a calculated motor run speed of 59.25 Hz at 60 Hz line frequency and 40.25 Hz motor run speed at 40.75 Hz line frequency, respectively. With this calculated run-speed and standard accelerometer-based vibration tools (spectrum, waveform, trend, etc.), machines being recorded remotely via externally mounted sensors with unpredictable speed changes are not any more difficult to analyze than standard, manually collected data.

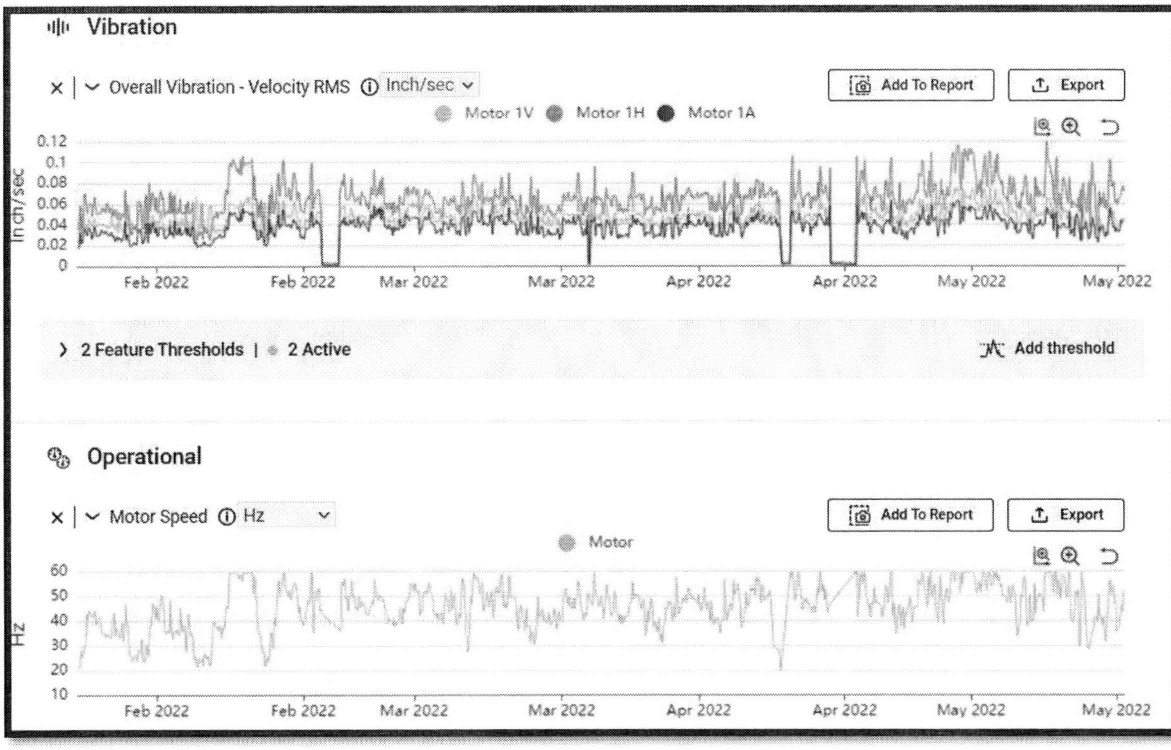

Figure 3: Velocity levels (in in/s RMS – top trend) at motor endpoint 1 show minimal change despited calculated speed (in Hz – bottom trend) varying between 20.67 and 59.29 Hz.

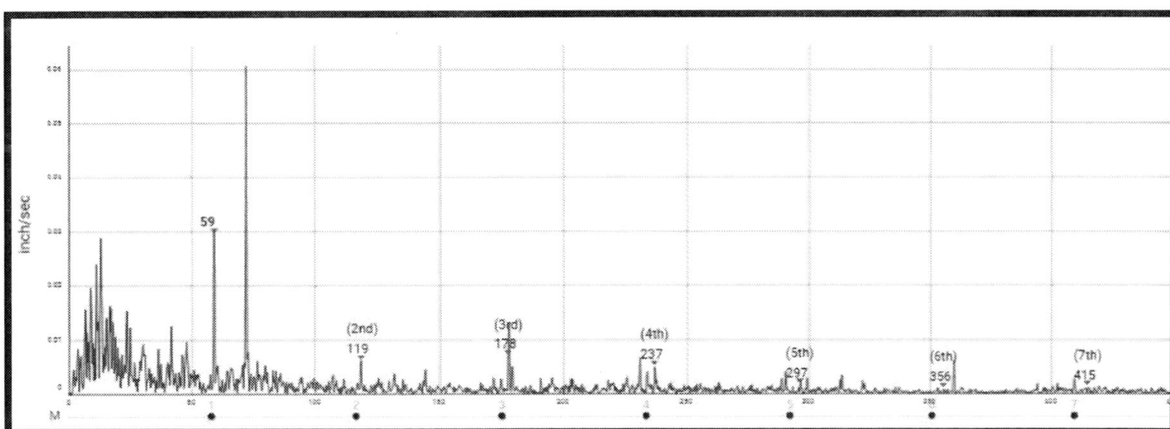

Figure 4: Motor 1V Velocity Spectrum (in in/s RMS) operating at 59.25 Hz motor run speed at 60 Hz Line Frequency.

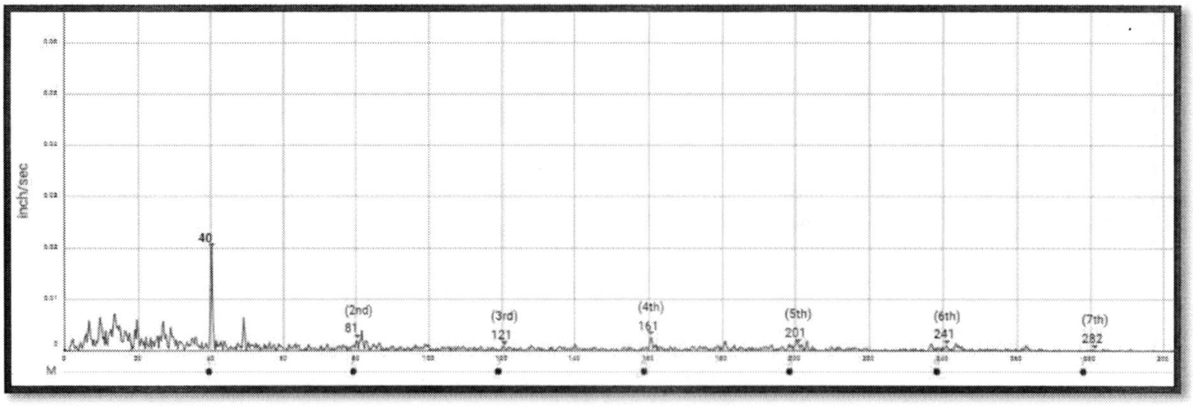

Figure 5: Motor 1V Velocity Spectrum (in in/s RMS) operating at 40.25 Hz motor run speed at 40.75 Hz Line Frequency.

OPERATIONAL CHANGES / RESONANCE:

The second most common scenario that magnetic data can be used to enhance remotely collected vibration data with unpredictable speed changes is the case in which the speed change(s) appear to have a strong correlation to a vibration increase. This can happen for many reasons, but the three general types are: 1. general operational changes (for example, product changeover), 2. vibration fault that worsens as the speed increases or decreases (for example, cavitation can increase at higher or lower speed operations depending on where the pump is operating on its respective pump curve), or 3. vibration fault that worsens only when the motor is operating in a certain range of speeds (for example, resonance).

It is not uncommon that additional data will need to be gathered to differentiate between types 1 and 2 that generally will require more information from the customer (either a production schedule, OEM supplied documentation, or similar) to correctly understand the operating context, but this would also need to be collected when vibration data is collected at the machine locally.

For simplicity visually, a resonance scenario is the easiest scenario to present. The data from Figure 6 is from a vertical pump at a water treatment facility. Due to the frequent demand changes of the water treatment process, small but frequent vibration and speed changes are commonplace. As seen in Figure 6 below, the velocity levels (top trend) have sharp, significant changes frequently while the motor's supplied line frequency (bottom trend) is only deviating a maximum of 12 Hz, with most changes being substantially less. Figure 7 shows a plot of the velocity (in in/s RMS) vs line frequency (in Hz) from the trend in Figure 6, which shows classic resonance behavior when line frequency is between 49.5 and 52.5 Hz line frequency. The actual line frequency was a more valuable turning value for this customer as the motor's VFD sets the line frequency based on system demand. Using this data, the customer was then able to lock out this problematic frequency range on the VFD, resulting in much lower and stable vibration levels.

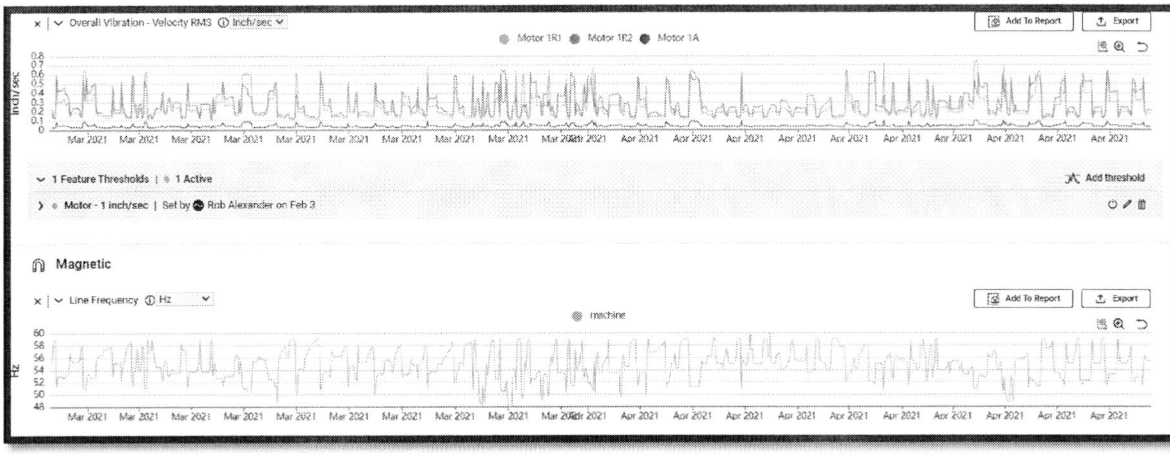

Figure 6: Velocity levels (in in/s RMS – top trend) at motor endpoint 1 show very distinct peaks every time the Line Frequency (in Hz – bottom trend) is between 49.5 and 52.5 Hz and then reduces to a lower vibration value.

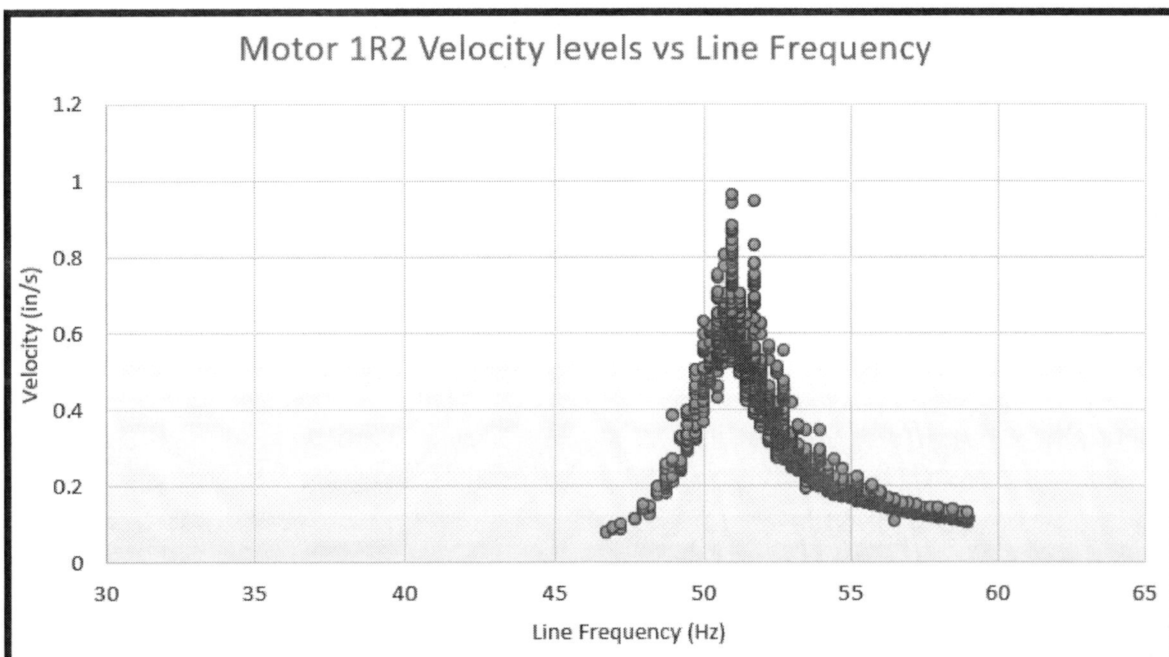

Figure 7: Velocity levels (in in/s RMS) vs Line Frequency (in Hz) plot of the trend data shown in Figure 6.

DETERMINING SLIP/LOAD CHANGES:

To ensure reliable motor performance during various loading conditions, the motor must be able to generate sufficient torque to drive the driven component(s) at the desired speed. For induction motors, as the load's torque requirement increases the speed decreases. This speed difference between the supplied rotating magnetic field known as synchronous speed and the rotor's actual turning speed is referred to as the slip speed as depicted in Figure 8 below.

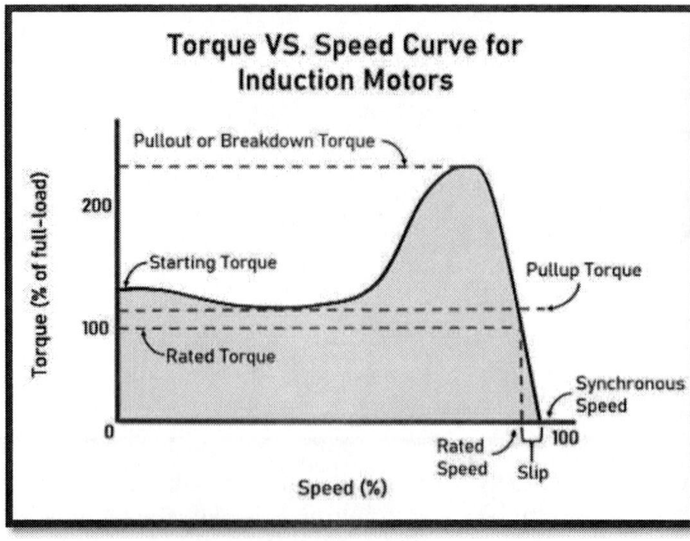

Figure 8: Torque (% of full-load) vs. Speed (in % of full speed) for induction motors where Slip is the difference between rotor speed relative to the synchronous speed.

The following example is from a 3HP motor driving a gearbox to a pulper feeding conveyor system. As seen in Figure 9 below, velocity levels (top trend) for the motor and line frequency (bottom trend) are stable (line frequency varies often, but in the same range) between December 2021 and March 2022. Figure 10 depicts the same line frequency trend from the previous figure, as well as slip that shows a significant increase in February that keeps occurring until the motor inevitably failed on March 17th as the customer was not able to replace the motor in time.

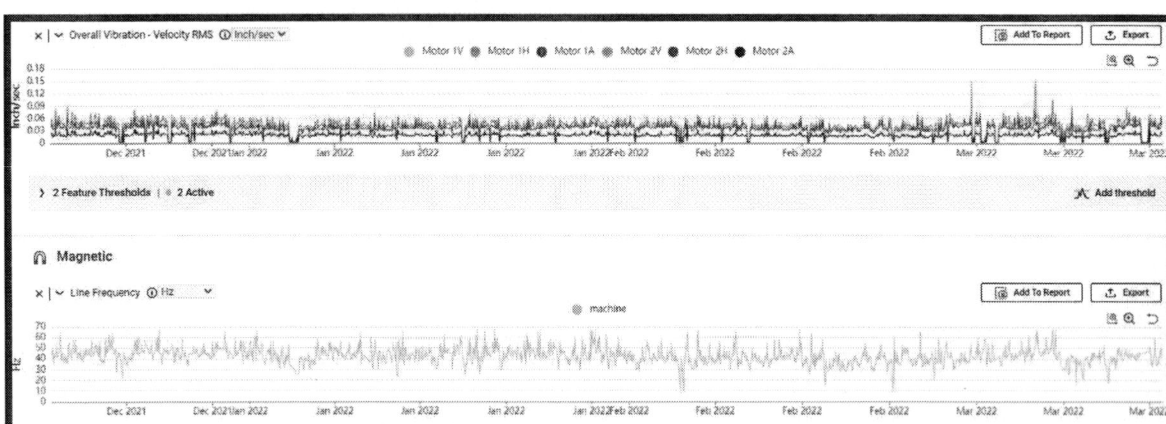

Figure 9: Velocity levels (in in/s RMS – top trend) at the motor endpoints is stable and Line Frequency (in Hz – bottom trend) while showing a lot of variation in magnitude, there is no evidence of a concern.

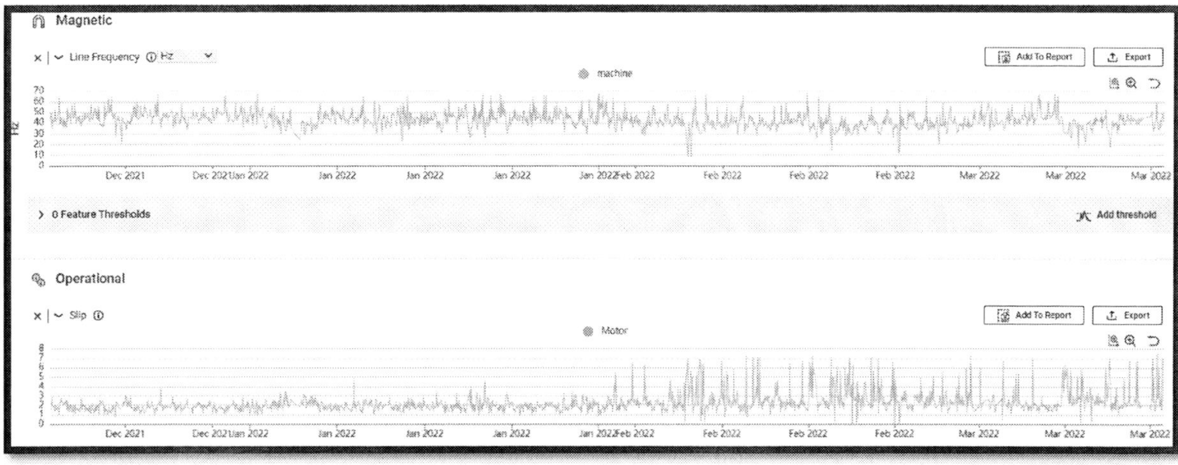

Figure 10: Line Frequency (in Hz – top trend) trend from the previous figure is stable while there is a significant increase in Slip (in %) starting in February that remained apparent at the machine until failure on March 17th.

ENDPOINTS FALLING OFF/MOVED/EXTERNAL EFFECTS:
When recording vibration data, ensuring high quality data is collected for every reading at a repeatable location is essential. In a traditional vibration route, this is typically done by marking with a permanent/semi-permanent ink at the exact locations on the machine data is supposed to be taken each time. Recording vibration data alone does not usually produce a reliable method to determine if there has been a change to the mounting condition of the sensor when data was last recorded. Like operational changes/resonance issues, augmenting the vibration data with magnetic data provides an easy way to leverage the correlation between these two data types and determine if there has been any change to the mounting position.

For the intent of explaining the general working principle, we will review magnetic field strength and vibration data for any change in the vibration data and magnetic data that occurred at the same time. From this discovery, there is a high probability that something in the mounting configuration has changed (sensor has been remounted in a different orientation or location, sensor has fallen off, sensor has been placed on a different machine or replacement machine, etc.) and will assess and recommend the mounting be adjusting as needed to continue recording data at the same position as previous data. With a standard hall sensor and accelerometer there are many different types of magnetic data and vibration data that can be compared simultaneously to increase the accuracy of detecting each type of the described phenomena.

Similarly, combining magnetic data with vibration data also allows us to easily identify situations in which vibration detected at the machine is not due to the machine being observed (for example, being transferred through the floor or common piping from another machine). This is done simply by observing a change in the vibration levels when the magnetic field strength signature is not indicating that the motor is turning. Figure 11 shows a slight reduction in motor vibration levels in late October and November 2021 in the trend after the pump was replaced between September and October 2021. Figure 12 then shows how easy it is to detect that the motor is not running despite only a minor reduction in acceleration levels when compared with actual run data. When the customer was alerted regarding this, it was discovered that they had added a secondary pump and common piping during the repair also and this vibration data was being transferred from a new pump we were not monitoring.

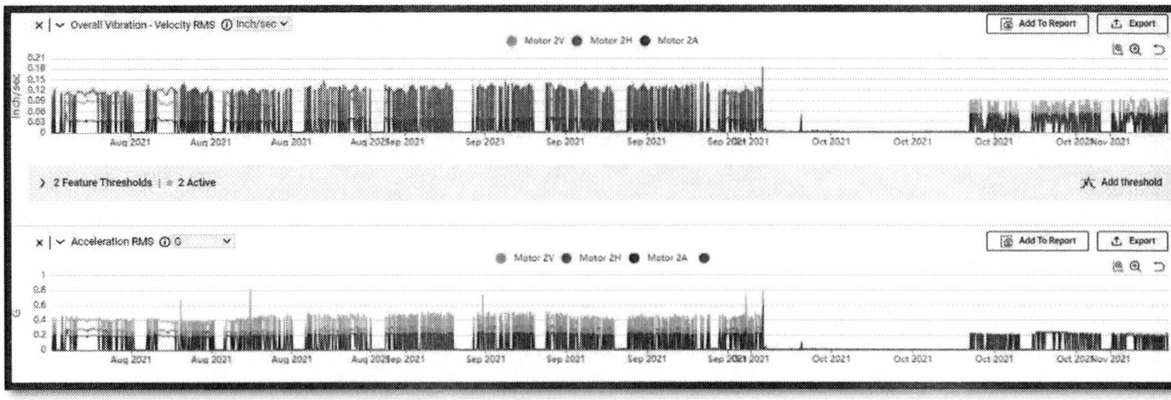

Figure 11: Velocity (in in/s RMS – top trend) and Acceleration (in G RMS – bottom trend) at the motor show a small reduction after the pump was replaced between Sep. 2021 and October 2021.

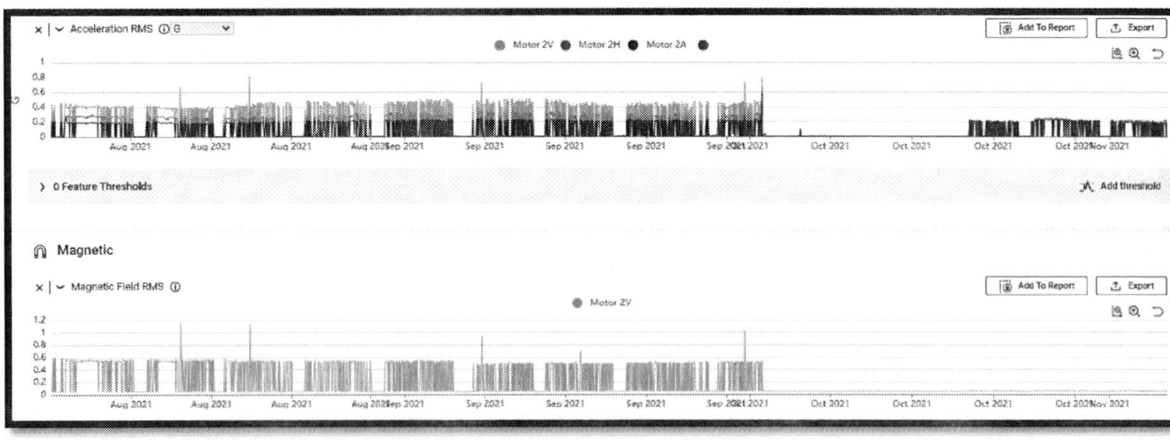

Figure 12: Acceleration (in G RMS – top trend) from the previous figure shows a minimal change after the pump was replaced and Magnetic Field (in Gauss RMS)

CONCLUSION:

The intent of this paper was to highlight how the simultaneous collection of magnetic data in conjunction with vibration data can be leveraged to strongly improve the ease and accuracy of vibration analysis, as well as provide additional machine health insights that would be difficult to detect with vibration data alone. With just this one additional piece of data, vibration monitoring on complex machines with multiple operational modes without being physically near the machine is made simple, as we can leverage a lot more context of the machine's operating condition that can be a challenge to collect simultaneously even when physical access to the machine is possible. As data processing, communication, and instrumentation technologies continue to mature, the ability to simultaneously receive data from many other relevant parameters such as temperature, pressure, and flow to leverage with vibration data easily to assess asset performance at an economical price appears inevitable.

BIBLIOGRAPHY:

Erik.Secula@nist.gov. (2019, December 6). *Hall Effect Measurements*. NIST; National Institute of Standards and Technology. https://www.nist.gov/pml/nanoscale-device-characterization-division/popular-links/hall-effect

FINITE ELEMENT REDESIGN TO MITIGATE STRUCTURAL RESONANCE

Author: Alex Tomsick, BSME

VIATC VIBRATION INSTITUTE ANNUAL TRAINING CONFERENCE & EXPO 2022 | MFPT ANNUAL CONFERENCE

2022 CONFERENCE PROCEEDINGS

VIATC TRAINING CONFERENCE & EXPO

SAVANNAH, GEORGIA · 2022

FINITE ELEMENT REDESIGN TO MITIGATE STRUCTURAL RESONANCE

Alex Tomsick, BSME
Red Wolf Reliability
912 Smith Field Dr. Suite 4
Fort Collins, CO 80524

Since graduating from Colorado State University with a Bachelor of Science in Mechanical Engineering, Alex Tomsick has worked for Red Wolf Reliability troubleshooting vibration and monitoring the performance and health of reciprocating equipment in many industries since May 2016. Specializing in high-end vibration troubleshooting and structural redesign using Modal, Operating Deflection Shape, and Finite Element Analysis techniques, Alex has helped customers across the United States mitigate damaging vibration. Certified as a Category III vibration analyst through the Vibration Institute, Alex Tomsick serves as a technical lead for Red Wolf Reliability.

ABSTRACT:
Resonance is an aggravating vibration condition that exists when a natural frequency of a system is coincident with a forced vibration frequency. When a resonant condition is excited, existing machine or structural vibration amplitudes are amplified leading to severe vibration amplitudes. Severe vibration amplitudes were noted in route-based condition monitoring data from two axial flow fans at fan running speed. Given the high and directional vibration amplitudes at fan running speed it was theorized a structural resonance or structural weakness was aggravating forced vibration amplitudes. Through the collection of operating deflection shape (ODS) and Modal data, it was found that the 1st bending mode of the fan structure was only separated from fan running speed by 8.4% and was being excited. To mitigate the resonant condition causing severe vibration amplitudes, a finite element model was created and validated with the field modal data. Once validated, structural support additions could be tested in the finite element model to shift the natural frequencies of the fan structure away from forced vibration frequencies. The recommended structural modifications have since been added to the fan structure, and the vibration amplitude at fan running speed decreased from 1.32 ips_{pk} at its highest to 0.01 ips_{pk}.

KEY WORDS: Bump test; finite element analysis; modal testing; operating deflection shape; resonance.

INTRODUCTION:

Two axial flow fans at a refinery had continually shown high vibration amplitudes at the motor and fan bearing locations as well as on the structure supporting each fan. Seen throughout route-based condition monitoring data, the highest vibration amplitudes measured on the motor and fan bearings were occurring perpendicular to the belt direction with the highest vibration amplitudes seen at fan running speed (498 cpm). Based on the extreme directional vibration amplitudes measured on motor and fan bearings, as well as the fan structure it was theorized that structural resonance or structural weakness was aggravating forced vibration amplitudes at fan speed.

OPERATING DEFLECTION SHAPE ANALYSIS:

The structure supporting the two axial flow fans consisted of two separate cantilever structures suspended from a heat exchanger. One cantilever structure supported each axial flow fan machine train consisting of a vertically mounted motor driving a vertical axial flow fan through a belt drive. Figure 1 shows a model of the original axial flow fan structure.

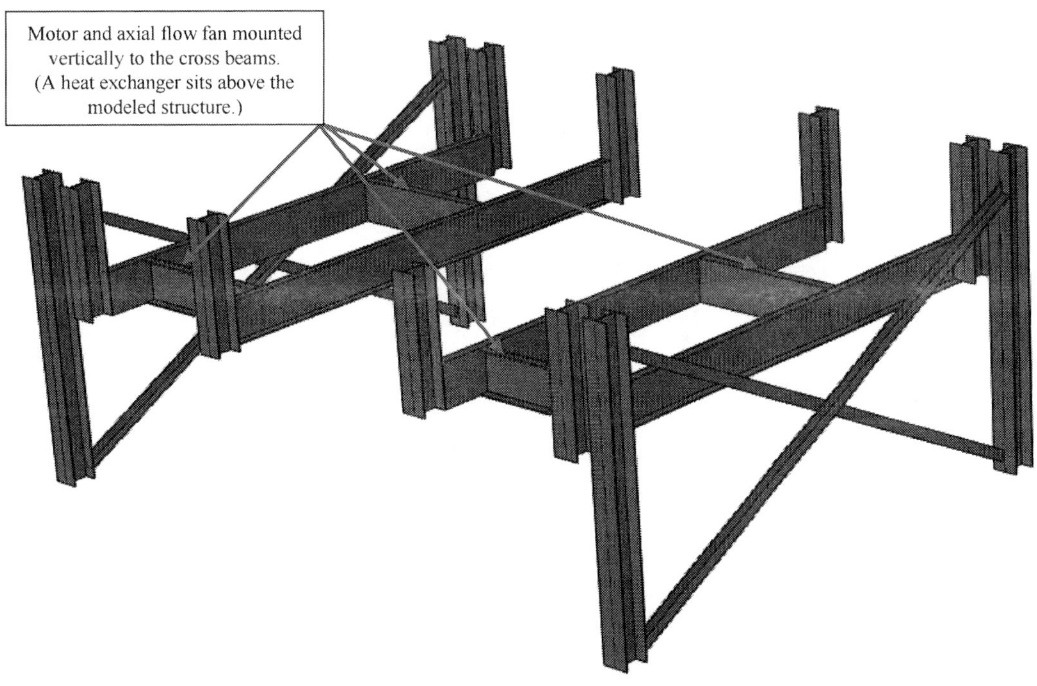

Figure 1: Model of the Original Structure

To better understand the forced response of the axial flow fan structure during the operation of both fans, operating deflection shape data was collected across the fan structure excluding the heat exchanger.

Figure 2 shows the undeflected model of the structure that was generated in the ODS animation software.

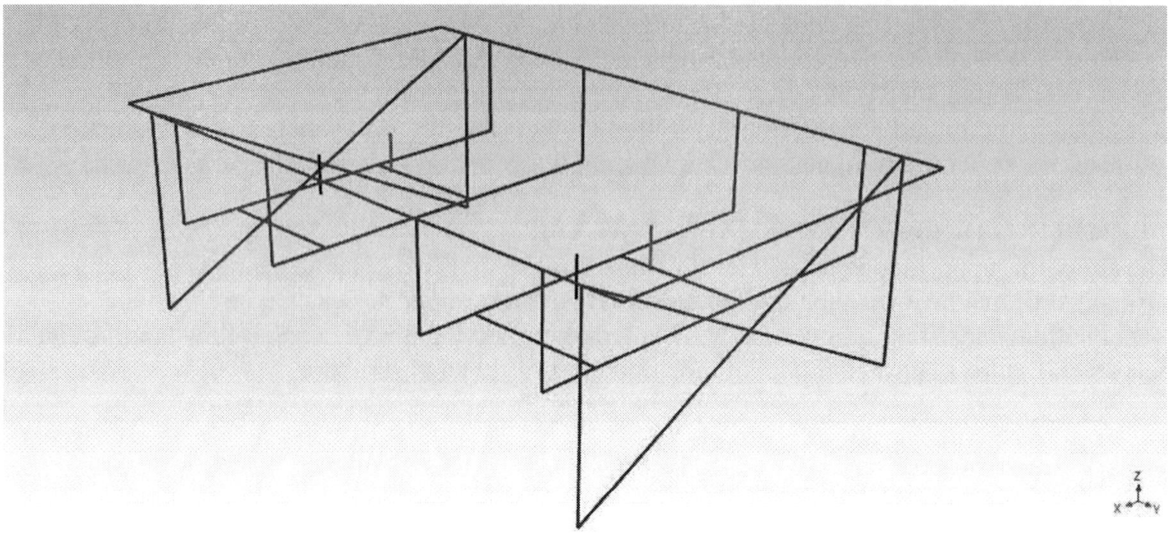

Figure 2: Undeflected ODS Model

ODS data were collected at the following locations:

- Structural steel beams (blue)
- Motor bearing locations (red)
- Fan bearing locations (black)

From the ODS data collected on the structure and bearing locations for both machine trains, it was found that the dominant vibration was occurring at fan running speed.

Figure 3 shows the operating deflection shape of the axial flow fan structure at 483 cpm from above.

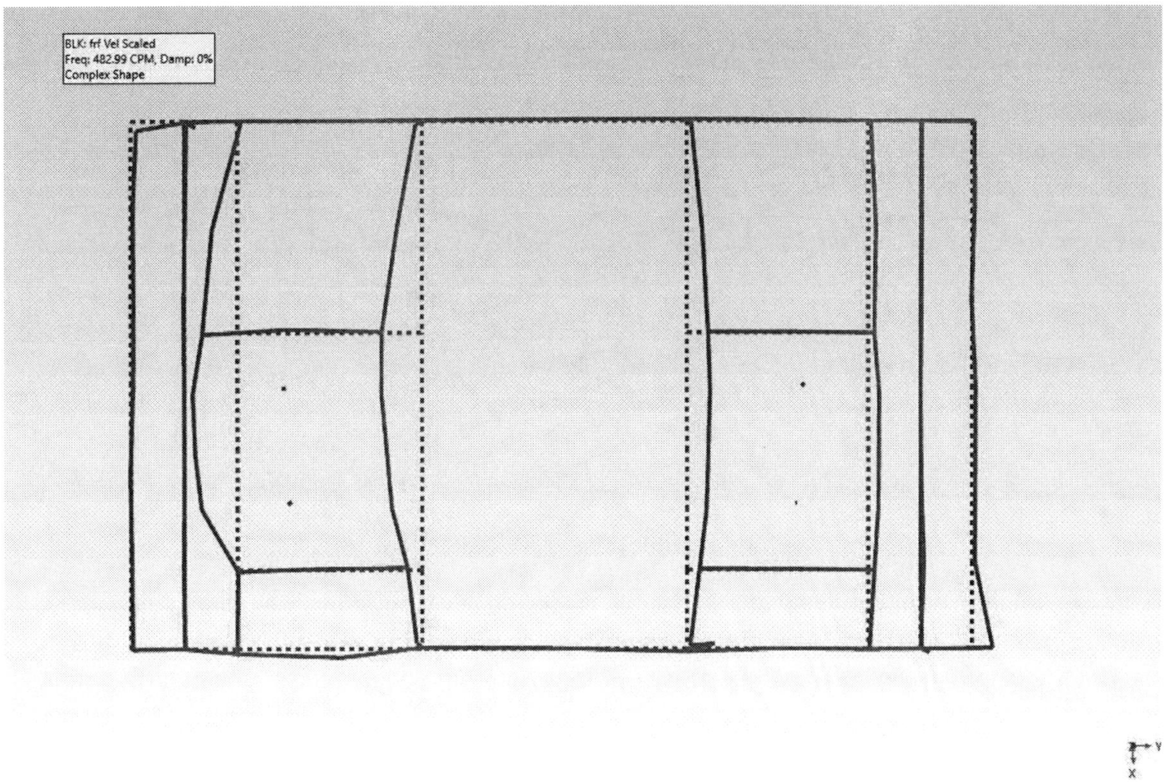

Figure 3: Structural Operating Deflection Shape at 483 cpm

The operating deflection shape of the tested structure exhibited the 1st horizontal bending mode of each cantilever structure. Given this, it was suspected this operating deflection shape was caused by excitation of a structural natural frequency of the cantilever support structure.

MODAL ANALYSIS:

To verify the observed deflection shape at 483 cpm was related to excitation of a structural natural frequency, modal data were collected on each cantilever structure. Given the suspected natural frequency near 493 cpm was related to the bending of the cantilever structure supporting each machine train, modal data were collected only on the cantilevered sections of the structure (red).

Figure 4 shows the modeled structure highlighting the locations tested during the modal collection.

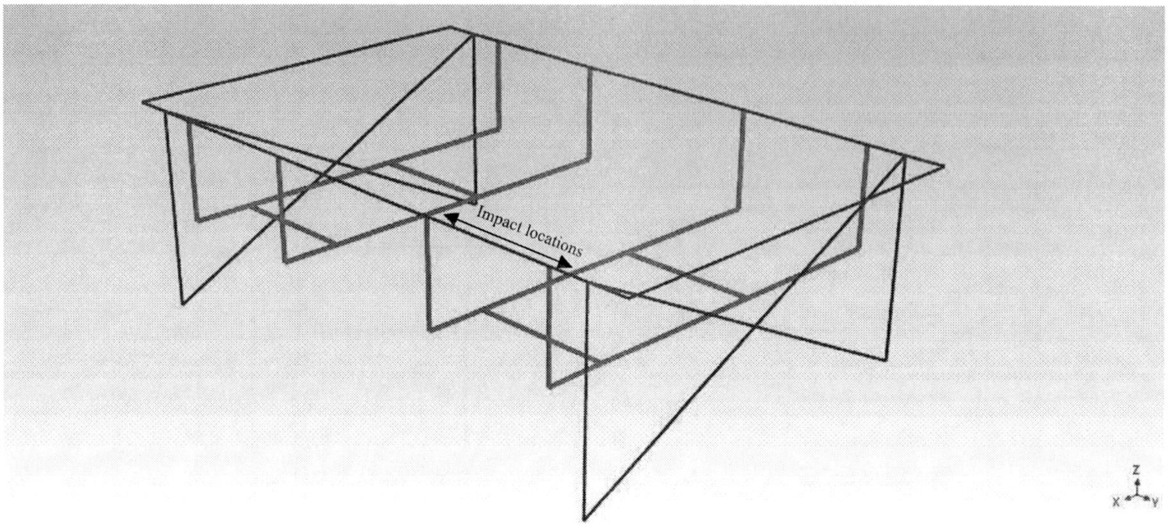

Figure 4: Undeflected Modal Model

To ensure the best structural excitation possible each cantilever structure was impacted and tested separately. Each cantilever structure was impacted in the middle of the C-channel beams that run the length of the cantilever portions of the structure perpendicular to the belt direction. From the modal data, it was found that the natural frequency related to the 1st horizontal bending mode of the cantilever structures was 540 cpm.

Figure 5 shows the overlaid frequency response functions (FRFs) measured on one of the cantilevered structures.

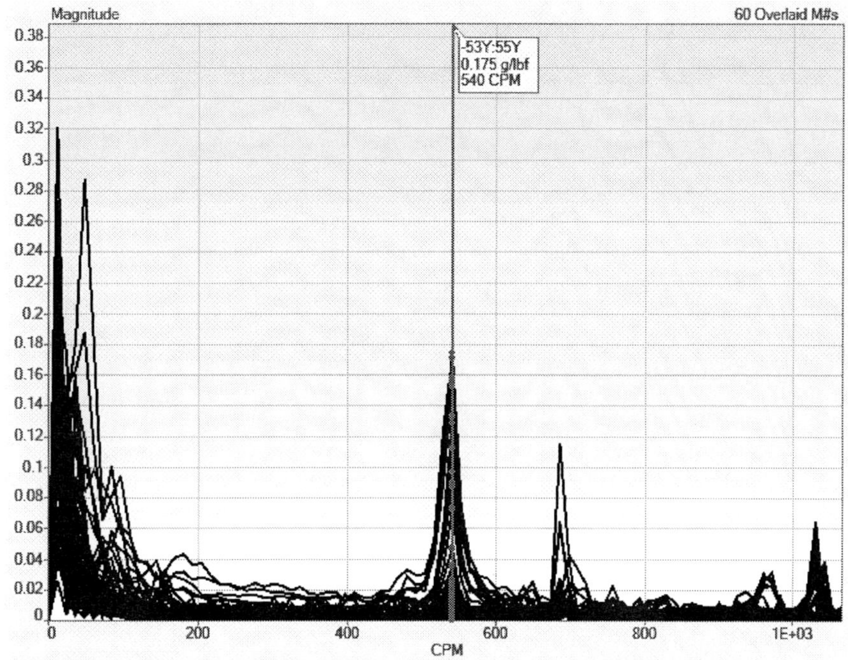

Figure 5: Overlaid FRFs Showing Measured Natural at 540 cpm

Figure 6 shows the 1st horizontal bending mode of the cantilever structures at 540 cpm as measured in the field modal data.

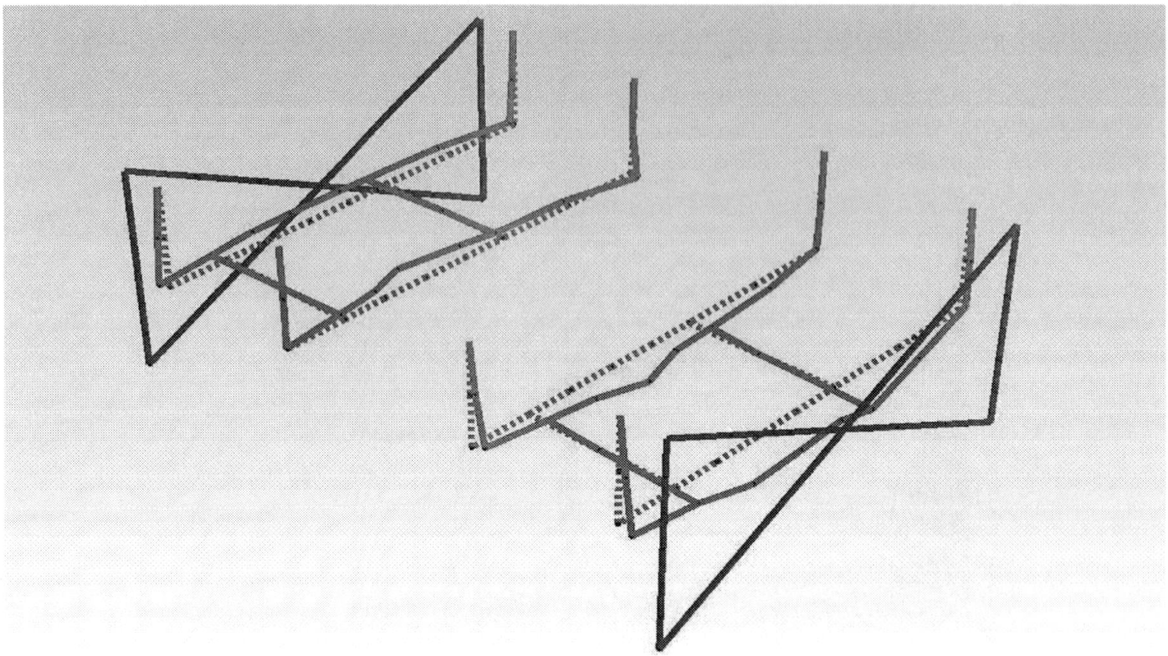

Figure 6: 1st Horizontal Bending Mode of Cantilever Structures at 540 cpm

Given the proximity of the 1st horizontal bending mode of the cantilever structures, and the observed excitation of this natural frequency found in the ODS analysis, a structural redesign of the cantilevered structure was recommended. Design modifications of the cantilevered structures were tested using finite element (FE) modeling techniques to shift the 540 cpm natural frequency away from fan running speed.

FINITE ELEMENT MODELING:
To create a finite element model, a CAD model of the structure was created. This model was based on the drawings and other information provided along with physical dimensions taken in the field for added detail and to account for any variations seen between the drawings and as-built dimensions. The CAD model was then imported into an FE modeling software. The structural steel members (angle iron, C-channel, and I-beams) were modeled using beam elements, while the motor, fan, and heat exchanger were modeled as lumped masses to simulate their weight and inertia.

When creating the FE model, the following boundary condition assumptions were made:

- All connections between structural steel members were modeled as rigid, by constraining all six degrees of freedom across connections.
- A fixed constraint was modeled at the bottom of the large vertical I-beams where the fan structure meets the bottom of the deck fixing the fan structure in place.

- The heat exchanger lumped mass was connected to the top of all vertical beams with rigid elements.
- The motor and fan lumped masses were connected to the long C-channel beams with stiff, three degree-of-freedom spring elements. The spring elements' stiffness in each direction was modified until the natural frequencies and mode shapes of the structure were consistent with the measured natural frequencies from the modal data.

Figure 7 shows the initial FE model.

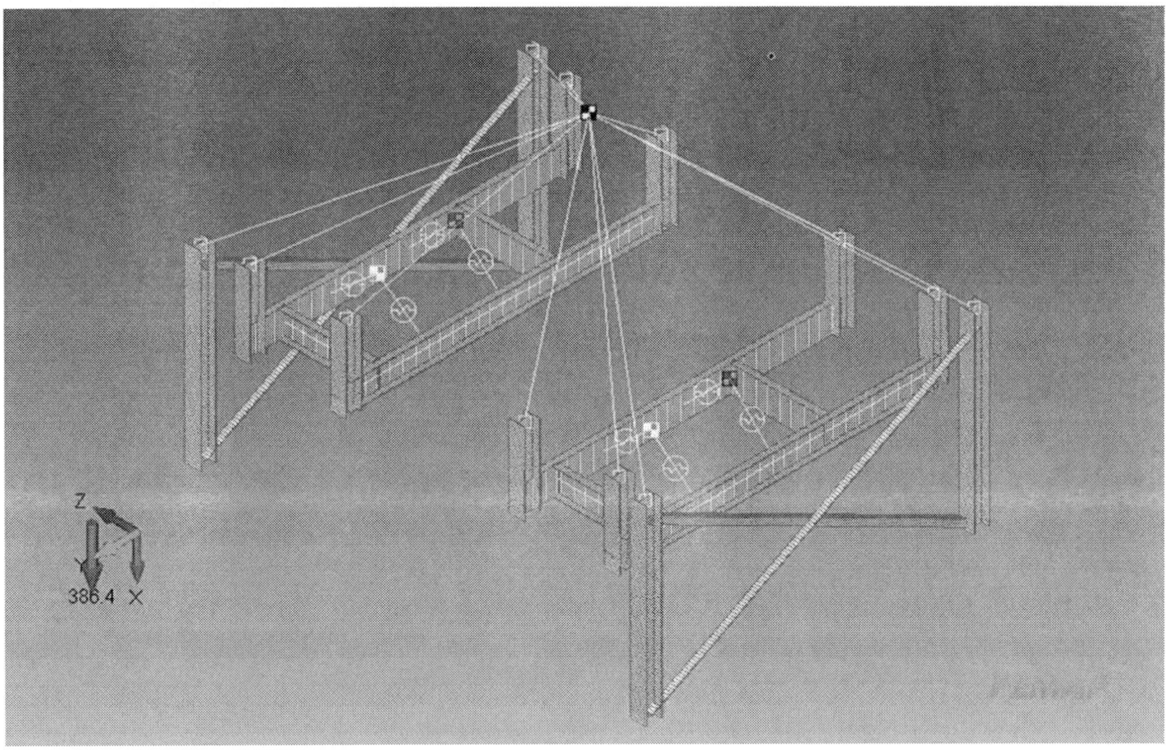

Figure 7: Initial FE Model

FINITE ELEMENT MODEL VALIDATION:

Based on the modal analysis, several natural frequencies were chosen to compare to the FE model. To validate the FE model, the output natural frequencies from the model must include the same mode shapes and natural frequencies as those found in the field modal data. Several natural frequencies from the field modal data were chosen to validate the FE model with the main natural frequency of interest being the 1st horizontal bending mode of the cantilevered structure at 540 cpm. The natural frequency corresponding to the 1st horizontal bending mode of the validated FE model was 10.48 Hz compared to the actual measured natural frequency of 9 Hz from the field modal data.

Figure 8 shows the 1st horizontal bending mode of the cantilevered structure at 10.48 Hz calculated from the FE model.

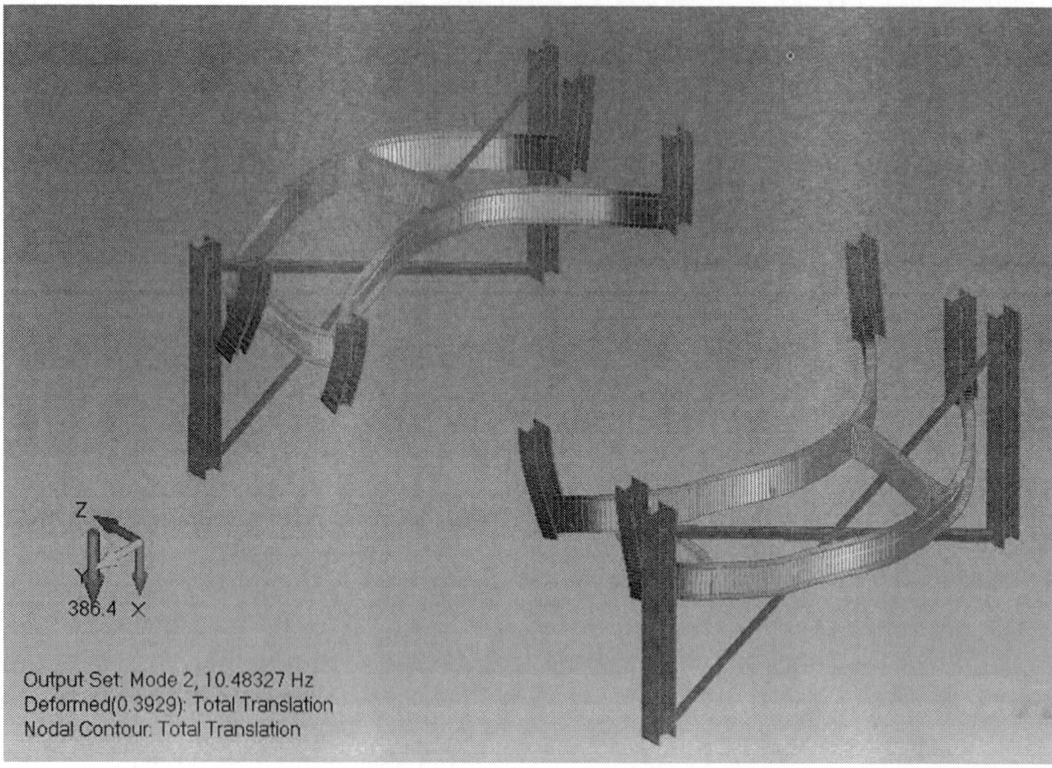

Figure 8: 1st Horizontal Bending Mode from the Validated FE Model

FINITE ELEMENT STRUCTURAL REDESIGN:

Using the validated FE model structural design modifications were implemented in the model to test how the modifications shifted the structure's natural frequencies. To ensure there is no interaction between the structural natural frequencies and the forcing frequencies of the machine trains design modifications were chosen to ensure a 25% separation between natural frequencies and forcing frequencies.

Table I: Expected Machine Forcing Frequencies

Description	Forcing Frequency (CPM)	Forcing Frequency (Hz)
Belt Frequency	305	5.08
Fan Running Speed	498	8.30
2X Fan Running Speed	996	16.60
Motor Running Speed	1765	29.42
2X Motor Running Speed	3530	58.83

Table I outlines the expected machine forcing frequencies generated by the machine trains. However, based on the ODS field data and data collected during route-based condition monitoring collections the only forcing frequencies that had historically caused high vibration amplitudes were fan running speed and 2X fan running speed. For this reason, it was decided that any design modifications must focus on maintaining at least 25% separation between structural natural frequencies and the machine forcing frequencies of fan running speed and 2X fan running speed.

Using the validated FE model, it was found that adding three additional steel members to the existing structure could ensure adequate separation between structural natural frequencies and the machine forcing frequencies of fan running speed and 2X fan running speed. Figure 9 shows a CAD model including the additional steel members shown in blue.

Figure 9: CAD Model Including Design Modifications

The additional steel members were added to the validated FE model using beam elements and constrained to the existing structure using rigid connections, constraining these connections in all six degrees of freedom. To match the modeled stiffness added by these cross braces, all connections when installed on the actual fan structure were welded connections. Table II shows how the addition of these structural members shifted the expected natural frequencies compared to the original measured natural frequencies from the field modal data.

Table II: Comparison of Natural Frequencies from Field Modal Data and the Modified FE Model

Description of Natural Frequency Mode Shape	Natural Frequency Measured in Field Modal Data (Hz)	Nearest Forcing Frequency (Hz) to the Measured Natural Frequency from the Field Modal Data	Expected Natural Frequency from the Modified FE Model (Hz)	Nearest Forcing Frequency (Hz) to the Expected Natural Frequency from the Modified FE Model
1st Horizontal Bending Mode of the Cantilevered Structure	9	8.3 (1X Fan Speed)	21.51	16.6 (2X Fan Speed)
2nd Horizontal Bending Mode of the Cantilevered Structure	28	29.42 (1X Motor Speed)	30.8	29.42 (1X Motor Speed)
1st Vertical Bending Mode of the Cantilevered Structure	42.2	29.42 (1X Motor Speed)	33.45	29.42 (1X Motor Speed)
1st Bending Modes of Vertical I-beams Near Fan	56.8	58.83 (2X Motor Speed)	58.42	58.83 (2X Motor Speed)

The design modifications added to the FE model increased the natural frequency associated with the 1st horizontal bending mode of the fan substructures far away from any expected forcing frequencies. This natural frequency was the highest concern due to its original proximity to the operating speed of the axial flow fans, which was causing a resonant condition of the structure. Also seen in the table, there were structural natural frequencies close to motor running speed and 2X motor running speed before and after the design modifications were added to the structure. Figure 10 shows the measured ODS response on the fan trains and the structure at each of the expected forcing frequencies.

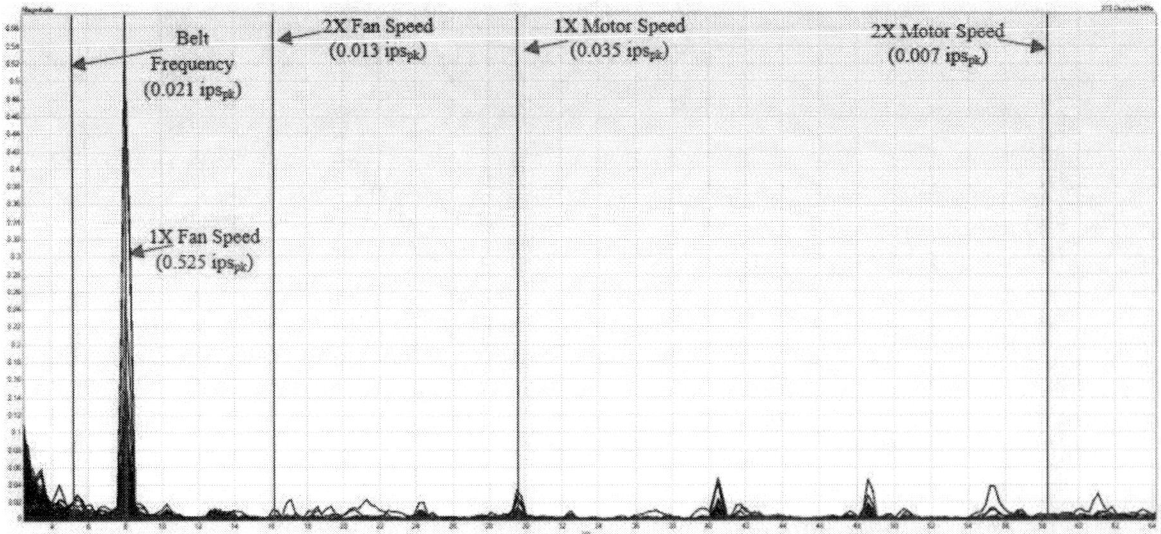

Figure 10: Measured ODS Response at Expected Forcing Frequencies

Based on the measured response at each machine forcing frequency in the ODS data the only forcing frequency that showed severe response was fan speed, even though there were originally natural frequencies found in the modal analysis at fan speed, motor speed, and 2X motor speed. Since there were not large amplifications in vibration amplitudes seen on the structure at motor speed and 2X motor speed in the ODS data, it was suspected there was not a high enough input force from the motor to excite nearby natural frequencies. For this reason, natural frequencies close to motor speed and 2X motor speed with the cross braces installed are less of a concern when compared to natural frequencies near fan speed and 2X fan speed.

RESULTS AFTER IMPLEMENTING DESIGN MODIFICATIONS:

After providing design recommendations to the end customer based on the FE modeling approach the design modifications were implemented on the existing structure and a large decrease in vibration amplitudes on both the motor and the fan structure was seen. At its worst, the highest vibration amplitude measured on either motor was 1.36 ips_{pk} perpendicular to the belt direction. After implementing the recommendations, the vibration amplitude at fan running speed at the same motor location was 0.01 ips_{pk}. Figures 11 and 12 show two spectra measured at the same location on one of the motors before and after implementing the design modifications.

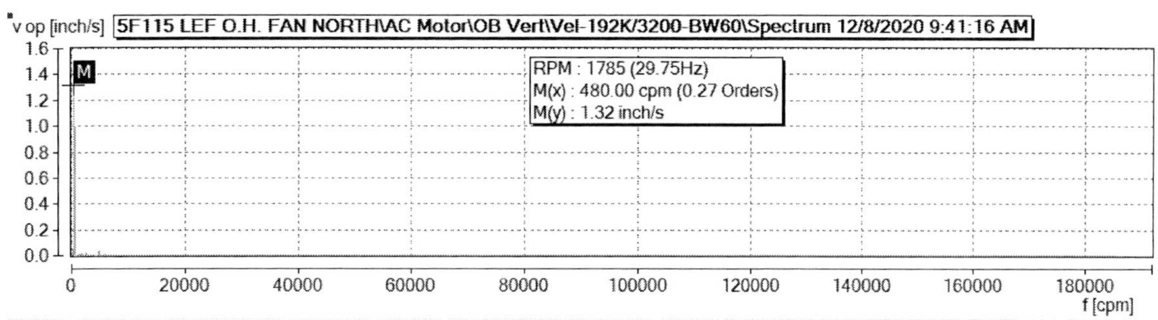

Figure 11: Vibration Spectrum Before Implementing Design Modifications

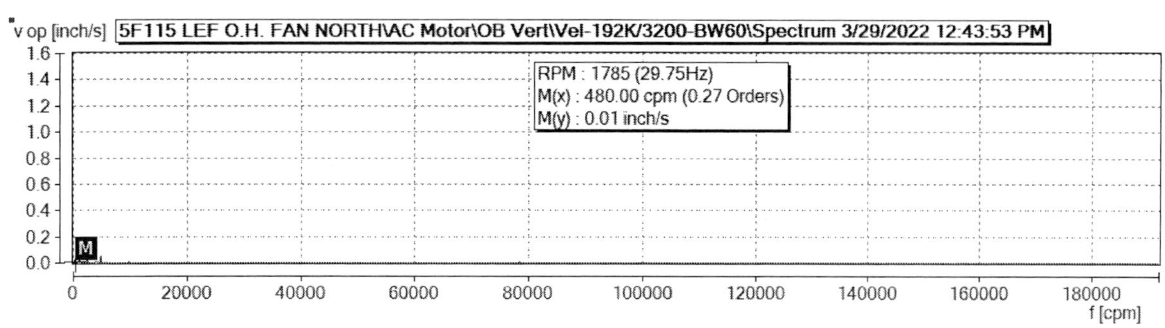

Figure 12: Vibration Spectrum After Implementing Design Modifications

CONCLUSION:

Given the extreme vibration amplitudes measured on the two axial flow fans throughout the data collection history gathered during route-based condition monitoring collections, the presence of a resonant condition was theorized. The 1st horizontal bending mode motion exhibited in the ODS animation at each axial flow fan's operating speed supported the theory of a structural resonance. Using the modal data collected on each cantilevered structure it was verified a structural natural frequency was present at 540 cpm. The measured field modal data could be used to validate the

FE model to ensure the boundary conditions used in the model were consistent with the real structure. Once validated with measured field data, FE modeling techniques could be used to rapidly test different design modifications to the cantilevered structures to shift natural frequencies away from machine forcing frequencies and mitigate future structural resonance. By using all of these techniques, design modifications to complex structures can be tested and verified quickly saving the end client time, money, and resources had they tested different structural changes on the existing structure.

INDUSTRIAL EQUIPMENT CASE HISTORIES: THINGS ARE NOT ALWAYS AS THEY SEEM

Author: Gregory E Henson

VIATC | TRAINING CONFERENCE & EXPO
SAVANNAH, GEORGIA · 2022

INDUSTRIAL EQUIPMENT CASE HISTORIES: THINGS ARE NOT ALWAYS AS THEY SEEM

Gregory E Henson
EMDS Electro-Mechanical Diagnostic Services
1550 E. Washington St.
Indianapolis, In

Greg Henson is a mechanical engineer graduate from Rose-Hulman Institute of Technology, with a level III certification from the vibration institute. He is a certified CMRP from the SMRP organization and has extensive experience in oil analysis, thermography and performance testing. He has 40 years' experience in the equipment reliability field with 20 years spent in the power industry and 20 years in a large pharmaceutical manufacturing plant. He is currently working as a part time consultant with Electro-Mechanical Diagnostic Services. .

ABSTRACT:
Vibration analysis and trending are valuable tools to diagnose the majority of equipment issues. However, sometimes vibration issues are not always as they seem. Without the use of past experiences with the equipment and discussions with the operators that use the equipment, a diagnosis based solely on vibration data may not be correct. The following case histories illustrate how misdiagnosis can occur if considerations to past issues are not included when tracking down the root causes of problems. The case histories will include a 400-ton chiller and its motor, a 30,000 CFM blower and its 2500 Hp motor.

KEY WORDS: Blowers; chillers; motors; oscillations

CASE HISTORY #1:

The first example of this is with a 685 Hp motor driving a 400-ton chiller. The equipment operates continually, cooling an exothermic reaction. The load on the chiller varies from 40% to 100% load. At the lower loads the chiller can go into a surge condition which will cause the entire machine to experience a brief vibration and noise condition.

THE USE AND ABUSE OF THE SINGLE DEGREE-OF-FREEDOM VIBRATION MODEL

Author: Robert J. Sayer, PE

VIATC VIBRATION INSTITUTE ANNUAL TRAINING CONFERENCE & EXPO 2022 | MFPT ANNUAL CONFERENCE

2022 CONFERENCE PROCEEDINGS

VIATC TRAINING CONFERENCE & EXPO
SAVANNAH, GEORGIA · 2022

THE USE AND ABUSE OF THE SINGLE DEGREE-OF-FREEDOM VIBRATION MODEL

Robert J. Sayer, PE
Owner, Applied Structural Dynamics, Westerville, OH
Past President, The Vibration Institute, Oak Brook, IL

Robert J. Sayer, P.E. is the Past President of the Vibration Institute and Owner of Applied Structural Dynamics (Westerville, Ohio). He has over forty-five years of experience as a mechanical and structural engineering consultant to a wide range of industries including pulp & paper, mining & metals, fossil-fuel & nuclear power generation, petrochemical, food processing, glass manufacturing, and aerospace & defense. Mr. Sayer provides analytical and experimental dynamic analysis of mechanical equipment for design projects, structural and mechanical failure studies, and process reliability and efficiency studies. Mr. Sayer holds engineering degrees from Ohio University (BSCE), Cleveland State University (MSIE), and Purdue University (MSCE). He had the honor and fortune of collaborating with the late Art Crawford, a pioneer of vibration analysis, on numerous projects. Mr. Sayer is a regular contributor at technical conferences and has published over 50 technical articles and co-authored 2 chapters in "The Simplified Handbook of Vibration Analysis".

ABSTRACT:
The single degree-of-freedom (sdof) spring-mass model has been used for years as a teaching tool at engineering schools and the Vibration Institute to illustrate concepts involving natural frequencies and the vibration response of structural-mechanical systems. Although there are few direct real-world applications of this model, it is still used in industry as an analysis/troubleshooting/design tool. Most real machines and structures are too complex to accurately predict vibration response using the sdof model. By extending the model to two degrees-of-freedom (2 springs – 2 masses), it can be applied to a significantly larger number of real-world applications. An example would be the response of a simple system with a dynamic vibration absorber attached.

The equations and calculations for the 2-dof model are significantly more complex than the sdof model but are still manageable by hand calculation. When the dynamic response of an even more complex machine and/or structure is needed, numerical techniques, such as finite element analysis (FEA), are available to provide an approximate solution. The computational intensity of these types of analyses requires the use of a computer.

KEY WORDS: Degree-of-freedom, dynamic vibration absorber, equivalent springs, finite element analysis, mechanical vibration, modal participation, tuned mass damper.

INTRODUCTION:

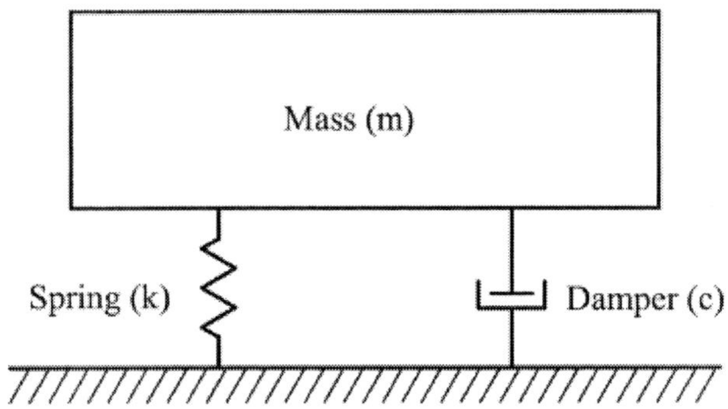

Figure 1: Sdof Model

The single degree-of-freedom (sdof) spring-mass model (Figure 1) is commonly used to demonstrate the dynamic response (vibration) of structural-mechanical systems to dynamic loading. The sdof model consists of a single rigid mass (m) supported by a single elastic spring (k). The motion of the mass is constrained in the horizontal direction and allowed only to move in the vertical direction.

Many real systems contain multiple elastic elements, each having its own stiffness. If a single degree-of-freedom model is to be used, then the stiffness of all the elastic elements must be replaced by a single equivalent stiffness. Figure 2 shows springs in series and parallel. The formulas for these cases are given in Equation 1 and Equation 2.

$$k = \frac{k_1 k_2}{k_1 + k_2} \text{ series} \quad \text{or} \quad \frac{1}{k_{eq}} = \frac{1}{k} + \frac{1}{k_2} + \ldots\ldots \qquad \textbf{Equation 1}$$

$$k = \frac{(a+b)^2}{\left(\frac{a^2}{k_2} + \frac{b^2}{k_1}\right)} \qquad \textbf{Equation 2}$$

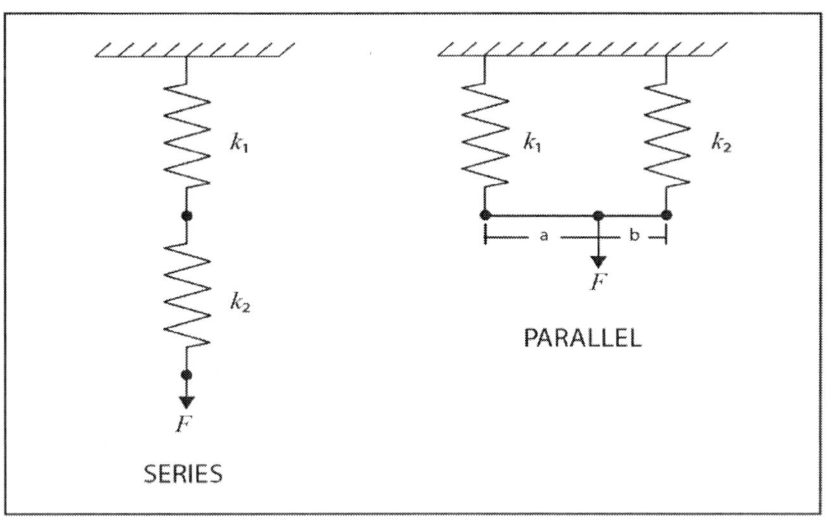

Figure 2: Equivalent Spring Types

For the special case of parallel springs where a = b, the equivalent spring stiffness is given by Equation 3:

$$k_e = k_1 + k_2 \qquad \textbf{Equation 3}$$

After accounting for all the components of the elastic supporting system as an equivalent spring, the damped natural frequency (f_n) of the sdof model can then be calculated using Equation 4.

$$f_n = (1/2\pi)[k_e/m]^{1/2}[1 - \zeta^2]^{1/2} \qquad \textbf{Equation 4}$$

where ζ = damping factor expressed as a percentage of critical damping.

The dynamic response (x) of the system when subjected to a dynamic force (F_{dyn}) applied at a frequency (f_d) is then calculated using Equation 5.

$$x = F_{dyn}/k_e[\{1-(f_d/f_n)^2\}^2 + \{2\zeta(f_d/f_n)\}^2]^{1/2} \qquad \textbf{Equation 5}$$

The sdof equation for natural frequency provides important insight into resonance. It shows that, generally the natural frequency increases as the stiffness is increased and decreases as the mass is increased. Equation 5 shows that the level of vibration is dependent upon the magnitude of the dynamic force, the proximity of the frequency of the dynamic force to the natural frequency, and the damping in the system.

The drawback of the sdof equations is that they provide an estimate for only one mode of natural frequency and that one hundred percent of the force, mass and stiffness participate in that mode. Most real systems have many modes of natural frequencies. The natural frequency of any mode is dependent upon the distribution of stiffness and mass. The amount of vibration is dependent upon where the dynamic load is applied in relationship to the mode shape and the proximity of its frequency to the natural frequency.

MODAL PARTICIPATION

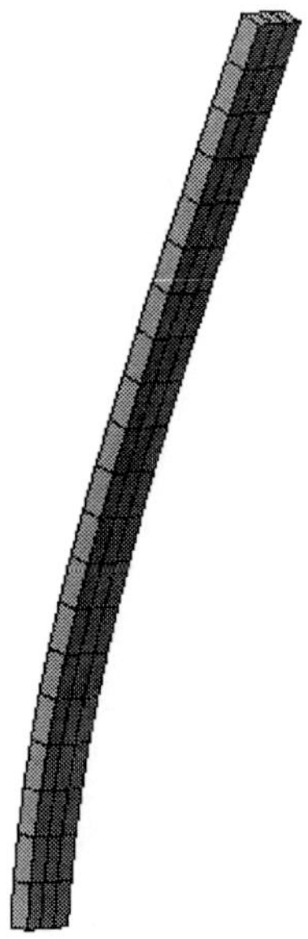

Figure 3: 1st Mode of Cantilevered Bar

The mode shape dictates how sensitive a natural frequency may be to a dynamic force. In multi-degree-of-freedom systems, it is not only the proximity of the frequency of the dynamic force to the natural frequency ($r = f_d/f_n$) that dictates the amount that a system vibrates, but the placement of that dynamic force in relationship to the maximum participation point on the mode shape associated with said natural frequency. If the dynamic force is applied at a rate close to the natural frequency and is applied at a location and in a direction corresponding to the maximum motion of the mode, then the vibration response will be a maximum. Conversely, if the dynamic force is applied at a location in which the mode shape does not have any motion, then there will not be any resonant amplification of response. This concept is referred to as *modal participation.*

The concept of modal participation is very important when modifying a design to change a natural frequency. Modifications should be made near locations of significant modal participation to maximize their beneficial effect. If a structural modification is made to a mechanical system at a location of zero modal participation, then it will not have any effect on the natural frequency associated with that mode.

To further illustrate this very important concept, consider as an academic exercise the first two modes of natural frequency for a simple cantilevered aluminum bar (3" wide x 40" long x1" thick) shown in Figure 3 and Figure 4. The bar is fixed at the base against motion in all directions.

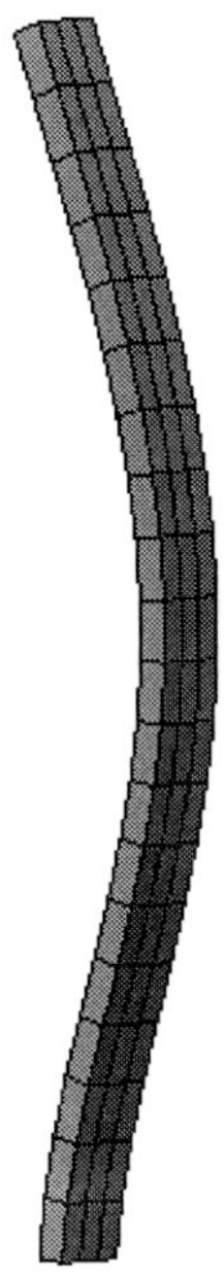

Figure 4: 2nd Mode of Cantilevered Bar

In the 1st mode, the entire bar sways in one direction, with the maximum motion occurring at the top. This is the principal flexural mode and has a natural frequency of 20.4 Hz. In the 2nd mode, the motion at the top is out-of-phase with the motion near the midpoint of the bar. This is a more complex mode and has a natural frequency of 127 Hz.

In the 2nd mode, there is a point, approximately 3/4 of the distance from the base to the top, where the mode shape is motionless. If a dynamic force were to be placed at this location, there would be absolutely no response to it, even if the frequency of the dynamic force equaled the natural frequency. This is because the mode shape does not participate for that particular dynamic force. In other words, there isn't any modal participation at that location, sometimes referred to as a node point of a mode. The 2nd mode would participate if the dynamic load were placed at any other location.

The 1st mode participates at all locations along its length, except at the base. However, the degree of participation varies, increasing dramatically from the bottom to the top of the bar. The 1st mode of natural frequency would be excited by a dynamic force placed at the node point of the 2nd mode.

The concept of modal participation is very important in vibration control. Placement of a stiffener at a nodal point in a mode shape will render the stiffener totally ineffective.

Since the sdof model is based upon the entire dynamic force, rigid mass and elastic spring participating in the dynamic response, it does not account for modal participation effect. The sdof model also provides only a single mode of natural frequency. Ignoring other modes during dynamic analysis of a real system can result in devastating errors in the prediction of expected vibration levels. Thus, although the single degree-of-freedom model can provide general insights into natural frequencies and dynamic response, the limitations of the model must be recognized when applying it to real systems.

NATURAL FREQUENCIES (2 DEGREE-OF-FREEDOM SYSTEMS)

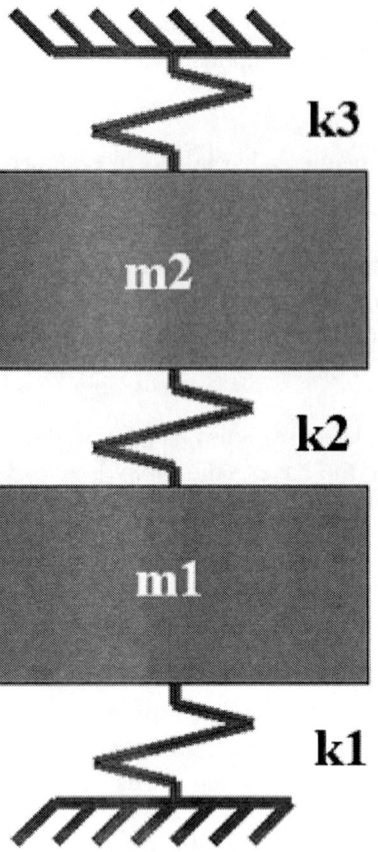

Figure 5: 2 DOF System

If it were determined that the mass of the structural component modeled as the elastic spring in Figure 1 was significant compared to the rigid mass and that there was significant flexibility in the support of the elastic spring, then a model with more than one degree-of-freedom might be required to analyze the system.

Figure 5 is a model of an undamped system with two degrees of freedom. It has two masses and three springs. Motion is still constrained to one direction.

The equations of motion for the transient solution are:

$$m_1(d^2x_1/dt^2) = -k_1x_1 + k_2(x_2 - x_1)$$
$$m_2(d^2x_2/dt^2) = -k_2(x_2 - x_1) - k_3x_2$$

The solution to these equations is cumbersome, but can be simplified by defining:

$$\alpha = (k_1 + k_2)/2m_1$$
$$\beta = (k_2 + k_3)/2m_2$$
$$\gamma = k_2/m_1$$
$$\varepsilon = k_2/m_2$$

The two natural frequencies for this spring-mass system can then be determined using Equation 6 and Equation7.

$$f_{n1} = (1/2\pi)\ [(\alpha + \beta) - [(\alpha - \beta)^2 + \Upsilon\varepsilon]^{1/2}]^{1/2}$$ **Equation 6**

$$f_{n2} = (1/2\pi)\ [(\alpha + \beta) + [(\alpha - \beta)^2 + \Upsilon\varepsilon]^{1/2}]^{1/2}$$ **Equation 7**

Each of these natural frequencies has a unique *mode shape* associated with it. The 1st natural frequency has a mode shape where the motions of the masses are in-phase with each other. The motions of the masses are out of phase in the mode shape for the 2nd natural frequency. The mode shape is important in vibration control.

EXAMPLE 1

Find the natural frequencies of an undamped 2 degree-of-freedom system that has two masses (m_1 = 10 lb-sec^2/in and m_2 = 2 lb-sec^2/in) and stiffness values of k_1 = 1000 lb/in, k_2 = 200 lb/in, and k_3 = 0 lb/in. These values for mass and stiffness, yield the following:

$$\alpha = (k_1 + k_2)/2m_1 = (1000 + 200)/2(10) = 60$$
$$\beta = (k_2 + k_3)/2m_2 = (200)/2(2) = 50$$
$$\gamma = k_2/m_1 = 200/10 = 20$$
$$\varepsilon = k_2/m_2 = 200/2 = 100$$

Inserting into Equation 6 and Equation 7, yields:

$f_{n1} = (1/2\pi)\ [(\alpha + \beta) - [(\alpha - \beta)^2 + \Upsilon\varepsilon]^{1/2}]^{1/2} = (1/2\pi)[(60+50) - [(60\text{-}50)^2 + (20)(200)]^{1/2}]^{1/2}$
$f_{n1} = 1.08$ Hz
$f_{n2} = (1/2\pi)\ [(\alpha + \beta) + [(\alpha - \beta)^2 + \Upsilon\varepsilon]^{1/2}]^{1/2} = (1/2\pi)[(60+50) + [(60\text{-}50)^2 + (20)(200)]^{1/2}]^{1/2}$
$f_{n2} = 2.10$ Hz

Note that the natural frequency of the individual spring-mass systems, if operating separate from each other would be:

$f_n = (1/2\pi)[k/m]^{1/2} = (1/2\pi)[1000/10]^{1/2} = (1/2\pi)[200/2]^{1/2} = 1.59$ Hz

The natural frequency of the individual spring-mass system falls between the lower and upper natural frequencies of the coupled system. Example 1 is the special case of a dynamic vibration absorber (DVA) that will be discussed later herein.

ROTOR WITH WHEEL ECCENTRIC RELATIVE TO BEARINGS:
The flexibility of an impeller is an important consideration in the rotor dynamic analysis of rotors with impellers mounted outside of the bearings (cantilevered) or impellers mounted between bearings, but not at the mid-span. Figure 6 is a single-inlet fan rotor where the impeller (fan wheel) is not located at mid-span between the bearings. It is offset to allow for the shaft to span the inlet duct that will be located on the left side of the impeller.

Figure 6: Single-Inlet Fan Rotor Designed to Operate Between Bearings

Figure 7 is a figure depicting another type of single-inlet fan rotor, similar to Figure 6, but with the fan wheel designed to be the mounted outside of the bearings. Equation 8 is a sdof textbook formula for the natural frequency of a cantilevered impeller, assuming the mass effects of the shaft are negligible compared to the overhung impeller.

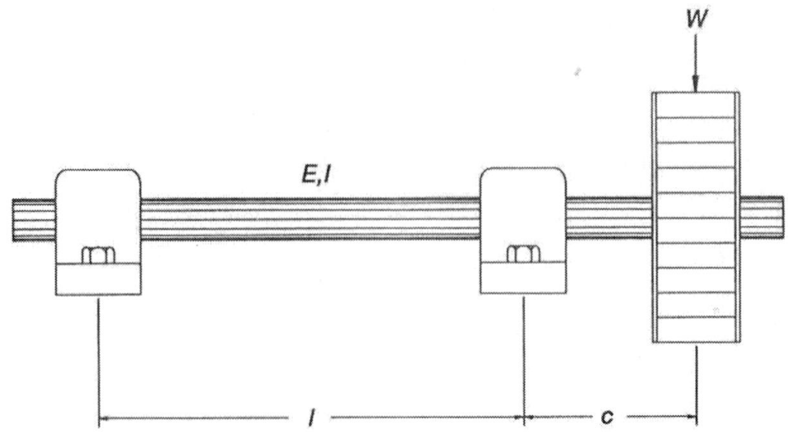

Figure 7: Single-Inlet Fan Rotor Dimensions

Natural Frequency = $f_n = (1/2\pi)[k/M]^{1/2}$ **Equation 8**

Where: k = flexural stiffness of the shaft = $3EI/(L + c)c^2$
M = mass of the fan wheel
L = span between bearings
c = overhung distance of center-of-gravity of fan wheel

Consider a rotor with a fan wheel weight (W) equal to 1,450 lbs, span between bearings (L) equal to 39.5 inches, overhang distance (c) equal to 14.7 inches, and the diameter of the shaft equal to 5 7/16 inches. The shaft is made of carbon steel (E = 29,600,000 psi). Substituting the known values into Equation 8 provides a calculated natural frequency of 46.4 Hz for the flexural mode of the rotor. Note that the mass of the shaft is neglected in Equation 8. Will the exclusion of the mass of the shaft result in an overestimation of the natural frequency?

The answer is yes, but it is not the only reason that the actual natural frequency will be overestimated using this textbook formula. The formula not only neglects the mass of the shaft, assuming it to be negligible compared to that of the fan wheel, but also treats the fan wheel as a rigid lumped component. In reality, a centrifugal fan wheel is a very flexible structural component. An assumption that the fan wheel is rigid compared to the shaft is an oversimplification that will lead to erroneous results. Since the fan wheel is a flexible structure, it results in another spring in the spring-mass system. Therefore, the rotor cannot be treated as a single degree-of-freedom system, for which Equation 8 is only valid.

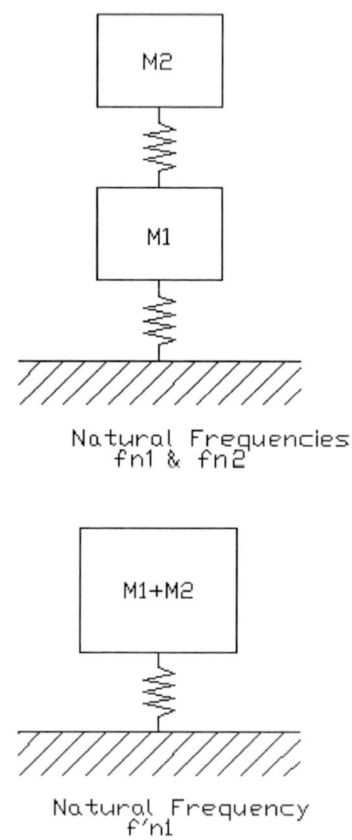

Figure 8: Spring-Mass Models

Figure 8 shows two separate spring-mass systems. Equation 8 is analogous to the lower system shown in the figure. The only spring is associated with the flexural stiffness of the shaft. The mass of the fan wheel is lumped along with the mass of the shaft. The actual condition is better represented by the spring-mass system in the upper system in the figure. The lower mass represents the effective mass of the shaft in the flexural mode. The lower spring represents the flexural stiffness of the shaft. The upper mass is the fan wheel, while the upper spring is the flexural stiffness of the back plate of the fan wheel.

Since the upper spring-mass system is a two degree-of-freedom system, it will have two natural frequencies. The motion of the masses will be in-phase with each other in the mode shape associated with the first natural frequency. The motion of the masses will be out-of-phase in the mode shape of the second natural frequency. The natural frequency of the single degree-of-freedom model will be somewhere between the natural frequencies for the modes in the two degree-of-freedom system.

If f'_{n1} is the natural frequency of the single degree-of-freedom system and f_{n1} and f_{n2} are the natural frequencies of the two degree-of-freedom system, then:

$$f_{n1} < f'_{n1} < f_{n2}$$

Figure 9 shows the mode shape for a finite element analysis (FEA) of the example single-inlet fan rotor. FEA [1] is a numerical technique that can account for the mass of the shaft and the flexural flexibility of the fan wheel.

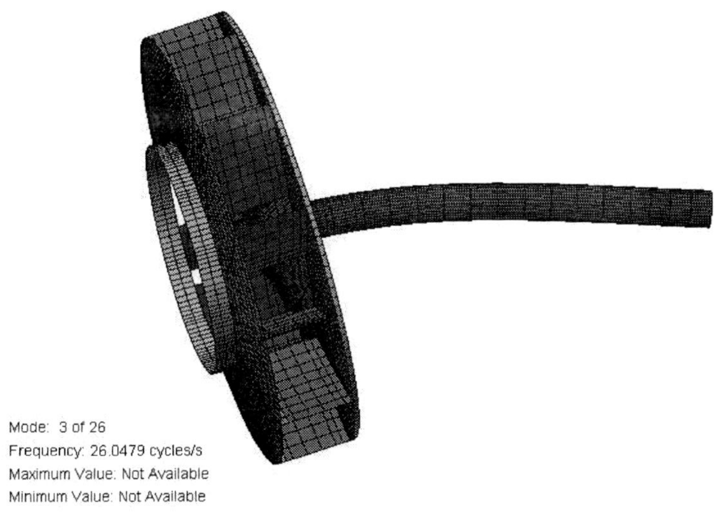

Figure 9: Mode Shape for Natural Frequency = 26.0 Hz

The mode shape in this figure is for the case where the flexural displacement of the fan wheel is in-phase with the flexural displacement of the cantilevered section of the shaft. The natural frequency estimated by the FEA is 26.0 Hz, which is much less than the natural frequency provided by Equation 8 (46.4 Hz). In this mode, there are two elastic elements in the system, one associated with the flexural stiffness of the backplate of the fan wheel and the other associated with the flexural stiffness of the shaft. Also, the mass of the fan wheel cannot accurately be considered a lumped value placed at the centerline of the shaft; its mass moment of inertia must be considered.

Figure 10 shows the mode shape for another natural frequency calculated using the FEA method. This mode shape is for the case where the flexural displacement of the fan wheel is out-of-phase with the flexural displacement of the cantilevered section of the shaft. The natural frequency estimated by the FEA is 107.3 Hz, which is much greater than the natural frequency (46.4 Hz) provided by Equation 8.

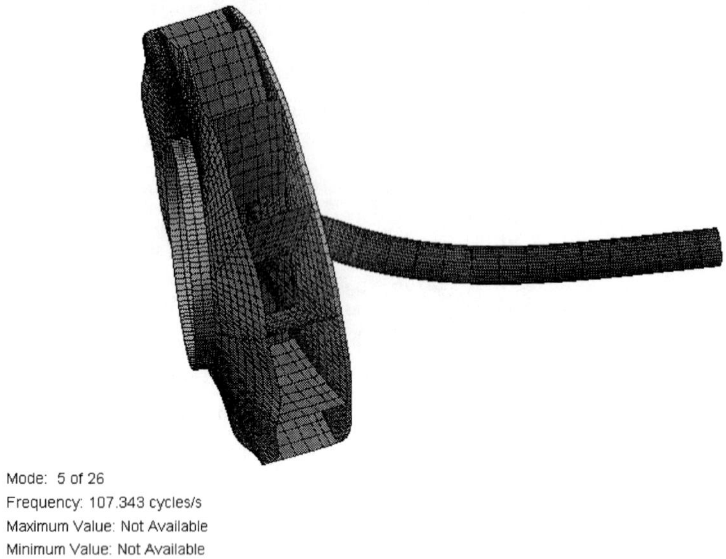

Figure 10: Mode Shape for Natural Frequency = 107.3 Hz

The significance of this analysis is that the textbook formula will always overestimate the natural frequency and provide an unrealistic estimate of the separation margin between the operating speed of the fan and its natural frequency. If the fan in the example above were to operate at 1800 rpm (30 Hz), the textbook formula provides a rather significant separation margin (f_o/f_n = 46.4/30 = 1.55). The more realistic FEA analysis would indicate a much lower separation margin (f_o/f_n = 26.0/30 = 0.87).

The textbook formula indicates that the principal "at rest" natural frequency of the rotor is 55% above operating speed, while the actual "at rest" natural frequency is closer to 13% below operating speed. The flexural stiffness of the backplate of the fan wheel will increase during operation due to centrifugal stress stiffening [2]. Therefore, it is not recommended that a fan wheel be design such that the "at rest" natural frequency of its principal flexural mode be less than operating speed. Designing single-inlet fan rotors using textbook formulae that treat the fan wheel as a lumped mass is not recommended.

2DOF APPLICATION (DVA/Tuned Mass Damper):
One method of vibration control is the tuned dynamic vibration absorber (DVA) which is a special case of tuning a natural frequency by adding stiffness, mass, and damping. It involves adding a spring-mass system to a machine or structure that alters its structural dynamic characteristics. A tuned dynamic vibration absorber (DVA) is a separate spring-mass system, that when added to the original spring-mass system, results in an additional natural frequency. If properly designed, the two natural frequencies of the combined system will straddle the natural frequency of the original system, one being lower and the other being greater.

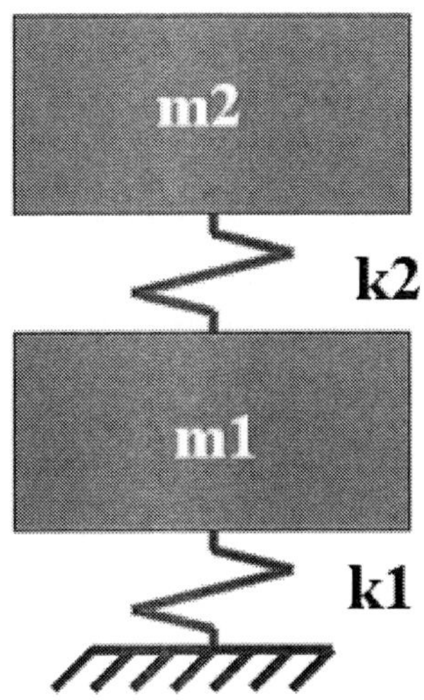

Figure 11: 2 DOF Spring-Mass System

Consider the undamped two degree-of-freedom system of Example 1. A smaller spring-mass system ($m_2 = 2$ lb-sec^2/in and $k_2 = 200$ lb/in) was attached to a larger spring-mass system. ($m_1 = 10$ lb-sec^2/in and $k_1 = 1000$ lb/in). The natural frequency of the original spring-mass system was calculated as 1.59 Hz. After attaching the smaller spring-mass system, the sdof model became a 2-dof model resulting in two natural frequencies at 1.08 Hz and 2.10 Hz

If a pure harmonic force ($F_{dyn} \sin \omega_f t$) is applied to the original mass (m_1), both masses will respond (vibrate). The calculation for the vibration of lower mass (X_1) and the upper mass (X_2) can be simplified by defining:

$\omega_1 = (k_1/m_1)^{1/2}$ = natural frequency of the lower spring-mass system and $\omega_2 = (k_2/m_2)^{1/2}$ = natural frequency of the upper spring-mass system, and

$X_0 = F_{dyn}/k_1$
$r_1 = \omega_f/\omega_1$
$r_2 = \omega_f/\omega_2$
$b = \omega_2/\omega_1$
$\mu = m_2/m_1$ = mass ratio
$D = b^2 r_2^4 - [1 + (1 + \mu)b^2]r_2^2 + 1$

Then non-dimensional displacement ratios $X_1/X_0 = (1 - r_2^2)/D$ and $X_2/X_0 = 1/D$.

EXAMPLE 2

Calculate the natural frequency of a 900 rpm motor that weighs 8,000 lbs supported on four springs, each having a stiffness of 46,000 lb/in. with negligible damping.

$k_1 = 4(46{,}000 \text{ lbin}) = 184{,}000$ lb/in
$m_1 = 8{,}000 \text{ lbs}/386.1 \text{ in/sec}^2 = 20.72$ lb-sec^2/in
$f_n = (1/2\pi)(k_1/m_1)^{1/2} = (1/2\pi)(4 \times 46{,}000/20.72)^{1/2} = 15.0 \text{ Hz} = 900$ rpm

The natural frequency of this supporting system is coincident with the operating speed of the motor. Vibration levels would be amplified considerably by this resonant condition. Investigate the new system natural frequencies if a DVA that weighed 1,200 lbs, having a spring constant of 27,625 lb/in was added to the motor.

$m_2 = 1{,}200 \text{ lb}/386.1 \text{ in/sec}^2 = 3.11$ lb-sec^2/lb
$k_2 = 27{,}625$ lb/in
$\alpha = (k_1 + k_2)/2m_1 = (184{,}000 + 27{,}625)/2(20.72) = 5{,}106.78$
$\beta = (k_2)/2m_2 = 27{,}625/2(3.11) = 4{,}441.3$
$\gamma = k_2/m_1 = 27{,}625/20.72 = 1{,}333.3$
$\varepsilon = k_2/m_2 = 27{,}625/3.11 = 8{,}882.6$

$$f_{n1} = (1/2\pi)\,[(\alpha + \beta) - [(\alpha - \beta)^2 + \Upsilon\varepsilon]^{1/2}]^{1/2}$$
$$= (1/2\pi)[(5{,}106.78 + 4{,}441.3) - [(5{,}106.78 - 4{,}441.3)^2 + 1{,}333.3(8{,}882.6)]^{1/2}]^{1/2}$$
$$= 12.37 \text{ Hz}$$

$$f_{n2} = (1/2\pi)\ [(\alpha + \beta) + [(\alpha - \beta)^2 + \Upsilon\varepsilon]^{1/2}]^{1/2}$$
$$= (1/2\pi)[(5{,}106.78 + 4{,}441.3) + [(5{,}106.78 - 4{,}441.3)^2 + 1{,}333.3(8{,}882.6)]^{1/2}]^{1/2}$$
$$= 18.18\ \text{Hz}$$

The original system natural frequency was right at operating speed. The dynamic vibration absorber resulted in two natural frequencies, one lower than and the other greater than operating speed. This provided a separation margin of over twenty percent.

Now determine the non-dimensional displacement ratio for the motor (X_1/X_0), which is similar to an amplification factor, and the displacement ratio (X_2/X_0) for the DVA at the operating speed of the motor. Since the operating speed is 15 Hz, the original natural frequency was 15 Hz, and the natural frequency of the DVA, by itself, is 15 Hz:

$$r_1 = \omega_f/\omega_1 = 1.0$$
$$r_2 = \omega_f/\omega_2 = 1.0$$
$$\mu = m_2/m_1 = 3.11/20.72 = 0.15$$
$$D = b^2 r_2^4 - [1 + (1 + \mu)b^2]r_2^2 + 1$$
$$= 1 - [1 + (1.15)] + 1 = -0.15$$

$$X_1/X_0 = (1 - r_2^2)/D = 0/D = 0.0$$
$$X_2/X_0 = 1/D = 1/-.15 = -6.66$$

The importance of this example is that, although theoretically the vibrational response of the motor has been totally removed ($X_1/X_0 = 0.0$), the DVA itself vibrates ($X_2/X_0 = -6.66$) out of phase with the dynamic force. The DVA must be designed not to fatigue during its useful life. It is important, not only to evaluate the vibration response of the system to which a DVA is attached, but also the DVA itself.

Figure 12 is the response curve for the case where the mass and stiffness of the DVA equals that of the original system. After adding the tuned DVA, the lower natural frequency is at 0.618x operating speed, and the upper natural frequency is at 1.618x operating speed.

The curve in Figure 12 should be considered an academic exercise, in that it is not probable that any mechanical system could be modified by adding another mechanical system of equal weight to it. Typically, the mass of a DVA is 15 – 20% the modal mass of the original resonant system, as in Example 2 ($\mu = m_2/m_1 = 0.15$). In these real cases, it is possible to implement a DVA that will reduce the vibration level of the original system. However, it most probably will not eliminate the vibration.

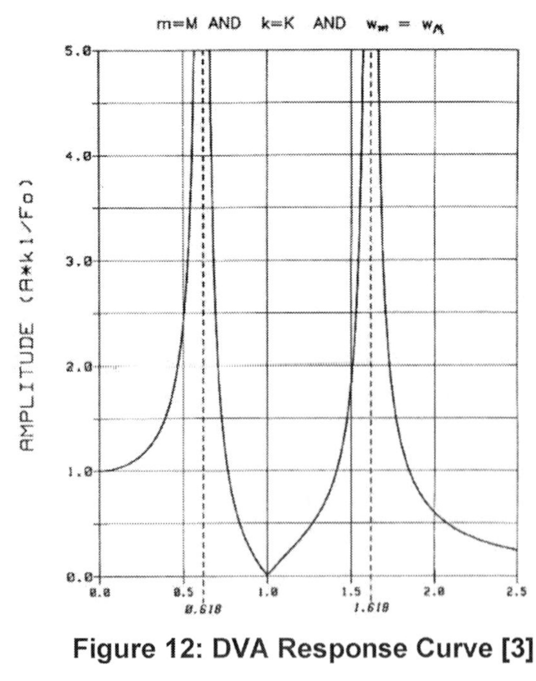

Figure 12: DVA Response Curve [3]

It is important to note that the DVA does not actually absorb vibration, as the name suggests, but modifies the modal characteristics of the original spring-mass system. It converts a single degree-of-freedom system into a two degree-of-freedom system. The system now has two natural frequencies, shifted to either side of the original natural frequency. If the frequency of the dynamic force changes, the vibration levels of the original system could increase dramatically due to the excitation of one of these new natural frequencies. Therefore, a DVA is typically not considered as a vibration control technique for machines that are operating with a variable frequency drive (VFD). Also, for mechanical systems with multiple dynamic forces, all operating at different frequencies, the use of a DVA should only be considered after evaluation of the response of the modified system to all the dynamic forces.

ROTOR-STRUCTURE INTERACTION

Typically, rotor dynamic response is determined by considering the flexibility of the supporting condition as equivalent springs. The mass of the supporting system is assumed not to participate in the rotor mode. Conversely, the dynamic response of the structural supporting system is typically determined by considering the equipment supported thereon as rigid lumped masses. The flexibility of the equipment is not considered.

In many cases, the modes of natural frequencies for rotors couples together with the structures that support them. In these cases, the stiffness and mass of the rotor and the structural support must be considered in any calculation for natural frequency and mode shape. The simple sdof model will not suffice since it neglects the mass of the support. Figure 13 is a three-dimensional view of the mode shape for a natural frequency wherein the rocking mode of the rigid foundation on the soil-structure interface couples together with the flexural mode of a rotor.

Figure 13: Rocking/Flexural Mode of Fd/Rotor

Figure 14 is an end view of the same mode shape that more clearly shows the rotor flexural – foundation rocking interaction. The participation of the rocking foundation is nearly the same as the participation of the flexing shaft. The mass of the foundation is significantly greater than the mass of the rotor. Treating the stiffness of the foundation as an equivalent spring in the rotor dynamic calculations can significantly overestimate the critical speed of the rotor. In this case, the natural frequency of the rotor can be increased by changing the structural dynamic characteristics of the foundation.

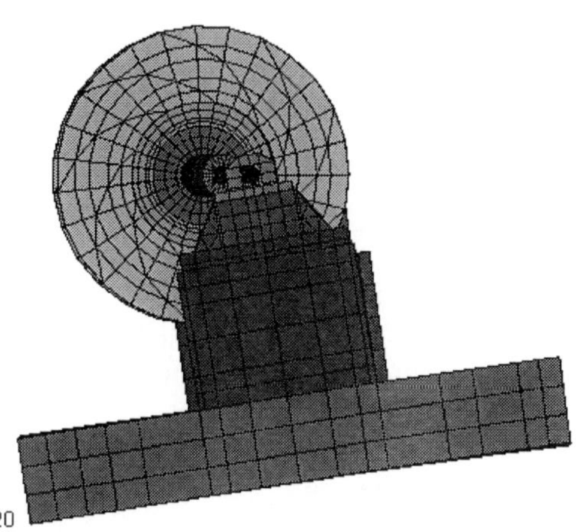

Figure 14: End View of Mode Sahpe

EXAMPLE 3

Consider the fan rotor shown in Figure 15 consisting of a shaft with a diameter equal to 16.25", with a span between bearings equal to 201.5". The weight of the fan wheel is 10,450 lbs. Each of the bearings sit on a pedestal which is supported by a W21x62 (21" deep, I = 1330 in^4) which spans between rigid walls. The distance of this span is 240". The weight of a single bearing and pedestal is 4,250 lbs. All materials are made from carbon steel. The pedestal is rigid compared to the shaft and support beams.

Figure 15: Fan Rotor Supported on Steel Beams

The cross-section area and mass of the shaft, mass of the fan wheel, and stiffness of the shaft are calculated below:

A = cross section area from shaft $= \pi d^2/4 = \pi(16.25)^2/4 = 207.4$ in^2

$m_{shaft} = AL\rho/g = (207.4)(201.5)(0.283)/386.1 = 30.63$ lb-sec^2/in

$M_{wheel} = W/g = 10{,}450/386.1 = 27.05$ lb-sec^2/in

The moment of inertia and flexural stiffness of the shaft are calculated as:

$I = \pi d^4/64 = \pi(16.25)^4/64 = 3{,}421.0$ in^4

$k_{shaft} = 48EI/L^3 = 48(29E06)(3{,}421.0)/(201.5)^3 = 582{,}000$ lb/in

The stiffness of the bearing and pedestal will be assumed as rigid compared to the shaft and supporting beam. The mass of each bearing and pedestal is:

$m_{bearing} = 4{,}250$ lbs$/386.1 = 11.0$ lb-sec^2/in

The combined stiffness of both W21x62 beams is:

$k_{beams} = 2(48EI/L^3) = 2\text{x}48(29E06)(1{,}330.0)/(240)^3 = 267{,}850$ lb/in

Using a single degree-of-freedom model, and neglecting the mass of the bearings, pedestal and support beams, the equivalent flexural stiffness of the shaft combined with the support beams is:

$k_e = k_{shaft}k_{beams}/(k_{shaft} + k_{beams}) = (582{,}000 \text{ x } 267{,}850)/(582{,}000 + 267{,}850) = 183{,}400$ lb/in

The natural frequency, using the single degree-of-freedom model is:

$$f_n = (1/2\pi)[k_e/(M + 0.5m_{shaft})]^{1/2} = 1/2\pi[183{,}400/(27.05 + 15.32)]^{1/2} = 10.5 \text{ Hz}$$

A numerical finite element analysis of this same structural-mechanical system yielded estimates of 9.3 Hz and 39.0 Hz for the first two modes of natural frequency associated with the flexural response of the rotor in the vertical direction. The mode shapes are shown in Figures 16 and 17.

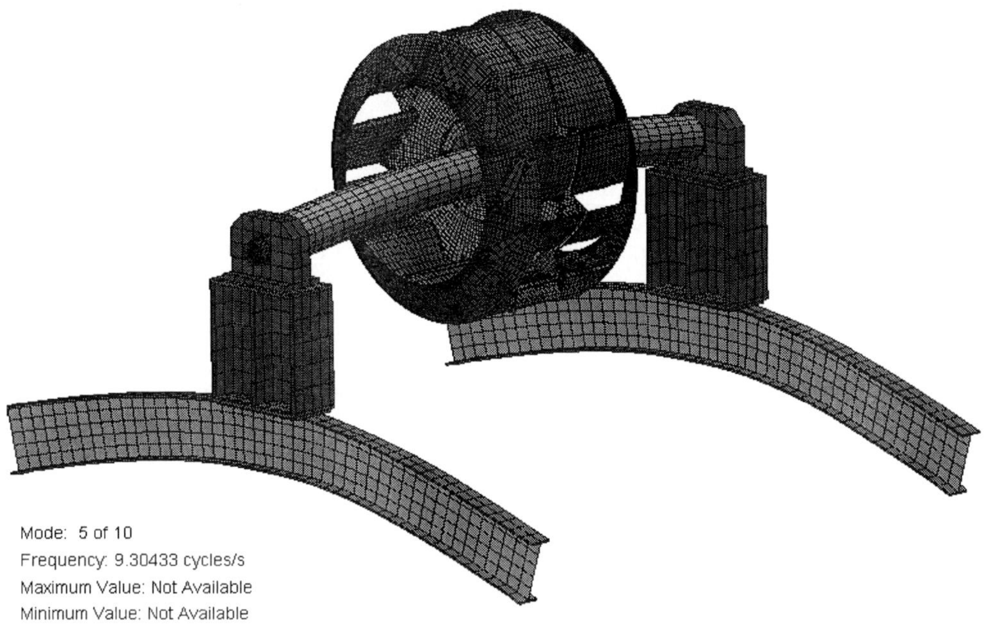

Figure 16: Mode Shape for Natural Frequency @ 9.3 Hz

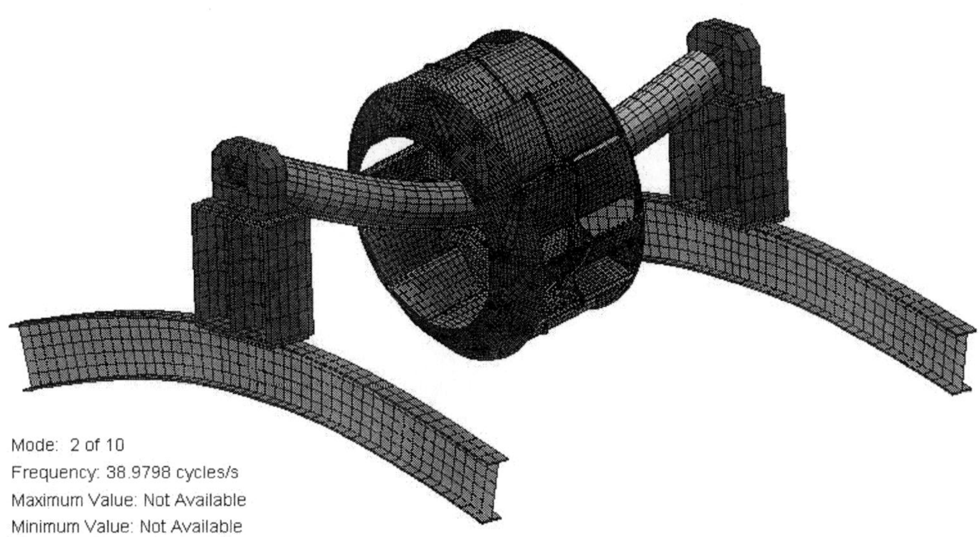

Figure 17: Mode Shape for Natural Frequency @ 39.0 Hz

In the mode shape for the lower natural frequency, the flexural motion of the shaft is in-phase with that of the structural beams. In the other mode shape, the flexure of the shaft and support beams are out-of-phase.

The natural frequency predicted by the single dof model was 10.5 Hz for the first mode compared to the 9.3 Hz estimated by the more accurate finite element method. This discrepancy (13%) may appear to be small, but it would account for a significant amount of any exclusion zone (separation margin) that may be specified for the rotor based upon the sdof calculation. The actual separation margin may be significantly less than the specified amount.

Since the sdof model can only provide an estimate for a single mode, it cannot provide an estimate for the mode shape of Figure 17.

Next, the problem was analyzed using the two degree-of-freedom equations (Equation 6 and Equation 7). This analysis technique yielded estimates of 9.0 Hz and 31.5 Hz for the two natural frequencies. The estimate for the 1st natural frequency is very close to that predicted by the FEA. However, there is a significant discrepancy in the prediction for the 2nd mode.

REFERENCES:

[1] Sayer, R.J., *"Finite Element Analysis: A Numerical Tool for Vibration Analysis"*, Proceedings of the National Technical Training Symposium and 27th Annual Meeting of the Vibration Institute, New Orleans, Louisiana, July 9-11, 2003.

[2] Sayer, R.J., *"Structural Dynamics of Centrifugal Fans"*, Proceedings of the National Technical Training Symposium and 34th Annual Meeting of the Vibration Institute, Chicago, IL, June, 2010.

[3] Leader, Malcolm, *"A Solution for Variable Speed Vertical Pump Vibration Problem"*, Proceedings of the Second International Pump Symposium, Houston, TX 1985.

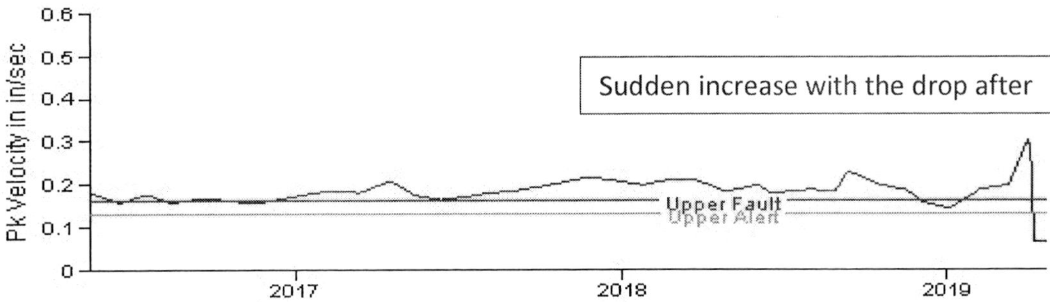

Figure 1

When the data was obtained around the first of April in 2019, the chiller was not loaded heavily, at around 60% load. As the trend data shows, this was a significant jump in the overall level. (Figure 1) When reported to operations, the comment was that the chillers had been surging and probably nothing to worry about. It was suggested to the customer that they should go to the machine and check it out, because it did not sound like a normal surge condition to me. (Figure 2)

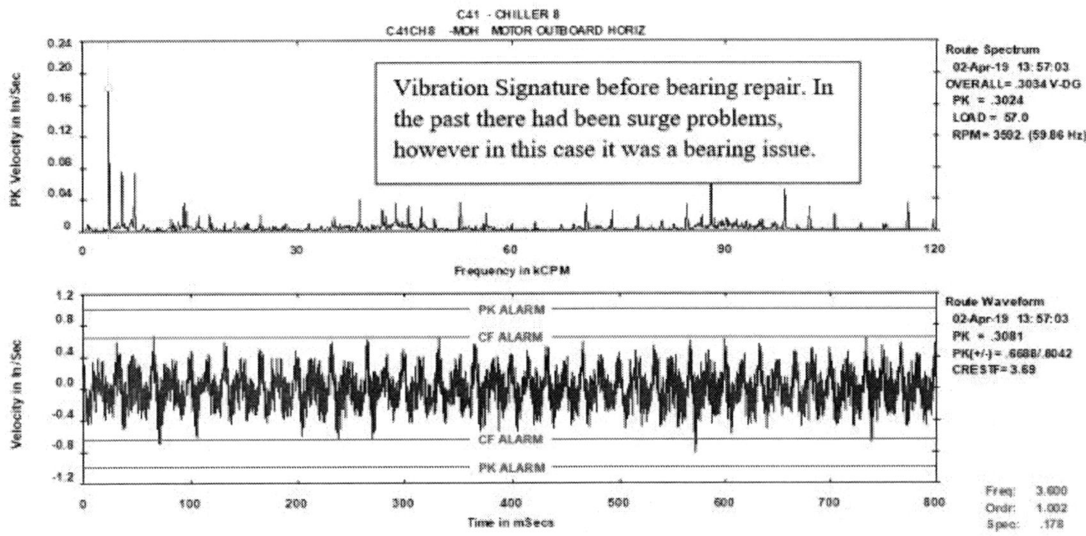

Figure 2

I left the customers site and later that day was informed that after listening to the machine they agreed the noise was not a normal surge condition and decided they had better pull the motor and send it out for inspection.

At the repair facility, the motor was inspected electrically and nothing was found. However mechanically the bearing fits were inspected and it was discovered that the fit between the bearing and the bearing housing was oversized by .004 inches. The bearing housing was refurbished to the proper fit and new bearings were installed. (Figure 3)

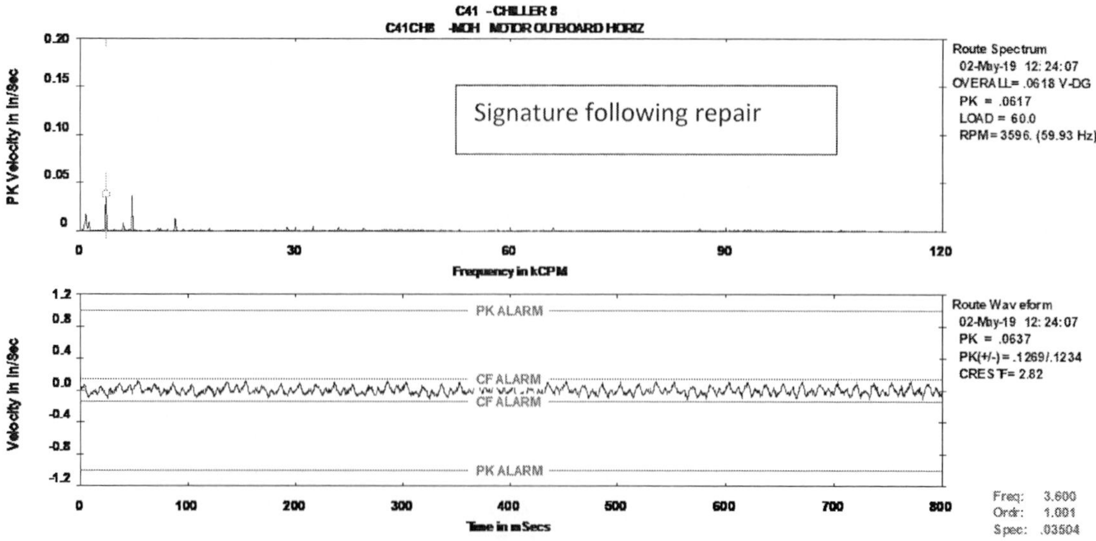

Figure 3

CASE HISTORY #2:

The second example is of one machine, but with both the blower and motor having issues. The machine is a 32,000 SCFM blower operating at 3600 RPM with 13 PSI discharge. This is a large volume low pressure machine. It is powered by a 2500 Hp motor.

The customer had experienced problems with the blower bearings previously, especially on the outboard end of the machine. The bearings would develop a spall area on the outer race and go into high alarm, within a matter of hours. As such the customer was very concerned about any readings on this machine that we felt might be of issue or concern. A contributing factor was that the customer had also had a similar blower that the inlet cone had partially broken and went into a wheel that did some internal damage and resulting in high vibration on the blower. The machine operates continually, unless rotated. There are five of these machines with four normally in service at all times. During colder weather, only three may be in operation.

When data was collected on this particular month, the exhaust from the motor seemed to be extremely warm. The customer was questioned if the air filters had been changed recently, as all of the exhaust temperatures felt warmer than normal. This was unusual since it was not that warm outside. It was discovered that the filters had not been changed for a slightly longer period than normal. What was also discovered was that when the filters were changed, the unit was not shut down. As was mentioned previously they had to operate usually four of these blowers for production requirements, so the filters were just changed on the fly.

From this information it was pretty well certain that the motor was probably full of dirt and build up that was most likely causing the vibration that was seen in the signature. (Figure 4)

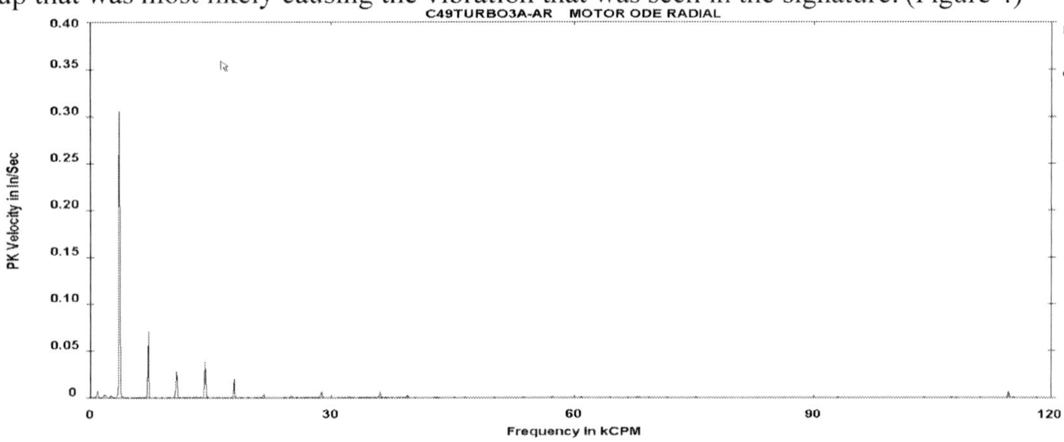

Figure 4

There were also some concerns around the vibration that was being seen on the blower. The blower has large ball bearings, and part of the data looked to be related to a possible bearing defect, but there was also data that looked more like an internal vibration issue. The customer decided to have additional testing performed before making any decisions on the course of action.

A detailed testing process was set up with triaxial accelerometers mounted on all four bearings. This allows data to be obtained in all three directions. The customer also had temperature monitoring on the bearings which could also be trended when the test was being ran. There were also pressure readings that could be observed on each individual stage of the blower, so pressure changes across the blower could be seen.

The testing procedure was designed so start up data along with operating data could be recorded. Since axial readings were higher than expected, it was felt that possibly the alignment was off and this could be causing some of the vibration seen on the motor or blower. The blowers normal operating condition was such that the motor and the blower had to grow into alignment. It would normally take over an hour before the blower would be at a steady state condition and vibration and temperatures had stabilized.

A 16 channel ZonicBook 618E IOtech vibration analyzer was connected to four 50mv/g triaxial accelerometers which were mounted on each of the four machine bearings. In addition to the accelerometers, 200 mv/mil proximity probes were mounted in the horizontal and vertical directions on the motor inboard bearing to evaluate absolute shaft vibration. A 100 mv/mil proximity probe was mounted axially to the compressor outboard bearing to monitor axial shaft movement. A Monarch SPS-5 photo tach was lined up axially on the motor outboard end to track motor rpm during operation. The machine had not been run in several days and data was collected from at rest position through the startup then data was continually monitored for two hours.

The motor provided the following information. The motor showed a 1x turning speed frequency that started at 0.25 in/sec and then after about 20 minutes into the run the vibration began to rise and started oscillating. The vibration amplitude would rise and fall and cycle through this rising and falling motion every 1½ to 3 minutes. Each cycle would return with greater amplitude increasing each cycle until the end of the test. The motor outboard bearing reached a maximum amplitude at 0.95 in/sec. (Figure 5)

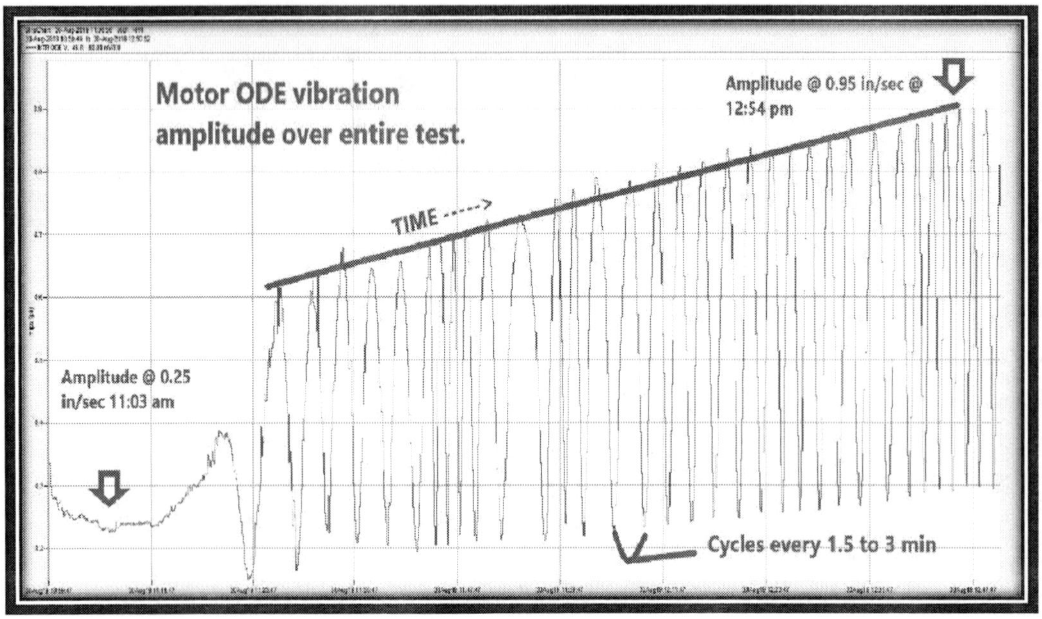

Figure 5

Along with the oscillating vibration, the motor casing phase angles were very unstable. These phase angles should remain fairly constant. The continued rotation of the phase angles on the motor usually indicates an internal rub. The oscillating vibration mentioned above is likely due to the rub moving in and out of phase with the motor's residual unbalance during operation. While the rub is in phase with the unbalance the vibration increases, the rub then moves out of phase with the unbalance and then the vibration decreases. This continued over and over once the machine temperatures began to climb. (Figure 6)

The proximity probe readings taken on the motor inboard side showed high vibration from the beginning and ended around 7.6 mils on the X probe and 6.7 mils on the Y probe. Slow roll data showed the runout to be around 2 mils on the proximity probes. With runout compensation considered the vibration levels were 5.2 and 4.2 mils on the X and Y probes. These levels are high for a 3600 RPM motor.

The compressor or blower showed some additional information. Vibration on the compressor inboard bearing slowly increased over time until the vibration exceeded 1 in/sec and the supervisory tripped the machine. There was no decrease or leveling out of this vibration, it began to rise and showed no signs of stopping until the machine was shut down. This vibration was greatest in the vertical direction and the source was undetermined. (Figure 7)

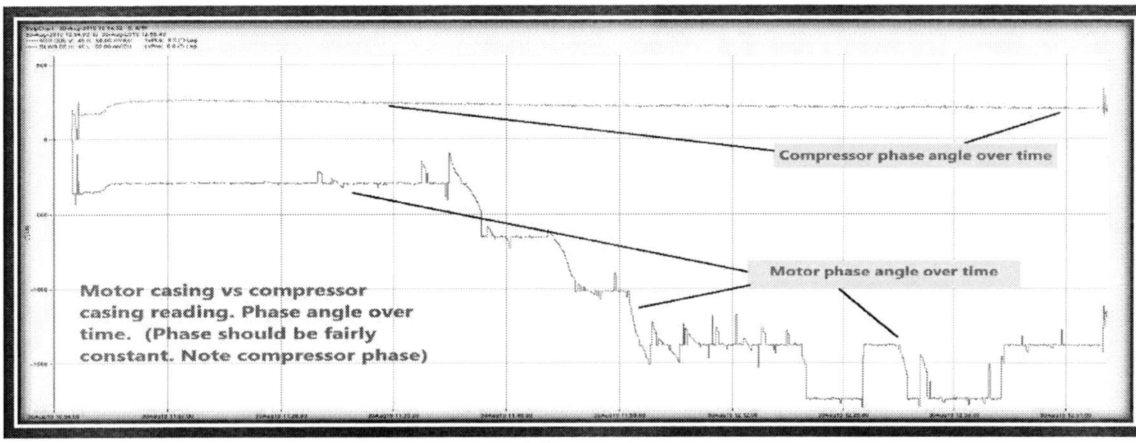

Figure 6

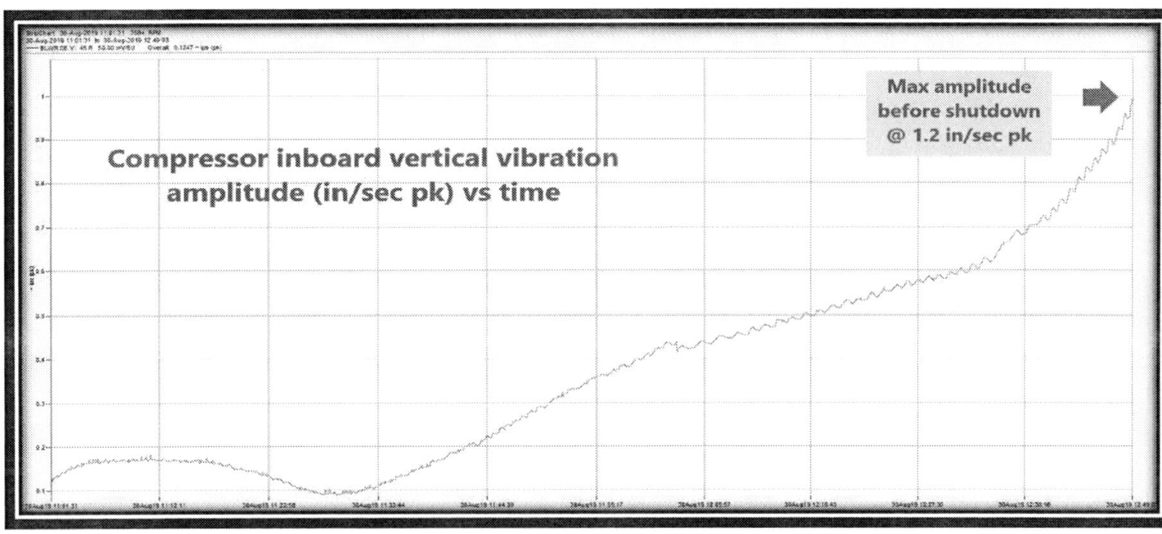

Figure 7

The proximity probe mounted axially on the compressor ODE bearing cap and was placed in line with the shaft to measure the shaft axial movement. It was determined that the axial movement was only 0.005" throughout the entire test period. Upon start up the probes measured 0.011" of movement towards the outboard bearing as the motor found magnetic center and then grew the 0.005" over the period of the test. Total float in that bearing measured was 0.016". Considering the length of the shaft and the temperature rise in this blower the 0.005" movement after startup did not seem to be enough axial movement.

Around 30 minutes after the start up the phase angle for the inboard vertical sensor shifted 180^{0}. This shift occurred over a 3-minute period. The horizontal sensor's phase angle remained constant as did all of the other sensors mounted on the compressor. What caused this phase shift is a mystery. The fact that the phase shift was only registered on the vertical sensor is even more puzzling. (Figure 8)

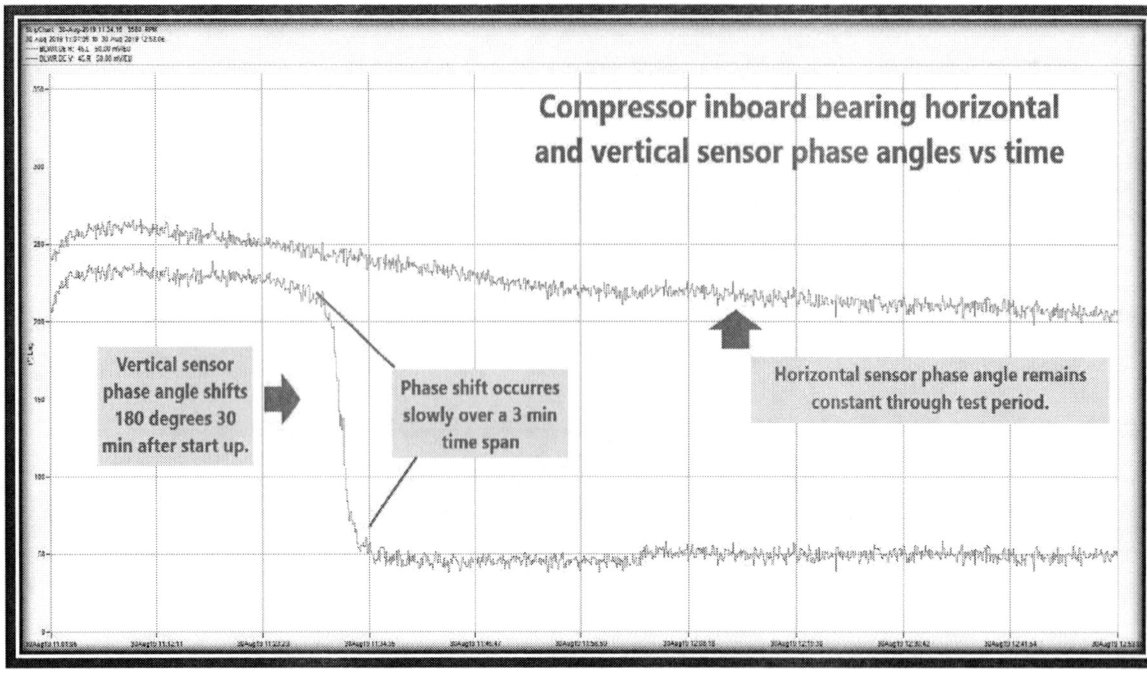

Figure 8

With the testing data, the customer decided to send both the motor and blower to a repair facility. The first tests ran on the motor were electrical tests which did not show any issues. The motor was disassembled and there was an excessive amount of dirt buildup on the inside of the motor. When the rotor was placed in a balance machine the magnitude of the residual unbalance was over 10 times the acceptable level on the one end and 5 times the acceptable amount on the opposite end. The bearing to housing clearances were checked and there was about .001" of excessive clearance.

The bearing to shaft clearances were also checked and the Inboard or drive end bearing had about .006 clearance and the outboard or non-drive end had .008" of clearance. The motor had a variety of issues that were not significant in themselves, but when added together resulted in high vibration levels.

The blower is difficult to inspect. This blower uses large 6320 ball bearings. The axial positioning of these bearings and spacers within the bearing housings has to be set correctly, or you can shift or push the motor off of electrical center, but this can also load the bearings differently and cause additional vibration problems if not set correctly.

To check internally, the blower has to be positioned on its end with the shaft in a vertical position. The sections, of which there are five, have to be individually unstacked, one stage at a time. The first stage or two did not show any significant abnormalities, but the third and fourth stages showed problems. Vanes on the individual wheels were found broken loose and had wedged against the next vane on both stages. The vanes had not broken and went into the adjacent wheel, but were not in the positions that they would have normally been in.

It appeared after further investigation, the rivets that hold the vanes in place had broken and further analysis indicated that the fit, between the vanes and the wheels were not tight and allowed more gap between the vanes and wheels, thus causing additional stress on the rivets that held the vanes in place. Additional inspection showed that one of the other stages on the blower also showed the same type of issue. The customer decided to replace all of the wheels on the blower. Repairs to both the motor and blower were completed and the machined was returned to service. Both the motor and the compressor ran well from both a vibration and temperature perspectives.

SUMMARY:

For the first case history, the operations group felt that it was probably just surging of the compressor. However, based upon a difference in sound, the resulting inspection found that there were mechanical issues causing the vibration instead of the suspected flow problem. In the second case, what started as a motor issue turned into problems with both the motor and blower. Vibration data does not lie, but it can mean different things and experiences from both operation and maintenance can help improve your abilities to properly diagnosis problems.

STUDY OF THE FRACTURE OF A LOAD CELL DUE TO IMPACT LOADING

Authors: Ashlyn Jackson
and John Havenar
and Austin Summer
and Seth Nowak
and Dr. Aniruddha Mitra

VIATC VIBRATION INSTITUTE ANNUAL TRAINING CONFERENCE & EXPO 2022 | MFPT ANNUAL CONFERENCE

VIATC | TRAINING CONFERENCE & EXPO
SAVANNAH, GEORGIA · 2022

STUDY OF THE FRACTURE OF A LOAD CELL DUE TO IMPACT LOADING

Ashlyn Jackson, Undergraduate Student
Georgia Southern University
Statesboro, GA 30458

Ashlyn Jackson is a senior mechanical engineering major at Georgia Southern University. She is a member of the Honors College at Georgia Southern, as well as the president of Pi Tau Sigma and Secretary of Tau Beta Pi, two honors engineering societies. Her main topic of research relates to material science and fracture mechanics.

John Havenar, Undergraduate Student
Georgia Southern University
Statesboro, GA 30458

John Havenar is an undergraduate mechanical engineering student at Georgia Southern University. His primary areas of interest are finite element analysis and computational fluid dynamics. His current research focuses on aeroelasticity modeling and computational analysis simplification.

Austin Sumner, Mechanical Engineer
Enercon
Kennesaw, GA 30144

Austin Sumner graduated Summa Cum Laude from Georgia Southern University with his Bachelor of Science in Mechanical Engineering. Through college, he completed all requirements to graduate from the honors college and held positions of leadership in both Pi Tau Sigma and Tau Beta Pi Engineering Honor Societies. He currently works for Enercon Services, with research interests in pressure vessel integrity, nuclear energy, and renewable energy production.

Seth Nowak, Graduate Student
Georgia Southern University
Statesboro, GA 30458

Seth Nowak graduated with his Bachelor of Science in Mechanical Engineering at Georgia Southern University in 2018. He is currently still studying at Georgia Southern University, pursuing his Master of Science in Mechanical Engineering as well as his Master of Science in Applied Engineering. In addition to this, he worked as a Graduate Research Assistant and is now a mechanical engineer at Aggero, LLC.

Dr. Aniruddha Mitra, Professor
Georgia Southern University
Statesboro, GA 30458

Aniruddha Mitra, PhD, PE is a Professor of Mechanical Engineering at Georgia Southern University.

ABSTRACT:
In this study, there are two loadcells used at the bottom of the legs of a pressure vessel. Being in an open environment, the pressure vessels are subjected to weather conditions that may increase external forces on the structures. When the wind speed is high enough, there is an increase in load transferred through the load cells, causing an impact loading to the load cells. In this research, a completely fractured and a partially fractured loadcell were analyzed for the possibility of fracture due to impact loads. Through surface imaging, the curved fractured surfaces, as well as the many small indentations on the contact point on the load cells, indicated failure over time, rather than from one single impact load. In addition to this, Finite Element Analysis was used to simulate the impact loads from the dropping of the leg onto the loadcell. These simulations revealed failure for the loadcell in two of the four cases simulated. From this research, the possibility of the fracturing of a loadcell due to a single impact load has been confirmed through finite element analysis. However, the results from the surface imaging demonstrated the likelihood of this failure resulting from multiple impact loads.

KEY WORDS: Finite element analysis; fracture; impact loading; load cell; pressure vessel; simulation; surface imaging

INTRODUCTION:
This research looks at the excess loading experienced by pressure vessels subjected to natural outdoor conditions such as wind. The forces placed on a pressure vessel by wind can result in a tilting or toppling effect [3]. This will further cause a detachment between the assembly of the bottom of the legs of the pressure vessel and the loadcell located below them, which is seen in

Figure 1. This causes an impact load to the top of the load cell when the pressure vessel begins to tilt or topple. This can damage both the legs of the pressure vessel and the load cells over time.

Figure 1. Assembly of load cell with pressure vessel leg.

This research came about due to the failure of two load cells made of 17-4 stainless steel as seen in Figure 2. These load cells are located at the bottom of each of the three legs of the pressure vessels and are used as load measuring devices [10]. One of these load cells has fractured into two pieces and the other load cell has a surface crack. The purpose of this analysis is to determine the possibility of these load cells fracturing due to a single impact load.

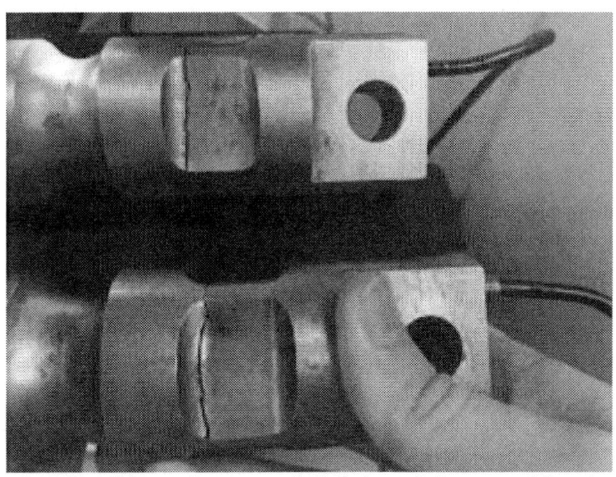

Figure 2. Load cell with surface crack (top) and fractured load cell (bottom).

SURFACE IMAGING:

Surface imaging can be used as a method of analyzing deformations or imperfections on a surface to conclude details about their origin [6]. For this research, a Keyence VHX-1000 and a Keyence VHX-7000 were used to image both the fractures and cracked load cell. Using a VH-Z20R and VH-500R lens, the microscope is able to obtain high-quality images revealing details that are hard to see with the naked eye [9].

ANSYS:
ANSYS is a program that allows users to model and simulate forces, stress, deflection, etc on objects, such as a load cell [9]. For this research, four explicit dynamic simulations were used to show the results of a free-falling object on the load cell model to demonstrate the effects of a single impact load. These ANSYS simulations also verified the possibility of the load cell failing due to an impact load.

PREVIOUS WORK:
Previously, wind loading experimentation has been conducted by Austin Sumner on these same models of pressure vessels with the use of a wind tunnel. To do this, scaled-down models of three different sized pressure vessels were 3D printed and placed in a wind tunnel [9]. A few of the variables considered were filled/empty conditions, orientation in the wind tunnel, distance from the center of the wind tunnel, and wind speed [9]. From this it was seen that with a wind speed of relatively high magnitude, the external force can cause the pressure vessel to either shake or tip over [9]. This verifies the possibility of a detachment between the leg of the pressure vessel and the load cell since tipping over would cause an imbalance of forces in each of the three load cells [9].

LITERATURE REVIEW:
Chavan, Kalaska, and Patil analyzed a model of a pressure vessel for wind loading effects. This was done with a non-linear finite element analysis method [1]. An improved model of the pressure vessel was created that changed the number of legs from eight to six and positioned them at an angle to the ground [1]. This was done because of the extreme precision that these types of pressure vessels that were being studied require [1]. The results of the finite element analysis revealed the safe conditions of wind loading for the pressure vessels [1]. It also showed that the improved model with the angled legs had better performance when compared to the original with eight vertical legs and also had less stress and deformation due to wind loading [1].

Mahandari and Kharisma looked at the effects of wind and earthquakes on a specific model of a pressure vessel. When studying the effects of wind loading, speed pressure and force coefficient were considered [7]. For the seismic activity portion, intensity and duration were taken into consideration [7]. Using ASME and ASCE codes, the factors previously mentioned were computed [7]. From these calculations, it was found that the pressure vessel under investigation did meet the standards for safety by remaining stable and having relatively minimal stress and deflection [7].

Infante and Silva studied fracture mechanics by conducting finite element analysis on a crack subjected to fatigue failure. For this analysis, three cases were used to vary the crack front geometry, material, and dimension [5]. Using ZENCRACK and ABAQUS, software able to analyze crack growth, the stress intensity factor was computed [5]. The simulations results were compared with typical existing methods for solving fatigue crack problems and were found to be reasonably similar [5].

Patil and Kadam reviewed a framed structure and studied its behavior when placed under an impact load. This was done with the use of the finite element method on six different models in ANSYS [8]. Varying the magnitude of the force, the loads were simulated to drop freely onto the framed structures from different heights [8]. These simulations showed several correlations with the increase in drop height [8]. As the drop height increased, the impact load, impulse, length of time of impact, and deflection also increased [8].

Hutiu et al. studied the effects of ductile and brittle fractures on the surfaces of different metals. This was done by using optical coherence tomography [4]. Three samples were used, including OL 37, OL 52, and an Sn-Sb-Cu alloy [4]. Using the results from the optical coherence tomography that determined the height profiles of the fractured surfaces, the type of fracture was recognized [4]. This method of analyzing fractured surfaces was found to be close in accuracy to other methods that are more expensive [4].

Dutta et al. reviewed methods of analyzing surface textures to determine what type of fracture occurred for different stainless-steel materials [2]. The samples used in this research consisted of AISI 304LN stainless steel that were subjected to tensile testing at different strain rates [2]. Following this, three methods of texture analysis including box-counting, gray level co-occurrence matric, and run-length statistical analysis [2]. Comparing the results from each of these, the method with the highest quality for the eight texture descriptors that were used was found to be the long-run emphasis and long-run high gray-level emphasis [2].

EXPERIMENTAL METHODS:

The first portion of this research consisted of analyzing the load cell for all forces present on the body. The free body diagram showing these forces can be seen in Figure 3. As seen in this figure, there are three lateral forces present directly on the contact points of the loadcell. This is where the bracket connected to the legs of the pressure vessel hold the load cell in place. From these predicted forces, the shear force diagram and bending moment diagram were made, as seen in Figures 4 and 5.

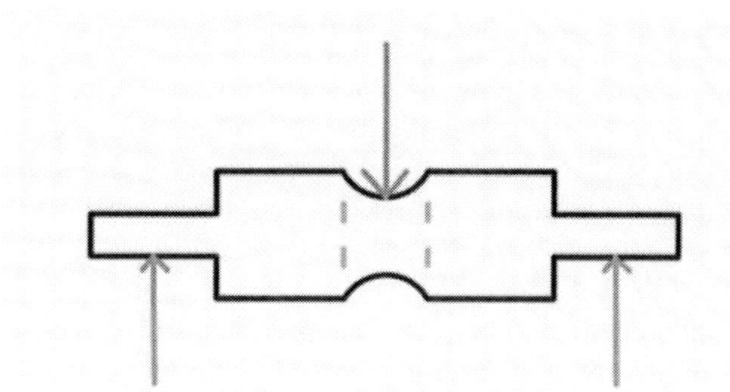

Figure 3. Free body diagram of load cell component (Sumner, 2020).

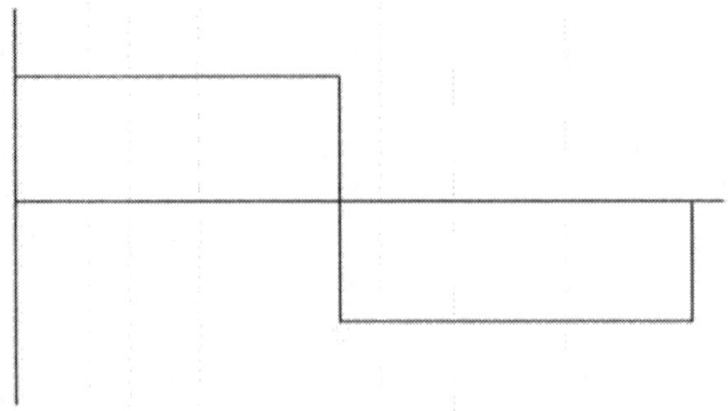

Figure 4. Shear force diagram of load cell component.

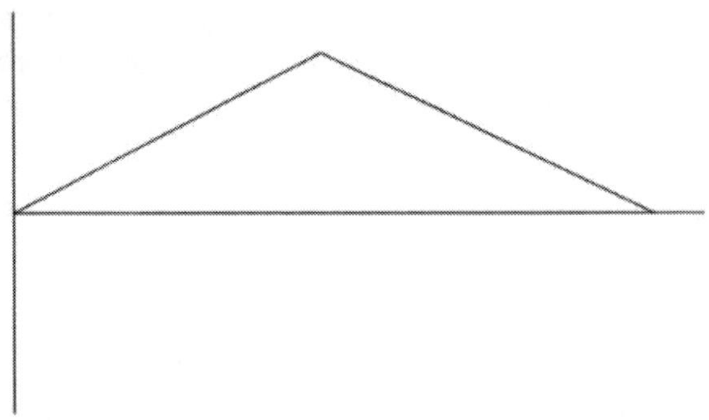

Figure 5. Bending moment diagram of load cell component.

A Keyence VHX-1000 and VHX-7000 microscope were used for further analysis of the load cell. Prior to the surface imaging, the microscopes were calibrated using the Keyence instruction manual [6]. The surfaces of the fractured load cell were cleaned with a wire brush then used for a 2D scan. The 2D scan images were only of small portions of the fractures surface, therefor they were combined into one image using the image stitching function. Next, the surfaces were scanned for a 3D image, using the 3D image stitching function. With careful measures, the smaller 3D images were stitched together to avoid errors concerning depth perception. Finally, the depth composition function was used on the fractures surfaces by first setting an upper and lower limit. The upper and lower limit were representative of the lens being focused on the highest and lowest points of the surface, respectively. The microscope then took pictures of the surfaces at small incremental distances between the upper and lower limit. From this, around 200 images were consolidated into one image that is focused on all areas of the surface.

This same fracture load cell as well as the non-fractured load cell contained a region subjected to contact from the central pin of the bracket. Analysis with the same microscope was also used for this surface. The upper and lower limits were reset for this region and the depth composition function was again used as well as 3D image stitching.

Finally, the last portion of the load cell that was imaged was the surface crack located on the un-fractured load cell. For this a 3D depth scan was used to compute the depth along the full length of the crack. In addition to this, a 3D image stitch was conducted for further analysis and observation.

The ANSYS portion of this research was used to show the results of a free-falling load on the load cell. For this, four explicit dynamic simulations were conducted. The material used was 17-4 PH Alloy Steel H1150. Keeping a load of 3135 lb., three of the trials used a drop distance from 0 to 1/8 in. in increments of 1/16 in. The fourth simulation used a load of 26000 lb. and no drop distance. Using high fidelity and a fine mesh resolution in the area of interest, which is the region near where the fracture will most likely occur, the results were computed.

RESULTS & DISCUSSION:

STRESS:

In Figure 6 below, the location of the fracture and surface crack on the load cells can be seen. It is important to note that there seems to be a trend on the location of failure of these load cells, indicated that they both fractured due to the same reason.

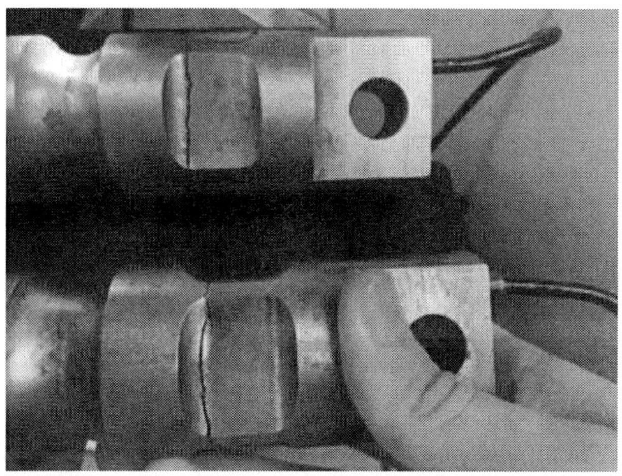

Figure 6. Location of failures on fractured and non-fractured load cell (Sumner, 2020).

It is also seen that the geometry throughout the load cell is not uniform, causing an inconsistent stress distribution. The region where the failure occurs is where the cross-section of the load cell is similar to an I-beam as shown in Figure 7. Using eqn. 1, the relationship between the moment of inertia, I, and the stress in the load cell, σ, can be used to determine why the fracture occurs in that location.

$$\sigma = \frac{Mc}{I} \tag{1}$$

As the cross-sectional area of the beam decreases, so does the moment of inertia. This further causes an increase in the stress according to eqn. 1. This area of higher stress is where the failure will occur, which is also at the point with the cross-sectional area of an I-beam.

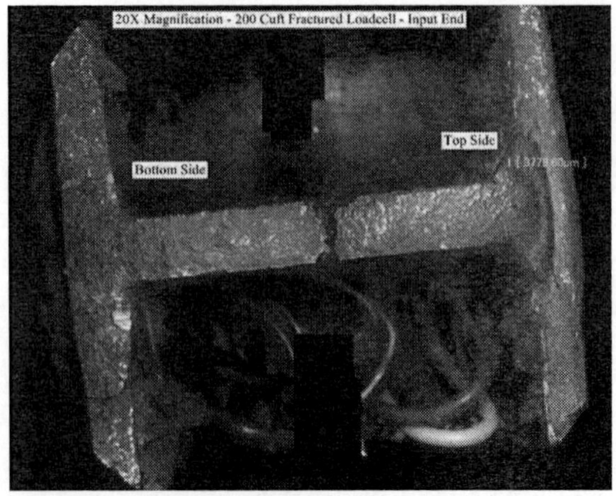

Figure 7. Cross-section of load cell where failure occurred (Sumner, 2020).

SURFACE IMAGING:
FRACTURE SURFACE:

Figure 8 shows the shape of the faces of the fractured surfaces of the load cell. From this image, two things can be taken away. First, it shows that the correlation between the location of the failure and the cross-sectional area. Second, it reveals that there seems to be a curvature in the surface. This is better seen by the depth profile in Figures 9 and 10. This curvature of the fractures surface implies that the failure most likely occurred over time rather than from one single impact load. This is further seen by the discoloration, only appears on one side of the surface. It infers that this side was exposed to environmental degradation longer than the side without discoloration due to a crack being present up to the point that the discoloration stops.

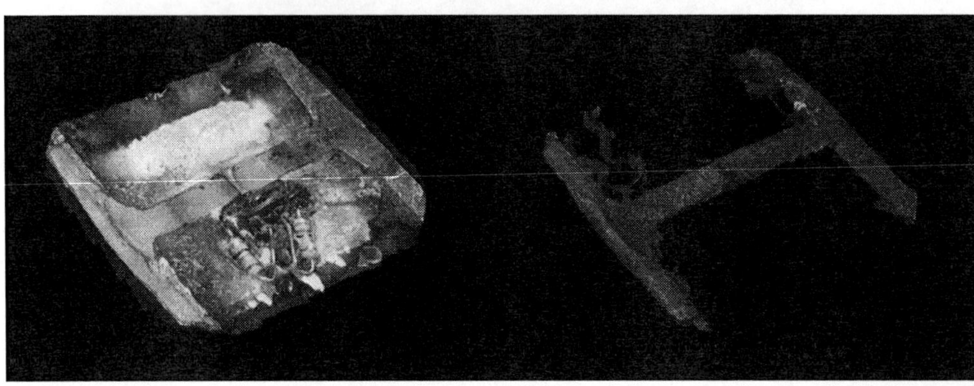

Figure 8. 3D image stitch of convex and concave face of fractured load cell (Sumner, 2020).

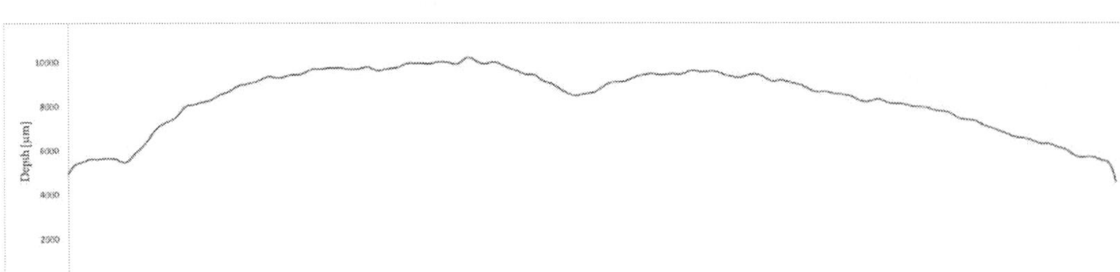

Figure 9. Depth profile of convex face of fractured load cell (Sumner, 2020).

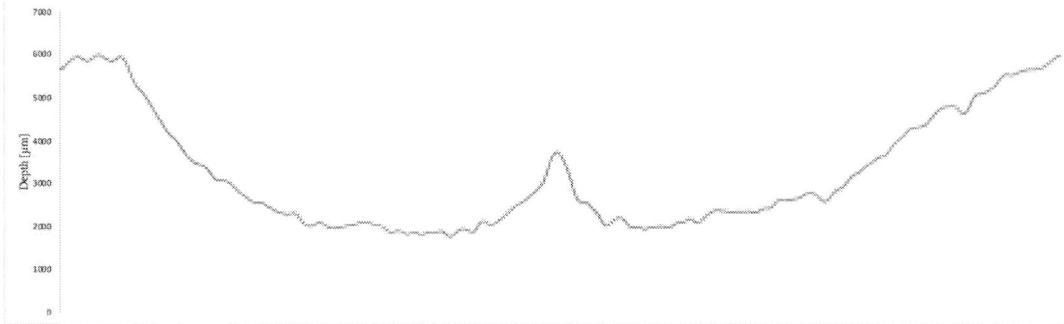

Figure 10. Depth profile of concave face of fractured load cell (Sumner, 2020).

SURFACE CRACK:
Figure 11 shows a 3D image stitching of the surface crack present on the unfractured load cell. This crack continues along the entire width of the load cell and also is seen to be located in the same spot that the fracture occurred in the other sample load cell. Figure 12 shows the depth profile at a point along the line of the surface crack, which is around 2.2 mm. Noting that the crack did not go all the way through the load cell, the possibility that the load cell fractured due to multiple impacts rather than one single impact is more likely.

Figure 11. Surface crack of unfractured load cell (Sumner, 2020).

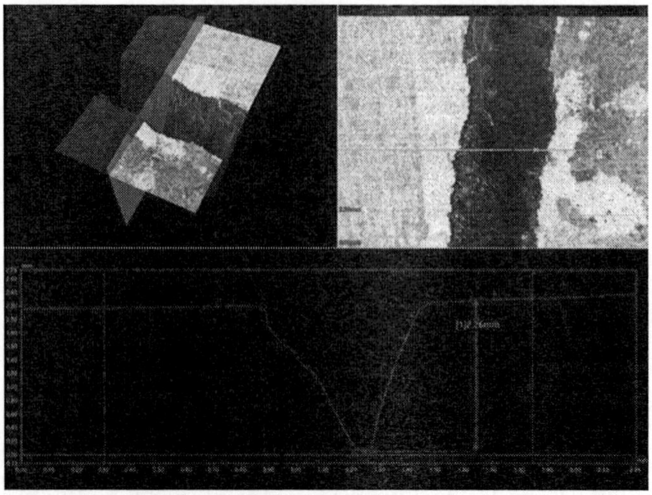

Figure 12. Surface crack of unfractured load cell and its depth profile (Sumner, 2020).

CONTACT REGION:

The contact region was the last portion to be analyzed using surface imaging. The region is at the midpoint of the load cell and can be seen in Figure 13 for the fractured load cell and Figure 14 for the unfractured load cell. This region is also the point at which the bracket of the pressure vessel assembly meets the load cell. Fig 15 shows some discoloration on the surface of the contact region for the fractured load cell. However, the surface of the contact region is best analyzed with the image in Figure 16, in which the results of a color depth profile are shown. The rough and bumpy surface is shown in this figure and reveals that the region consists of small indentions. These indentions are most likely the result of repeated impacts by the pressure vessel bracket.

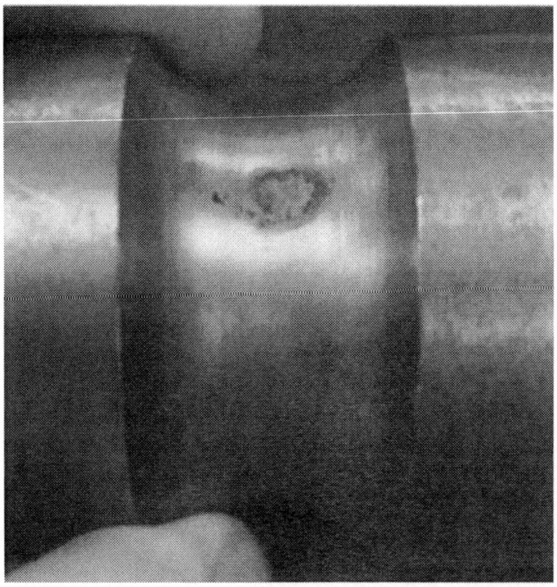

Figure 13. Contact region on fractured load cell (Sumner, 2020).

Figure 14. Contact region on unfractured load cell (Sumner, 2020).

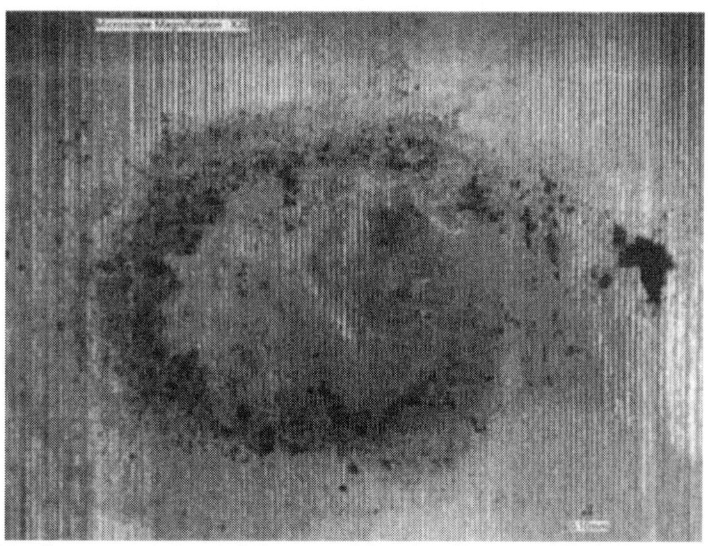

Figure 15. Contact region of fractured load cell, magnified (Sumner, 2020).

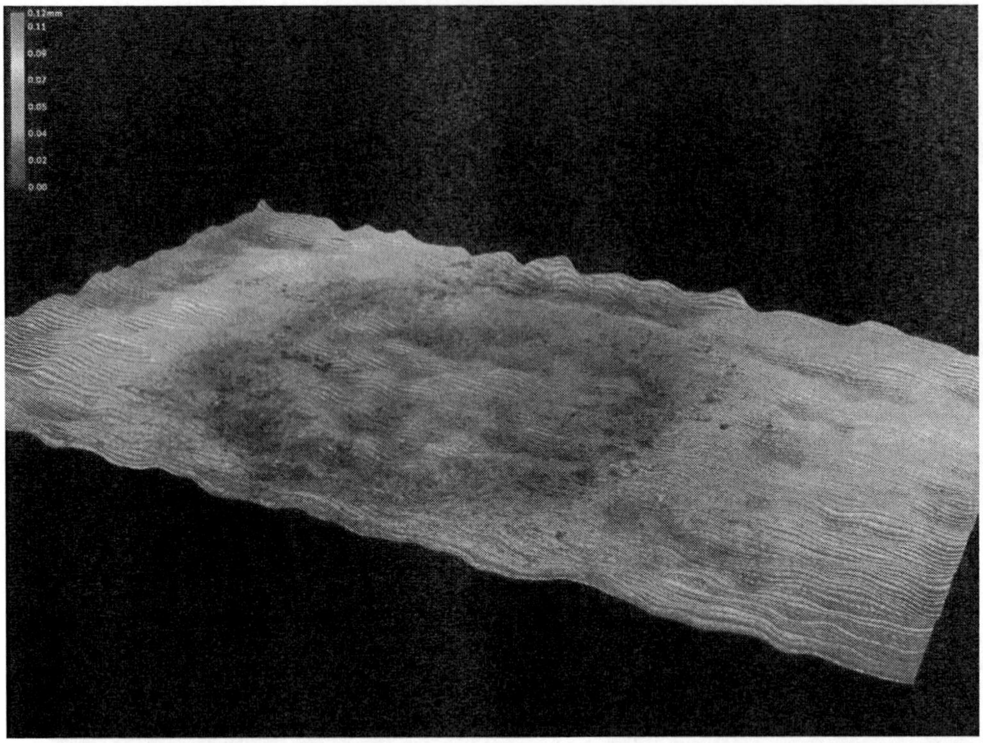

Figure 16. Color depth profile of contact region for fractured load cell (Sumner, 2020).

FINITE ELEMENT MODEL:

The simulations in Figures 17, 18, 19, and 20 were found using finite element analysis in ANSYS 2020. The purpose of these simulations is to demonstrate the possibility of the load cell fracturing to a single impact loading. Figures 17, 18, and 19 use a load of 3134 lb, but the drop height is varied in increments of 1/16 in., starting at 0 in. All three simulations show a concentration of stress near where the sample load cells fractured. The only simulation that resulted in fracturing was the when the drop height was set to 1/8 in., therefore increasing the impact force. This simulation shows the case of the load cell fracturing due to a single impact loading is possible. The fourth simulation, in Figure 20, increases the load to 26000 lb. and has no drop height. In this simulation, it can be seen that the load cell again results in a fracturing due to impact loading. However, this simulation also confirms one other thing. Even if the bracket does not detach from the load cell, a sudden spike in loading could cause the load cell to fracture if the magnitude is high enough.

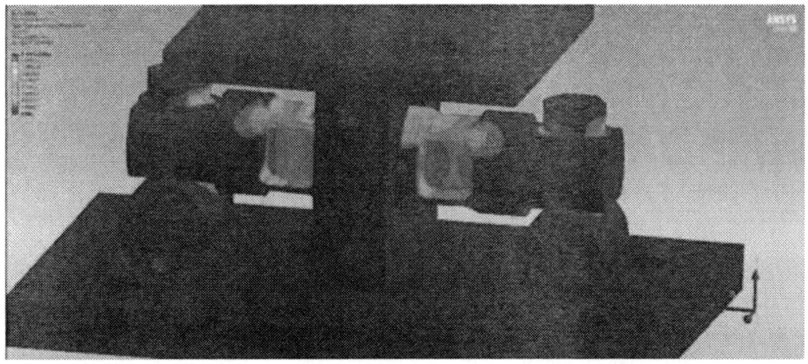

Figure 17. Impact load of 3134 lb. with no drop height.

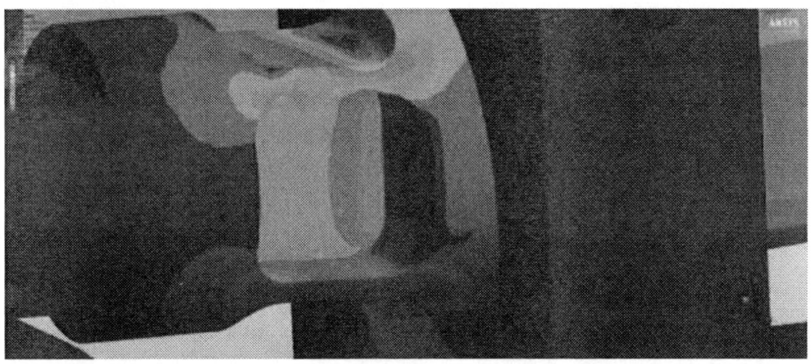

Figure 18. Impact load of 3134 lb. with drop height of 1/16 in.

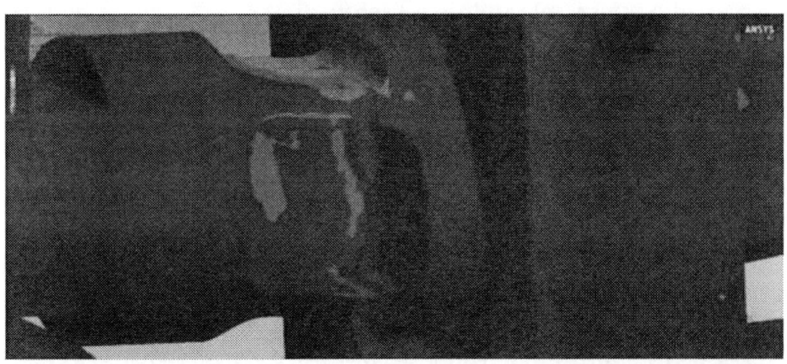

Figure 19. Impact load of 3134 lb. with drop height of 1/8 in.

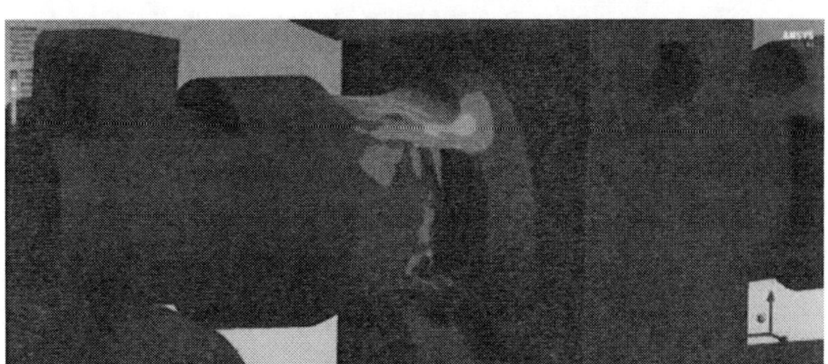

Figure 20. Impact load of 26000 lb. with no drop height.

CONCLUSION:
Using finite element analysis, it was found that the load cell failing due to a single impact loading is possible. The load cell in the simulations fractured when subjected to an impact load of 3134 lb. and a drop height of 1/8 in as well as with an impact load of 26000 lb. and no drop height. In addition to this, clues found from surface imaging also indicated the likelihood of failure due to impacts loading. Many small indentations were found on the contact region between the bracket and the load cell, indicating that the bracket impacted the load cell many times. The curvature of the fractured surfaces as well as the discoloration that appeared on only part of the surface also indicated that the load cell was subjected to multiple impact loadings and failed over time. Therefore, while the simulations verified the possibility of the load cell fracturing from one larger single impact, it was more likely that the load cell failed over time due to many smaller impact loads.

For continuing this research, the implications of earthquake effects on the pressure vessel would need to be evaluated just as wind loading was. To do this, a shake table will be utilized to experimentally determine the forces on a pressure vessel when being subjected to seismic activity. From the data obtained using the shake table, dimensional analysis can be used to compute what the actual forces for a life-size pressure vessel would be.

ACKNOWLEDGMENTS:
First, I would like to thank Georgia Southern University's Mechanical Engineering Department for the funding provided for this research as well as for allowing us to use their facilities and equipment. I would also like to thank Dr. Brian Vlcek, for allowing us to utilize his surface imaging lab, which played a major role in this research.

REFERENCES:
[1] Chavan, S. S., Kalaska, S. H., & Patil, P. B. (2013). Effect of Wind Load on Pressure Vessel Design by Using Non- Linear Finite Element Analysis (FEA). International Journal of Engineering Research & Technology, 2(12), 2566–2571.
[2] Dutta, S., Das, A., Barat, K., & Roy, H. (2012). Automatic characterization of fracture surfaces of Aisi 304LN stainless steel using image texture analysis. Measurement, 45(5), 1140–1150. https://doi.org/10.1016/j.measurement.2012.01.026
[3] Haque, S., Nowak, S., Callaghan, R., Mukerjee, A., Prasad, R., Rahman, M., & Mitra, A. (2019). A Case Study of Structural Industrial Pressure Vessel Under Wind Load. Proceeding of the 2019 Conference for Industry and Education Collaboration.
[4] Hutiu, G., Duma, V.-F., Demian, D., Bradu, A., & Podoleanu, A. G. (2014). Surface imaging of metallic material fractures using optical coherence tomography. Applied Optics, 53(26), 5912–5916. https://doi.org/10.1364/ao.53.005912
[5] Infante, V., & Silva, J. M. (2010). Case studies of computational simulations of fatigue crack propagation using finite elements analysis tools. Engineering Failure Analysis, 18(2), 616–624. https://doi.org/10.1016/j.engfailanal.2010.09.027
[6] Keyence Digital Microscopes - VHX-7000 Series. KEYENCE. (n.d.). Retrieved March 30, 2022, from https://www.keyence.com/products/microscope/digital-microscope/vhx-7000/
[7] Mahandari, C. P., & Kharism, A. A. (2015). Wind and Earthquake Loads On The Analysis of a Vertical Pressure Vessel For Oil Separator. Proceeding Seminar Nasional Tahunan Teknik Mesin XIV.

[8] Patil, P. S., & Kadam, S. R. (2014). Investigation of Dynamic Behavior of Framed Structure Under Impact Load. International Journal of Emerging Engineering Research and Technology, 2(3), 79–86.
[9] Sumner, A. T. (2020). Dynamic Loading Experimentation and Surface Imaging of Pressure Vessel Loadcell Fracture (thesis).
[10] Wolfendale, P. C. (1986, February 25). Load Cell.

BIBLIOGRAPHY:

[1] Liu, D., & Pons, D. (2018). Crack propagation mechanisms for creep fatigue: A consolidated explanation of fundamental behaviours from initiation to failure. Metals, 8(8), 623. https://doi.org/10.3390/met8080623
[2] Mirzadeh, H., Najafizadeh, A., & Moazeny, M. (2009). Flow curve analysis of 17-4 ph stainless steel under hot compression test. Metallurgical and Materials Transactions A, 40(12), 2950–2958. https://doi.org/10.1007/s11661-009-0029-5

PIPING VIBRATION ASSESSMENT

Authors: Gyorgy Szasz, PhD
and Scot McNeill, PhD, PE
and HyeongUk Lim, PhD

VIATC TRAINING CONFERENCE & EXPO
SAVANNAH, GEORGIA · 2022

PIPING VIBRATION ASSESSMENT

Gyorgy Szasz, PhD Scot McNeill, PhD, PE HyeongUk Lim, PhD
Stress Engineering Services, Inc.
13610 Westland East Blvd, Houston, TX 77041

Dr. Szasz is a Staff Consultant at Stress Engineering Services, Inc. He has 25 years of experience with hands-on vibration testing, measurement and analysis. He has performed an extensive amount of field work globally in various power plants (including hydro, fossil and nuclear), refineries, pipeline stations, LNG facilities, offshore platforms, etc. He is a certified category IV vibration analyst, member of ASME, president of the VI board of directors, and a former member of the ASME Technical Committee on Sound and Vibration.

Scot McNeill is a Principal at Stress Engineering Services in Houston, Texas. He has extensive experience with flow induced vibration of fixed equipment in upstream, midstream and downstream applications. Scot has contributed to industry standards, such as: the Energy Institute Guidelines and API 579. Scot received his B.S. and M.S. degrees in engineering mechanics from the University of Wisconsin at Madison and his Ph.D. in mechanical engineering from the University of Houston.

Dr. Lim is an Associate at Stress Engineering Services, Inc. He has worked on a variety of oil & gas projects that apply numerical modeling and data analytics/science. He has published a number of journal articles regarding structural vibration/reliability for offshore structures. His technical proficiency includes Monte Carlo simulation, surrogate modeling, reliability analysis, and uncertainty quantification. He is an Editorial Board member of Journal of the Korean Society for Advanced Composite Structure.

ABSTRACT:
A great body of knowledge exists to support rotating machinery vibration analysis and a large number of vibration analysts are comfortable evaluating rotating machinery. Assessing piping vibrations is still more art than science in a lot of cases. This paper describes the steps that go into performing piping vibration assessments in a formal manner, ranging from data acquisition to advanced modal analysis and stress computation based assessments. The paper also provides piping vibration case studies to illustrate Level 1 and 2 assessments. The described approach is largely based on industry guideline documents co-authored by Stress Engineering Services with other contributors.

KEY WORDS: Assessment; broadband; fitness-for service; Level 1; Level 2; Level 3; narrowband; piping; random; screening; vibration;

INTRODUCTION:

Solving a piping vibration problem begins with recognizing that there is a problem in the first place. Sometimes the problem is obvious, as a related failure occurs. Failures of small bore connections (SBCs) are far more common than large bore lines, large bore piping vibration problems, however, often manifest themselves in the form of damaged piping supports. Still, it is not a foregone conclusion that large bore piping can never fail due to vibration. It is likely however, that the amplitudes necessary to fail large bore piping will be noticed visually or otherwise by operators, and a vibration study will be initiated before the failure occurs. One notable exception is thin wall large bore piping, which when subjected to high frequency excitation can fail rapidly due to shell mode vibration.

Whether the vibration study is initiated due to a failure or operator observations, the first step in the assessment is the collection of appropriate vibration data. The quality of the collected data will have a significant impact on the assessment results, therefore, no effort should be spared in making sure that the best possible data is gathered within the constraints always present when working in the field.

DATA ACQUISITION:

Collecting piping vibration data can present many challenges ranging from accessibility and sensor mounting issues to data collection time and environmental concerns. One of the first things that has to be decided is whether the process generating the vibration is stable enough to allow the data to be collected in a survey fashion, where the data is collected sequentially by visiting one measurement location after the other. If the process is such that vibration levels change over the time it takes to perform a vibration survey, the data must be collected simultaneously at all locations. Even if the data can be collected surveys style, it is often desirable the reference channel is maintained so phase can be calculated later and this adds a degree of difficulty since at least one long cable has to be carried along to each measurement location.

Survey style data collection can be performed using a portable data collector (Figure 1, left), but monitoring typically requires a multi-channel analyzer/recorder (Figure 1, right). Multi-channel monitoring requires far more setup time of each location that has to be instrumented and long cables have to be stretched from each location to a central spot where the data acquisition system is located.

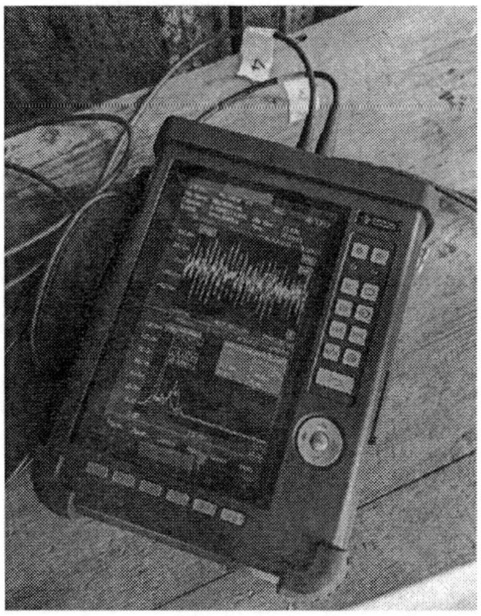

Figure 1: Sample Data Acquisition Hardware Options

If the piping is made of carbon steel and it is reasonably low temperature typically the sensors can be mounted magnetically; for stainless steel piping or for high temperature piping other solutions have to be found. These solutions may include the application of mounting brackets (Figure 2, left) or mounting blocks to the piping. In select cases mounting brackets can be strapped on the outside surface of insulation (Figure 2, right). Although this practice is generally not recommended, reasonably accurate data can be gathered for very low frequency (visible) vibration.

Figure 2: Frame (left) and Strap (right) Mounted Accelerometers

Problem classification we will be discussed in the next section, but some idea of the piping vibration type must be known at the time of data collection since for example random vibration requires much longer data sets than periodic vibration. This is to ensure that valid statistics can be computed from the recorded time histories. Note that this also implies that time histories have to be recorded at every location. Averaged frequency spectra are useful but insufficient for piping vibration assessments since certain statistical parameters can only be calculated from the time history.

Measuring high frequency vibration requires more stringent sensor mounting methods than low frequency. In some cases special sensors may be required to collect the data such as a laser vibrometer. Recently other optics based vibrations and sensing technologies have become available such as video vibration amplification, which is highly suitable for vibration frequency measurement and determining optimum sensor location by revealing the operating deflection shape.

PROBLEM CLASSIFICATION:

Piping vibration can be classified into a variety of categories such as periodic and random, single frequency or multifrequency, broadband versus narrowband, stationary versus non-stationary, high frequency or low frequency, and so on. If the waveforms have a recognizable repeating pattern, the vibration is periodic (Figure 3, top), in all other cases the vibration is random (Figure 3, bottom).

The randomness can manifest itself both in terms of magnitude and frequency. Periodic vibration can be single frequency or multi frequency. Random vibration can be single frequency, multi frequency, narrowband or broadband. Narrowband means that the vibration is limited to a range of frequencies, while broadband means that vibration frequency covers a larger range in the frequency spectrum. Stationary vibration has statistics that remain constant with time while the statistics of nonstationary vibration, such as its RMS amplitude keep changing. Generally speaking, vibration above 300 Hertz is considered high frequency.

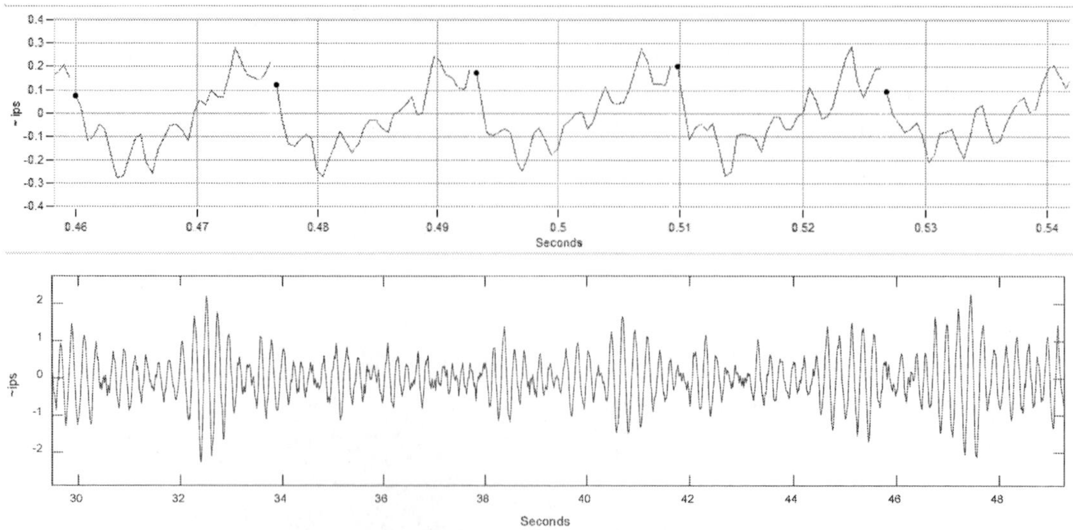

Figure 3: Sample Periodic and Random Waveforms

Knowing the type of vibration occurring will have an impact on how the data is collected and also how it's assessed. Even in the case of the most basic assessments at level one screen some industry guidelines differentiate between periodic and random vibration as well as small bore versus large bore piping.

LEVEL 1 – GENERIC SCREENING

Some basic screening criteria do not take the piping configuration into account. They are intended to be conservative enough to apply to all possible piping configurations. These criteria include those from the energy institute (EI) [3] and Engineering Dynamics, Inc. (EDI) [2]. These methods were primarily intended for evaluating single frequency vibration in a resonance condition. Recent work extended this method to evaluating random piping vibration as well [5].

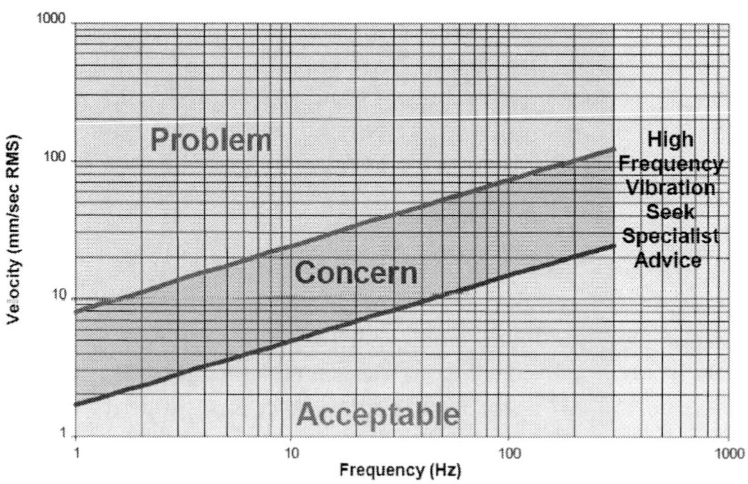

Figure 4: EI Vibration Screening Curves

The extension was based on the overall root mean square (RMS) velocity and mean zero up-crossing frequency, plotted as a single point on the screening curve chart (Figure 4) for vibration screening assessment of piping response between 1 and 300 Hz, instead of having to rely on a dominant peak in the spectrum.

A more comprehensive approach is currently being developed for future inclusion in the API 579 FFS Standard [4]. Note that Level 1 assessments are not universally applicable. Due to the assumption made when they were developed certain situations call for higher level assessments such as vibration with high kurtosis. Therefore, while kurtosis is not necessary to perform a level 1 assessment, it is worthwhile checking it to make sure that the assessment is valid. Kurtosis is a statistical measure of how pointy or flat the peak of the distribution curve associated with a random data set is. Data sets with high kurtosis tend to have more outliers, while data sets with low kurtosis tend to have a lack of outliers.

LEVEL 2 ASSESSMENT:

A Level 2 assessment is an extension of the Level 1 assessment in that it provides a less conservative allowable vibration velocity level to be compared to the highest overall vibration measured in the piping system. It takes into account the piping characteristics as well as the process conditions, but stops short of requiring the calculation of the stresses in the piping system. This means that Level 2 assessments can still be performed using spreadsheets, without dedicated stress analysis software. One such method is the VM2 approach from American Society of Mechanical Engineers (ASME) [1]. An improved version of that approach is currently being developed for future inclusion in the API 579 FFS Standard [4]. A sample table of the main inputs needed for a Level 2 assessment are shown in Table 1. This is clearly a lot more information than what was needed at the Level 1 assessment.

Table 1. Inputs to a Level 2 Piping Vibration Assessment

Quantity	Symbol	Units	Circuit 1	Circuit 2	Circuit 3
Design life		years	50	50	50
Pipe material	-	-	304/304L	304/304L	304/304L
Pipe material ultimate tensile strength	UTS	ksi	75	75	75
Normal operating temperature	T	deg F	650	650	650
Pipe Young's Modulus at operating temp.	E	psi	2.49E+07	2.49E+07	2.49E+07
Pipe outside diameter	OD	in	6.625	8.625	10.75
Pipe wall thickness	t	in	0.28	0.322	0.375
High low misalignment	h	in	*0.070*	*0.081*	*0.094*
Density of pipe	r_p	lb/in3	0.287	0.287	0.287
Density of contents	r_f	lb/in3	*0.0280*	*0.0280*	*0.0280*
Insulation thickness	t_i	in	3	3	3
Insulation density	r_i	lb/in3	8.68E-03	8.68E-03	8.68E-03
Highest vibration level measured	V_m	in/s RMS	0.82	1.03	0.56
Effective frequency	f_e	Hz	4.3	5.7	4.8
Skew	Sk	-	0	0	0
Kurtosis	Ku	-	3.5	3.6	3.4
Dominant frequency 1	f_d1	Hz	4.1	5.0	4.3
Dominant frequency 2	f_d2	Hz	N/A	N/A	N/A
Reference Young's modulus	E_ACS	psi	2.83E+07	2.83E+07	2.83E+07
Weld joint non-propagating stress range	S_0	psi pk-pk	7740	7740	7740
Smooth bar fatigue curve endurance limit	S_e	psi pk	13600	13600	13600
Reference RMS stress allowable value	S_ref	psi RMS	631	631	631
Parameter for non-gaussian, high kurtosis signal	η	-	0.018	0.021	0.014
Calculated crest factor from kurtosis	CF	-	5.86	6.08	5.62

Two key additional inputs that are difficult to obtain are the locations of the high stress and the stress concentration factors (SCFs) at the high stress location. If the high stress location is at a weld, welded joint fatigue curves apply to the analysis, while if the high stress location is in a fitting or pipe, smooth bar fatigue curves apply to the analysis. In order to remain conservative, typically both scenarios have to be checked, assuming that the high stress location could be at either place.

Stress concentration factors are easiest to derive from stress intensification factors (SIFs) since those are readily available from publications such as the AMSE B31J standard. For welded joints the SCF = SIF, while in smooth bar situations SCF = 2*SIF. This difference stems from the fact that SIFs are determined relative to an unmachined butt weld which has a peak stress index of 2 [6]. When non-standard components appear in the piping circuit, sometimes a conservative assumption can be made for the SCF, or a Level 3 assessment with local component FEA becomes necessary.

LEVEL 3 ASSESSMENT:

When a Level 3 assessment is performed, the stresses in the actual piping system must be estimated, which generally requires the construction of a computer model of the piping and dedicated stress analysis software. Beam element based tools such as CASEAR II (C2) can be adequate as long as there aren't any non-standard piping components in the circuit such as manifolds or trunnions (dummy legs) welded to elbows. Such components will often require the construction of a local FE model in a higher-level analysis software package such as ANSYS or ABAQUS.

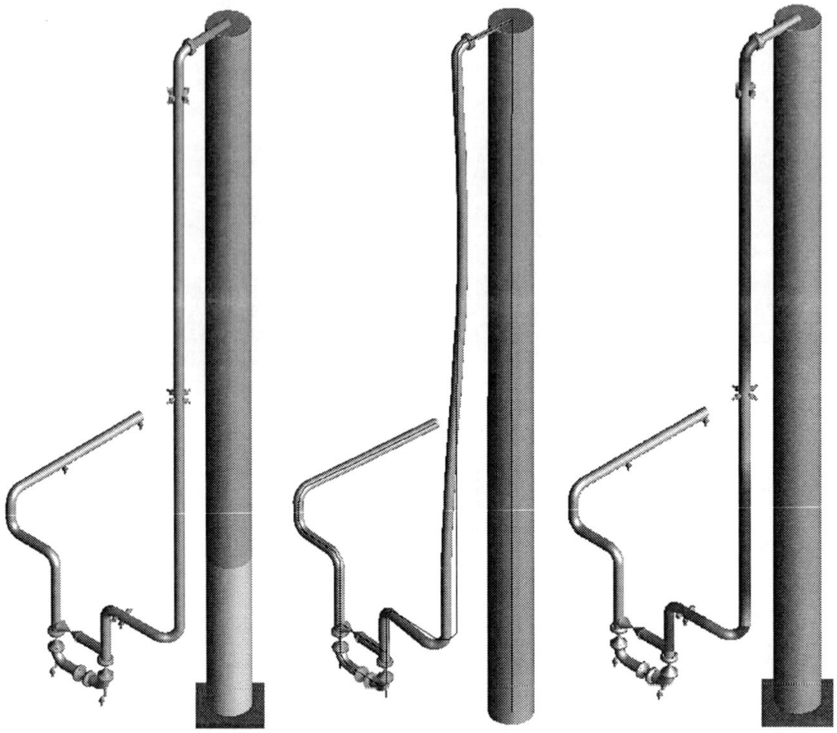

Figure 5: C2 Piping Model (left), Mode Shape (middle), Dynamic Stresses (right)

The computer model that is constructed needs to be benchmarked against the collected data. This means that sufficient and appropriate data must be available to compute the operating deflection shape of the piping. It is generally assumed that the operating deflection shape matches one of the vibration modes or a combination of the modes of the pipe, and for random excitation this is a reasonable assumption. When collecting piping vibration data it is often best to collect it in such a way that phase information can be derived even if it is not necessary for the level 1 and level 2 assessments. When multi-channel monitoring is performed phase is always available however only referenced surveys, mentioned previously, allow performing this task later.

As with the L2 assessment, it matters where the high stress location is. C2 for example can automatically handle some but not all stress concentrations. As alluded to previously, SIFs at weld locations can be used directly, SIFs in fittings need to be doubled. If there is a weld in a straight run of piping that is shown to be a high stress location, a misalignment related SCF has to be applied while post processing the data. A sample C2 model is shown in Figure 5.

When the piping vibration is single mode, the Level 3 assessment produces an even more refined allowable for the overall vibration measured. For multi-mode vibration the assessment gets considerably more complex. The vibration may need to be evaluated one mode at a time, and the resulting stresses combined at the end using the square root of sum of the squares approach. Matching multiple piping vibration modes with a computer model is exceedingly difficult, typically a compromise has to be made in order to be able to stay within budget and schedule constraints.

MITIGATION:

If the piping vibration fails all three levels of assessment, the vibration must be mitigated. Mitigation is not the central focus of this paper, however, it is still worth mentioning that the options for mitigation (in order of preference) are:

- Reducing or eliminating the excitation
- Reducing the response
- Increasing the fatigue resistance of the piping

Even though the preferred method is excitation reduction, the most frequently employed method is the adding supports to reduce the response. This is mainly due to cost and operational constraints. Reducing excitation may require expensive changes to the system such as replacing the piping with larger diameter piping, installing elbows with larger turn radii, installing accumulators, switching pumps to a variable frequency drive, etc. Meanwhile, reducing response typically includes adding supports, mass, and damping for resonance mitigation or vibration energy dissipation. Corrective actions for increasing fatigue resistance may include using fatigue resistant fittings, replacing fabricated tees with forged tees, grinding butt weld caps flush, etc.

CASE STUDY 1 - LEVEL 1 ASSESSMENT:

In this section we provide a case study for a Level 1 assessment performed for metering skid piping in a facility. Vibration measurements were performed on a metering skid, containing two coriolis flow meters. Vibration data were collected for various flow rates; a hand-held, battery-powered vibration analyzer was used to collect vibration data. A single-axis accelerometer at a fixed position served as a reference and a tri-axial accelerometer is moved from location to location (multiple locations for each flow rate).

After data collection, the vibration levels were calculated and compared to "Concern" and "Problem" levels in the EI vibration screening curves. It was found that at a particular flow rate, vibration levels exceeded the concern level at a couple of locations near the flow meters. A sample table prepared for the Level 1 assessment is shown in Table 2. In the table, the location marked as "4" was a location where the measured RMS velocity exceeded the concern level; and repeated measurements confirmed high vibration at that location. The mean zero up-crossing frequency (used as the effective frequency where the vibration levels were evaluated on the screening curve) is denoted as 'fz' in the table.

Table 2. EI Vibration Screening Results for Level 1 Piping Vibration Assessment

Location	Direction	Vrms	fz	Concern	Problem	Average	Kurtosis	CrestFactor	Direction
[-]	[-]	[in/Sec]	[Hz]	[in/Sec]	[in/Sec]	[in/Sec]	[-]	[-]	
1	Ch Ref	0.047	177.486	0.763	3.731	0.009	2.861	3.909	
1	Ch2	0.203	147.284	0.701	3.421	0.003	1.657	2.139	N
1	Ch3	0.128	152.131	0.712	3.467	0.009	1.702	2.197	Vert
1	Ch4	0.068	162.816	0.730	3.580	0.004	2.059	2.680	E
2	Ch Ref	0.050	177.280	0.764	3.732	0.008	2.951	3.701	
2	Ch2	0.205	194.714	0.795	3.901	0.004	1.888	2.631	N
2	Ch3	0.143	244.589	0.882	4.358	0.010	3.042	3.685	Vert
2	Ch4	0.094	172.887	0.755	3.689	0.008	2.314	3.027	E
3	Ch Ref	0.056	183.663	0.776	3.801	0.004	2.843	3.917	
3	Ch2	0.192	332.290	1.018	5.045	0.008	2.312	3.325	N
3	Ch3	0.464	223.674	0.842	4.173	0.010	1.781	2.330	Vert
3	Ch4	0.373	165.386	0.732	3.612	0.003	1.642	2.244	E
4	Ch Ref	0.053	181.312	0.764	3.778	0.006	2.926	3.600	
4	Ch2	0.757	182.294	0.766	3.782	0.007	2.112	2.804	W
4	Ch3	1.136	167.325	0.744	3.627	0.004	1.659	2.159	Vert
4	Ch4	0.433	147.229	0.692	3.416	0.007	1.787	2.238	N
4	Ch Ref	0.179	63.067	0.478	2.272	0.001	6.257	5.118	
4	Ch2	1.306	169.767	0.740	3.657	0.008	1.665	2.125	W
4	Ch3	1.536	173.853	0.750	3.698	0.008	1.811	2.179	Vert
4	Ch4	0.442	159.505	0.719	3.548	0.003	1.917	2.608	N
5	Ch Ref	0.047	184.630	0.774	3.809	0.005	2.926	3.772	
5	Ch2	0.693	153.345	0.714	3.481	0.006	2.207	2.745	N
5	Ch3	0.398	225.192	0.851	4.191	0.005	2.743	3.391	Vert
5	Ch4	0.387	133.336	0.667	3.260	0.005	2.712	3.645	E
6	Ch Ref	0.049	187.568	0.780	3.836	0.001	3.009	4.491	
6	Ch2	0.224	513.971	1.252	6.218	0.004	2.816	3.819	N
6	Ch3	0.219	313.507	0.995	4.909	0.005	2.349	3.153	Vert
6	Ch4	0.261	203.542	0.806	3.992	0.001	1.993	2.636	E

Figure 6 shows the EI vibration screening curves prepared for Location "4"; it is seen that vibration data is close to periodic with a fundamental frequency of 146 Hz and several harmonics (marked as 2X, 3X, and 4X). Thus, it can be concluded that when vibration levels were high, the nature of the vibration switched from predominantly random to pseudo-periodic, dominated by a fundamental frequency and several harmonics. The signal characteristics can also be seen in the velocity time history collected from another location of concern. The response has mildly varying amplitude as shown in Figure 7.

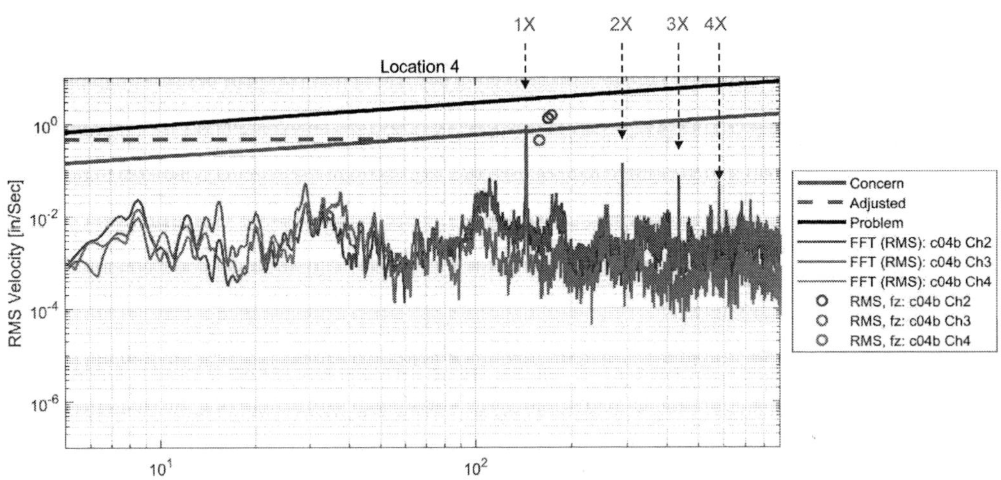

Figure 6: EI Vibration Screening Curves in a Location near Flow Meter

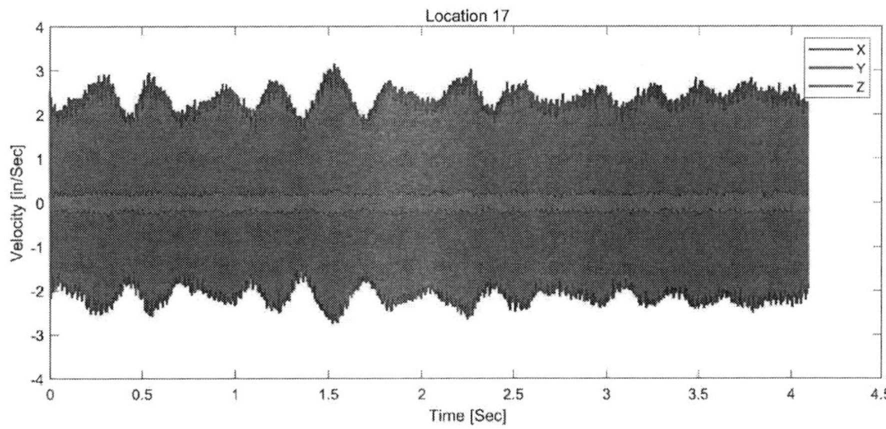

Figure 7: Measured Vibration (Velocity) Time History in a Location near Flow Meter

The cause of the high vibration in this facility was diagnosed as likely internal Flow-induced Vibration (FIV). Vortex shedding over a thermowell or dead leg coupled with acoustic resonance can lead to strong vibrations. In this instance there were thermowells between the two flow meters, and calculated vortex shedding/acoustic resonance frequencies indicated that it could be the cause of the vibrations. To estimate vortex shedding frequencies and acoustic mode frequencies, the following information was needed:

- Pipe isometric drawings or field measurements for skid piping
- Fluid properties at the operating conditions
    - Speed of sound
    - Temperature, pressure, density
- Geometry of any flow disruptions
- Branch pipe sizes, thermowell or other probe details

CASE STUDY 2 - LEVEL 2 ASSESSMENT

Tall columns in refineries and chemical plants often have long piping running up and the down their side to deliver or withdraw process fluids and vapors. If the support configuration for these lines is primarily based on static loads, excessive vibration is sometimes exhibited at the bottom elbow where the line departs the column. These vibrations can easily exceed the Level 1 assessment criteria. In one such instance a Level 2 assessment was performed which takes the piping and process properties into account, and is summarized in Table 3. The Level 2 vibration limits may not end up being significantly higher than the Level 1 limit and therefore the piping may require a Level 3 assessment. That was the case for the line assessed in Table 3, however, the piping did pass the Level 3 assessment, which raised the allowable to 0.94 in/s RMS after dynamic stresses were estimated using a C2 piping model.

Table 3. Level 2 Piping Vibration Assessment Calculations

Intermediate Calculations

Quantity	Symbol	Units	Circuit 4
Calibration Factor	C_x	-	1.81
Correction Factor for Piping Arrangement	C_a	-	1.43
Correction Factor for Frequency	C_f	-	1
Correction Factor for Concentrated Masses	C_cm	-	1
Weight of Pipe per Unit Length	w_p	lb/in	4.91
Weight of Fluid per Unit Length	w_f	lb/in	0.87
Weight of Insulation per Unit Length	w_i	lb/in	2.32
Correction Factor for Distributed Masses	C_dm	-	1.28

Allowable Velocity (Weld Joint)

Quantity	Symbol	Units	Circuit 4
Eccentricity	d_m	in	0.15
Weld Root Misalignment SCF	K_mr	-	1.52
Weld Toe Misalignment SCF	K_mt	-	2.16
Dimensionless Velocity Factor	C_T	-	7.17
Stress to Velocity Factor	K_v	psi/ips	1022.1
Temperature correction factor for stress	f_MT	-	0.979
Allowable RMS stress	Sa_rms	psi	841.8
Allowable RMS velocity	Va_rms	in/s	0.82

Allowable Velocity (Smooth Bar)

Quantity	Symbol	Units	Circuit 4
Stress Intensification Factor (elbows)	SIF_el	-	6.40
Stress Intensification Factor (tees)	SIF_T	-	5.61
Stress Concentration Factor	SCF	-	6.40
Dimensionless Velocity Factor	C_T	-	21.27
Stress to Velocity Factor	K_v	psi/ips	3031.0
Temperature correction factor for stress	f_MT	-	0.979
Allowable RMS stress, smoth bar	Sa_rms	psi	1522.6
Allowable RMS velocity, smoth bar	Va_rms	in/s	0.50

Final Result

Quantity	Symbol	Units	Circuit 4
Highest vibration level measured	V_m	in/s RMS	0.68
Allowable RMS velocity	Va_rms	in/s	0.50
Assessment result for measured vibration			**FAIL**

CONCLUSION:

The following conclusions can be drawn from the previous sections:

- Piping vibration data must be collected with the assessments to be performed later in mind.
    - Always record time history data for the full length of the data collection.
    - Try to ensure that phase data can be reconstructed later if a Level 3 assessment becomes necessary, i.e. use a reference channel when doing surveys.
- Data collection may become challenging due to access, material type and or temperature. Plan ahead and bring extra hardware.
- Level 1 assessments are the simplest however they are not always applicable and are designed to be very conservative.
- Calculate time history based statistics to verify applicability of Level 1 assessments.
- Level 2 assessments are less conservative however, they are based on several assumptions, and may not be applicable to non-standard components.
- Level 3 assessments are the most complex and require the most knowledge of the piping as well as analysis skills. Actual complexity varies with the problem; however, all piping vibration problems can be handled with this type of assessment.
- If mitigation becomes necessary, evaluate the option of reducing excitation, before moving onto reducing response by adding supports.

REFERENCES:

[1] ASME OM-3 (2009), "Vibration Testing of Piping Systems"

[2] Wachel, J.C., Morton, Scott J., Atkins Kenneth E., "Piping Vibration Analysis"; Proceedings of the 19th Turbomachinery Symposium; Turbomachinery Laboratory, Texas A&M University, College Station TX, 1990.

[3] "Guidelines for the Avoidance of Vibration Induced Fatigue Failure in Process Pipework," Energy Institute, London, 2/e, 2008.

[4] "Fitness-for-Service," API 579/ASME FFS-1, 2007.

[5] Breaux, L.E., McNeill, S.I., Szasz, G., "Fitness-for-Service Assessment of Piping Subject to Random Vibration," Proceedings of the Pressure Vessel and Piping Conference, July 17-21, 2016, Vancouver, BC, Canada.

[6] ASME Boiler and Pressure Vessel Code, Section III, Part NB